America as Second Creation

D1367971

America as Second Creation

Technology and Narratives of New Beginnings

David E. Nye

The MIT Press
Cambridge, Massachusetts
London, England

© 2003 Massachusetts Institute of Technology

Set in Sabon by Graphic Composition, Inc. Printed and bound in the United States of America.

Library of Congress Cataloging-in-Publication Data

Nye, David E., 1946–
America as Second Creation: Technology and Narratives of New Beginnings / David E. Nye
    p.   cm.
Includes bibliographical references and index.
ISBN 0–262–14081–0 (hc.: alk. paper) – 978-0-262-64059-6 (pb.: alk. paper)
1. Frontier and pioneer life — United States. 2. Frontier and pioneer life — United States — Historiography. 3. Technology — Social aspects — United States — History. 4. Technology — Social aspects — United States — Historiography. 5. Land settlement — United States — History. 6. Land settlement — United States — Historiography. 7. National characteristics, American. 8. United States — Discovery and exploration. 9. United States — Colonization. 10. United States — Historical geography.   I. Title.

E179.5 N94   2003
978'.02—dc21

2002029564

10 9 8 7 6 5 4

. . . nature builds us no house or temple, spins no dress. She writes no poetry, composes no music, presents us with no forms of intercourse. Having given out forms enough to beget activity in human taste, she scants her work that we may go on and exert a creative fancy for ourselves.

The wild forests are cleared away, the green slopes are dressed and laid out smiling in the sun, the hills and valleys are adorned with beautiful structures, the skins of wild beasts are laid aside for robes of silk or wool. In a word, architecture, gardening, music, dress, chaste and elegant manners—all inventions of human taste—are added to the rudimental beauty of the world, and it shines forth, as having undergone a second creation at the hand of man.

—anonymous, "Taste and Fashion," *New Englander and Yale Review* 1 (1843), no. 2

Contents

Acknowledgments

Research and writing require many resources and much advice. Important early support for this work came in 1997, when The MIT Press offered me a contract on the basis of several conference papers, an article, and an outline. Syddansk Universitet and its rektor, Henrik Tvarnø, further supported the project with a timely teaching-free semester in 1998. During a Leverhulme Trust research fellowship, Leeds University provided office space and an undisturbed environment. A year later, the Carlsberg Foundation enabled me to spend the Michaelmas term at Churchill College, Cambridge University.

Some background materials for this work came from my university library, but much of the research was done in Britain, at the John Reylands Library in Manchester, Leeds University's Brotherton Library, the Cambridge University Library, and the British Library in London. I also benefited greatly from short visits to the Winterthur Library in Delaware, the University of Washington's library in Seattle, the Wilson Library at the University of Minnesota, the Northwestern University Library in Evanston, Cornell University's Olin Library, and the various libraries of the Massachusetts Institute of Technology. Many scholars helped me think about this project, and I learned more from their comments and encouragement than I suspect they realized. Among them were Thomas Hughes, Mick Gidley, Tony Badger, Bob Shulman, Robert Friedel, Lindy Biggs, Julie Roy Jeffrey, Merritt Roe Smith (who shared several books with me), and particularly Leo Marx (who showed me why chapter 2 was necessary). An anonymous reviewer provided a helpful reading of the manuscript, as did Marty Melosi, who drew my attention to works in environmental history that enriched the final result, and Cecelia Tichi, who pointed me toward useful sources.

In a more general sense, I owe a debt to particular publications that made this one possible. My intellectual debts are more fully suggested by the bibliography. Yet I do want to single out Leo Marx, who in his classic *The Machine in the Garden* links a wide range of literary texts to the central question of what the industrial revolution meant to a nation that identified itself with nature and with agrarianism, and John Kasson, who in *Civilizing the Machine* examines the closely related political question of how Americans sought to reconcile new technologies with the ideology of Republicanism. In addition to their work, the multi-volume projects of Louis Hunter on energy, John Stilgoe on landscape, and D. W. Meinig on cultural geography established the background for the present study.

Several institutions encouraged my early forays into the topic by inviting me to speak. These included the School of English, Leeds University; The Science Museum, London; the Program in Science, Technology, and Society at the University of Trondheim, Norway; the Department of American Studies, Amsterdam University; the English Department of Copenhagen University; the Center for American Studies, Syddansk University; Cornell University's Science, Technology, and Society Department; and MIT's Program in Science Technology, and Society. I presented chapter 2 at the University of Bayreuth in October 2001, chapter 3 in Graz at the annual meeting of the European Association of American Studies in April 2000, portions of chapters 7 and 8 at a meeting held by the Danish Association for the History of Technology in May 2001, and chapter 11 at a conference held at MIT's Dibner Library in April 2000.

Most of all, I thank Larry Cohen, who has been my patient and encouraging editor for five volumes with The MIT Press. For two decades I have relied on his judgment and valued his brief, penetrating comments. In a publishing world increasingly driven by finance and characterized by ephemeral contacts, our long association has been my extremely good fortune. Larry has been remarkably tolerant about changes and delays. His understanding for the kind of history that I have pursued has been an indispensable support, and I am proud to dedicate this work to him.

America as Second Creation

Introduction

In the American Beginning

In the beginning there were no technological creation stories about the New World. When Europeans invaded the edges of the newly discovered continent of North America, the stories they told to make sense of their world came from their homelands. Their communities existed as extensions of the French, Dutch, British, and Spanish empires. Political thought in the colonies during the seventeenth and eighteenth centuries focused on relations to and conflicts between these great European powers. Economic thought was strongly marked by the mercantilist questions of tariffs, trade restrictions, and the division of labor between mother countries and colonies. Religious thought reflected the European struggle between Catholics and Protestants as the Reformation and its aftermath led to migrations of conflicting groups and doctrines into North America. An independent nation called the United States of America was not the imagined outcome of any of this. There could be no American narratives until the revolutionary period, and even then it would take a generation to develop stories that could replace the colonial sense of European political and religious origins with nationalist stories about secular, American origins.

In the American beginning, after 1776, when the former colonies re-imagined themselves as a self-created community, technologies were woven into national narratives. A few assumed particular prominence, among them the axe, the mill, the canal, the railroad, and the irrigation dam. This book is about those technologies and the stories that clustered around them. It is about an American story of origins, with America conceived as a second creation built in harmony with God's first creation.

Americans constructed technological foundation stories primarily to explain their place in the New World, not to understand the technologies. A new machine acquired social meaning when placed in a context and used for some purpose. A small number of technical experts may have understood a particular steam engine as the practical expression of experiments and equations; however, the passenger on a steamboat or a train comprehended it first as a novel experience of noise, power, and movement, and later as a dynamic part of a larger narrative about American expansion and progress, but probably never in terms of thermodynamics. This book will examine not the public's immediate experience of new technologies but the stories that wove new technologies into the history of settling the nation.[1] In popular literature, speeches, advertisements, paintings, and many other forms, European-Americans invented foundation stories as they entered new regions of North America and used powerful technologies to transform these spaces into familiar landscapes.

Every society constructs narratives to make sense of its existence, to explain how its people came to live in a particular place. Native Americans' stories of origin provide an instructive counterpoint to those of European-Americans. Native Americans' stories express their sense of oneness with the land. The first people are said to have emerged out of the local earth or to have come into the world through the intervention of spiritual beings. The Hopi believe that their ancestors came out of the earth at particular sacred locations.[2] One group points to a spot at the junction of the Little Colorado and Colorado Rivers, at the northern end of the Grand Canyon.

Anthropologists have found that on average the Navajo regard one place per 26 square miles as sacred. While some of these are man-made sites, such as Anasazi ruins, most are natural features—"mountains, hills, rock outcrops, canyons, springs and other bodies of water, natural discolorations on rocks, areas where certain plants grow, mineral deposits, isolated trees, places where rocks produce echoes, air vents in rocks, sand dunes, flat open areas, lightning-struck trees and rocks."[3] These sites are apprehended not as isolated but as integral parts of the larger landscape. When anthropologists ask about places, the replies are often organized as stories, and these tales link sacred spaces both to rituals and to the central myths of Navajo society. Though the details differ widely, most other

Native American tribes have also been found to weave their culture into the landscape in detailed narratives linked to specific locations.[4]

For those who arrived after Columbus, neither ancient sacred places nor local stories of origin were possible. Instead, the new Americans constructed stories of self-creation in which mastery of particular technologies played a central role. The Native Americans' self-conception was inseparable from the first creation of the world; former Europeans had to project a second creation. They started without a detailed knowledge of the land itself, and they could not imagine away their belated arrival. Instead, they constructed stories that emphasized self-conscious movement into a new space.

One cluster of stories consisted of frontier epics about the hardships of pioneers and their conflicts with Native Americans. The central technologies were firearms, particularly the rifle and the six-gun, and the stories were not only about defeating the Native American but also, as Richard Slotkin has shown, about the psychological transformation of settlers who experienced "regeneration through violence."[5] Such stories defined the self against an alien "other." Perhaps the earliest popular example was the captivity narrative, which began to circulate in the seventeenth century. It told the story of a white person captured by Native Americans and forced to adopt their way of life. One of the first of these colonial best sellers was Mary Rowlandson's, written after she had been ransomed.[6] The tale of regeneration through violence did not reach full articulation until the second half of the eighteenth century, however. Disseminated through speeches, songs, dramatic performances, and popular literature, it was embodied in such heroes such as Daniel Boone and Davy Crockett. Buffalo Bill Cody, whose Wild West Show toured the United States, was a later avatar of this story. As Richard White notes, the Wild West Show "told of violent conquest, of wresting the continent from the American Indian peoples who occupied the land."[7] Characteristically, whites were depicted as having been attacked and forced to defend themselves.

Tales of regeneration through violence represent only one tradition. Narratives about settling the land after Native Americans had been swept aside also deserve study. These foundation stories are about how "Americans" transformed a wilderness into a prosperous and egalitarian society. Their dramatic action focuses on transforming an uninhabited, unknown, abstract space into a technologically defined place. They valorize particular man-made objects, such as the American axe or the railroad, that made this transformation possible. Second-creation stories depict not heroic founders so much as generic first settlers. They express in secular form the beginnings of a new social world, and they establish the ideal ground rules of the society.

These stories about the foundation of new American communities, which became prominent at the end of the eighteenth century, were unlike the stories settlers and explorers told during the colonial period. To be sure, explorers and the first immigrants possessed tools and were often proud of their settlements, but neither their tools nor the transformation of the land were at the center of their narratives.[8] When Europeans began to colonize the New World, some imagined themselves to be returning to the bower of paradise.[9] Others saw themselves as conquerors of a pagan land in need of Christian redemption.[10] Early settlers in New England believed they had embarked on what Perry Miller called an "errand into the wilderness" to create a better Christian community than had been possible in Britain. The Puritans came to understand their relocation in biblical terms: they were a new chosen people, and they had come to a new Promised Land.[11] This way of thinking about America by no means disappeared in 1776, and derivatives of such ideas can be traced to the present. But after the Revolution, and particularly in the nineteenth century, Americans developed another way to understand their settlement of the western hemisphere: as the technological transformation of an untouched space. This story was projected back in time as well as forward into the immediate future.

The versions of the technological foundation story that emerged and circulated during the nineteenth century were literally about creating society by applying new technologies to the physical world: the axe, used to create the log cabin and the clearing; the mill, the center of new communities; canals and railroads, used to open western lands to settlement; irrigation, which converted "worthless" desert into lush farmland. Each narrative described a process of community creation. In each, the dominant society understood the narrative to be a valid account of its historical emergence. Each reduced a complex set of actions and experiences to an apparently simple assertion of facts. Though often the story was not literally true or was at least partially false, it was told with conviction and was widely accepted as valid.

The preference in the United States for particular technological narratives is inseparable from the country's geography, history, and pattern of population growth. Because the dominant English-speaking culture moved from the Atlantic seaboard westward, these narratives emerged in a sequence based on an east-to-west perception of geography. They started in eastern woodlands, where the axe and the mill provided the means of appropriating the land, and moved into the Ohio and Mississippi Valleys, where steamboats, canals, and railroads accelerated the pace of development. Later, settlers in arid regions that required irrigation created another variant of the foundation narrative. Such stories, addressed primarily to the English-speaking population, provided both justifications for and after-the-fact vindications of their settlement. Foundation narratives explained and validated expansion as white Americans cut down the forests, drained the swamps, plowed the plains, and irrigated the deserts.

As every gardener knows and as every colonist soon learned, the tool precedes the garden. If some seventeenth-century locations looked like Eden, it was because Native Americans had burned off the underbrush or had shaped the landscape with hoes, digging sticks, hatchets, and other tools. Groups of colonists soon realized how necessary the axe, the plow, the sawmill, the grist mill, and other technologies were to reshaping the land to their own form of agriculture. They used these tools to transform the space of America—to drain lands if they were wet, to water them if they were dry, to log them if they were wooded, to plow and to plant new crops.[12]

The axe, the mill, the canal, the railroad, and the dam stand at the center of stories about how European-Americans naturalized their claim to various regions of the United States. This book is about how Americans narrated their place in the world and imagined their position in history in technological terms. It focuses primarily on narratives rather than on the technologies themselves. Americans constructed stories that emphasized how particular tools and machines enabled them to inhabit new places and to create new communities. Importantly, they expressed a belief that these technologies would enable them to preserve their egalitarian difference from the Old World. For example, in contrast to tales of regeneration through violence, second-creation stories presented the American axe not as a weapon but as an instrument of peace used by individual pioneers to carve a civilization of independent farmers out of the wilderness. Water-powered American factories were not expected to re-create a European proletariat; they were expected to preserve a rural nation while making it more productive and prosperous. The steam engine was not expected to undermine Republicanism; it was expected to enhance it. The canal and the railroad were not expected to monopolize transportation; they were expected to increase competition and hasten development. The irrigation systems of the far West were not expected to concentrate economic and political power; they were expected to extend family homesteading into the desert. These narratives sought to explain how Americans could use new machines to transform the land while establishing communities similar to those they had known before. In these progressive narratives, machines were dominant yet democratic, transformative yet conserving. The narratives naturalized the technological transformation of the United States so that it seemed an inevitable and harmonious process leading to a second creation that was implicit in the structure of the world.

The foundation stories that white Americans have told about the reconstruction and habitation of a new space are secular stories about what in many cultures are religious matters: how a group comes to dwell in a particular space, and how it wields power to transform the land and make a living from it. Had the United States been static during the nineteenth century, it would only have been necessary to define this narrative (chapter 1), specify its ideological underpinnings (chapter 2), and analyze the narratives of the axe (chapters 3 and 4) and the mill (chapters 5 and 6). But American foundation stories were rewritten periodically as social and economic relations changed. Two variants might have been sufficient for a small society commanding a few technologies that enabled it to persist in one location. But because white American society grew rapidly, because it moved into new forms of terrain, and because it mastered new forms of power, the early forms of the foundation narrative required periodic revision. Newer foundation stories focused on canals and railroads (chapters 7 and 8) and on irrigation (chapters 9 and 10). This does not mean that the first stories disappeared. Stories of isolated families on the frontier carving homesteads out of the wilderness with an axe, and of communities growing up beside a sawmill and a grist mill, have persisted to the present day, alongside the later narratives. Yet, as chapter 11 makes clear, the ideological assumptions that underlay second-creation stories had collapsed by about 1900. The concluding chapter considers the relationship among such stories, conservation's recovery narratives, and the idealization of wilderness.

1

Narrating the Assimilation of Nature

Alexis de Tocqueville remarked that white Americans observed the rapid disappearance of the Native American with equanimity and naturalized it as part of an inevitable process. The "Anglo-American race," he wrote, "fells the forests and drains the marshes; lakes as large as seas and huge rivers resist its triumphant march in vain. The wilds become villages, and the villages towns. The American, the daily witness of such wonders, does not see anything astonishing in all this. This incredible destruction, this even more surprising growth, seems to him the usual progress of things in this world. He gets accustomed to it as to the unalterable order of nature."[1] Yet how could one reconcile the contradiction between Native American decline and white American growth? How could radical change be a part of an "unalterable order"?

Most nineteenth-century Americans believed in a deceptively simple story in which the natural world was incomplete and awaited fulfillment through human intervention. Being incomplete, the land needed technological improvements that would express the pattern latent in it. The transformations Americans envisioned were thought of less as violations of nature than as useful improvements. John Greenleaf Whittier contrasted the American and British views of such matters:

When the rail-cars came thundering through his lake country, Wordsworth attempted to exorcise them by a sonnet; and were I not a very decided Yankee, I might possibly follow his example, and utter in this connection my protest against the desecration of Pawtucket Falls, and battle with objurgatory stanzas these dams and mills. . . . Rocks and trees, rapids, cascades, and other water-works are doubtless all very well; but on the whole, considering our seven months of frost, are not cotton shirts and woolen coats still better? As for the spirits of the river, the

Merrimac Naiads, or whatever may be their name in Indian vocabulary, they have no good reason for complaint; inasmuch as Nature, in marking and scooping out the channel of their stream, seems to have had an eye to the useful rather than the picturesque. After a few preliminary antics and youthful vagaries up among the White Hills, the Merrimac comes down to the seaboard, a clear, cheerful, hard-working Yankee river. Its numerous falls and rapids are such as seem to invite the engineer's level rather than the pencil of the tourist; and the mason who piles up the huge brick fabrics [of mills] at their feet is seldom, I suspect, troubled with sentimental remorse or poetical misgivings.[2]

Rather than protest against the creation of an industrial landscape, Whittier appealed to the practicality of producing clothing in factories, and he read the landscape for nature's intentions. The river had been created not to be picturesque but to be useful. There was no need for sentimental misgivings about building dams; the sites themselves "seem to invite the engineer's level." For Whittier, and for most Americans of the middle of the nineteenth century, the river was waiting to be dammed; similarly, the prairie was waiting to be farmed, the woodlands to be cut down, and the desert to be irrigated. In this view, Americans used new technologies not to overrun nature but to complete the design latent within it. The second creation, though man-made, was in harmony with the first.

Nineteenth-century Americans repeatedly told themselves stories about the mastery and control of nature through technology in which radical transformations of the landscape were normal developments. Ralph Waldo Emerson, in his essay "Wealth," argued: "Men of sense esteem wealth to be the assimilation of nature to themselves, the converting of the sap and juices of the planet to the incarnation and nutriment of their design. Power is what they want, not candy—power to execute their design, power to give legs and feet, form and actuality to their thought; which, to a clear-sighted man, appears the end for which the universe exists, and all its resources might be well applied."[3] In this vision, which was by no means limited to Emerson, the land exists for a purpose, and humans are expected to facilitate "the end for which the universe exists."

Emerson's self-reliance was teleological, aiding an inevitable organic process. In the larger scheme of things, it did not matter that Emerson's "clear-sighted man" was by Emerson's own admission a "monomaniac" whose thoughts were riveted to one project. Emerson had seen "the men of the mine, telegraph, mill, map and survey—the monomaniacs who talk up their project in marts and offices and entreat men to subscribe—how did our factories get built? How did North America get netted with iron rails, except by the importunity of these orators who dragged all the prudent men in?"[4] In his formulation, unimpeded exercise of the imagination might encourage single-minded obsessions, but these were necessary to prod the citizenry into action. 'Monomania' is an unsettling word, but it seems to be balanced by 'prudence'. A century and a half later these remarks suggest less sanguine observations, however; "monomaniacs" with an aggrandizing view of nature are still lobbying in state legislatures and in Congress.

The persistent desire to assimilate nature to a second technological creation was the central feature of technological foundation stories.[5] In each case, popular narratives explained how Americans were using new tools and machines to assimilate nature. These stories described the creation of new social worlds, ranging from frontier settlements to communities based on irrigation. In each case, a new form of society based on successful exploitation of a new technology became possible. The stories were central to the new nation's perception of history and geography, which is to say its perception of time and space.

Technological foundation stories also provided a framework for the individual's "pursuit of happiness." They were narratives of abundance that emerged during the period of Enlightenment and were actualized during industrialization. To put it another way: These stories emerged when new machines, notably railways and textile mills, exceeded the power of humans, draft animals, or simple grist mills. A surplus of mechanical force was taken to be axiomatic, making possible new landscapes, boomtowns, sudden profits, personal success, and national progress. For most Americans, the foundational belief in naturally abundant power described (and was inseparable from) a laissez-faire ideology in which the self-reliant individual had only to exert himself in order to rise in the world. This story flourished in the nineteenth century in many forms. It continues to resonate powerfully with the American public despite the increasing awareness of environmental limits to growth.[6]

A technological narrative is selective. It singles out particular objects while deemphasizing or even deleting others. Outside the confines of training manuals and how-to books, the stories told about tools and machines are not cluttered by technical details. Foundation narratives appear not only in the work of novelists and poets but also in travelers' accounts, in newspaper stories, in editorials, in political speeches, in diaries, in world's fair exhibits, in letters, in legal cases, in history books, in media events, in pageants, in paintings, in advertisements, and in popular songs. These narratives within politics, commerce, and popular culture form a common storehouse that poetry and fiction explore, amplify, play with, and sometimes subvert.

Second-creation stories can be found in the loose fabric of a newspaper or in the midst of a speech, a travel book, or a diary. These often short narratives articulate the common understanding of technologies in the creation of society and serve as a discursive bridge between social history and fiction. They refer to verifiable things and events, but they have begun the process of translating them into literature. For example, a wide range of earlier stories about steamboating on the Ohio and Mississippi Rivers drawn from newspapers, travelers' accounts, and local histories place Mark Twain's writings within the larger discourse about the settlement of the Mississippi Valley.[7] In this context, the publication of *Life on the Mississippi* in the 1870s marks the final stage in the development of a set of second-creation narratives focused on steamboats.[8] During the nineteenth century, similar narratives developed based on the axe, the mill, the canal, the railroad, and irrigation. In each case, the dominant narrative was inseparable from building a house or founding a community in the wilderness.

Technological foundation narratives appealed to most but not all Americans. Their popularity arose from their *apparent* ability to explain historical events and fuse them with cultural values. They structured a selected set of events in a way that appealed to many white middle-class Americans as a reasonable account of their history. During the nineteenth century, however, a complex narrative system emerged that extended, defended, amended, or contradicted the foundational, progressive formulation.[9] Many of these alternative narratives, or "counter-narratives," were written by or addressed to groups that had been silenced in or absent from the original formulation.

Although no fixed structure lies beneath the many technological creation stories, there are recurrent features. My purpose is not to establish a "deep" structure underlying hundreds of individual examples, nor is it to suggest an idealized form that such stories ought to "live up to." Rather, it is to suggest what a technological foundation story may contain and how it unfolds:

- A group (or an individual) enters an undeveloped region.
- They have one or more new technologies.
- Using the new technologies, they transform a part of the region.
- The new settlement prospers, and more settlers arrive.
- Land values increase, and some settlers become wealthy.
- The original landscape disappears and is replaced by a second creation largely shaped by the new technology.
- The process begins again as some members of the community depart for another undeveloped region.

The order and the meaning of these elements vary with the author and with the audience. In subsequent chapters I will explore this narrative variety; I will not focus on how completely any individual instance realizes the pattern that is abstracted from the many individual cases and presented as an introduction, not a conclusion. Not every story contains all the elements, their order may vary, and the meaning given to the whole changes over time. However, a few generalizations can be made:

- A foundation narrative is usually not about an individual hero.
- Often it is told in the passive voice and emphasizes the technology.
- In such cases, it is the technology—the axe, the mill, the canal, the railroad, or the irrigation ditch—that "causes" the chain of events.
- Though an individual or a corporation is acknowledged to have initiated the process or to have profited from it, the story is presented as a typical case of what "inevitably" will take place.[10]
- The narrative is less a story about a hero than an example of a developmental process. It is an exemplary tale of progress in which human will is conflated with natural forces (as in Emerson's essay, where individual entrepreneurship merges with men's "assimilation of nature to themselves").
- These stories are about "the end for which the universe exists, and all its resources might be well applied."[11]

The foundation narratives were progressive and optimistic. Not merely descriptions written after the fact, they were stories that encouraged settlers to give up a familiar life, move westward, and put new lands into production. They gripped the imagination and convinced people to leap into the unknown.

A foundation story had to appear to be both a sober matter of fact and a promise of betterment. It had to seem a verified part of the past as a guarantee of its probable reenactment in the immediate future. Because a story had to seem repeatable, its action emphasized not the struggles and triumphs of an extraordinary individual but the movement of a people as a whole. The narrative of second creation was about the unfolding of "destined" processes. Therefore, although the foundation narrative was a story of national transformation, the state's role was reduced to a guiding influence. If the state were to play a decisive role, then the repeatability of the narrative would come into question, and it would become a tale about politics rather than a matter of manifest destiny. The narrative was, therefore, implicitly based on laissez-faire economics and a whiggish sense of history—assumptions that a majority of nineteenth-century Americans shared.

Whiggish history, manifest destiny, and technological conquest all are open to questioning today. As Michael Adas notes, "evidence of scientific and technological superiority has often been put to questionable use" and has "legitimized efforts to demonstrate the innate superiority of the white 'race' over the black, red, brown, and yellow."[12] Chapters 3, 5, 7, and 9 explore how mastery of tools and machines became fundamental to the dominant technological creation stories. Chapters 4, 6, 8, and 10 examine counter-narratives based on the same events but seen from the viewpoint of those whom new technologies disturbed or displaced, such as Native Americans and early environmentalists. Counter-narratives resist or reimagine technological change and seek to ground identity not in machines but in other cultural artifacts or values.

Hayden White has argued that differences between historical interpretations arise from contrasting techniques of encoding facts within a larger design. A story achieves its meaning through "the destructuration of a set of events (real or imagined) originally encoded in one tropological mode and the progressive restructuration of the set in another tropological mode." When a narrative such as the technological creation story has "become encoded by convention, authority, or custom," it is attacked through "a process of decodation and recodation." The explanatory force of this attack depends on "the contrast between the original encodation and the later one."[13] Native Americans, farmers, fishermen, striking workers, and environmentalists constructed counter-narratives cast in a different figurative mode that emphasized conflict rather than the harmonious unfolding of events. Whereas second-creation stories treat the land as empty space, ignoring the original inhabitants, the counter-narratives are told from the viewpoint of the indigenous community and/or emphasize the ecological effects of technological change. The ways in which the foundation narrative can be recoded are many and can be extremely complex. Though it has no ideal form, the following inversion of the earlier example suggests how a dominant story might be challenged:

- Outsiders enter an existing biotic and/or human community.
- They acquire its land and assets by force or legal trickery.
- They possess powerful new technologies.
- They begin to use these technologies to transform the landscape, undermining the existing community's way of life.
- The existing community and the new one come into conflict.
- The new community wins.
- Additional settlers arrive and complete the transformation of the landscape.
- The original community loses population and goes into decline.
- Its people become marginal and disappear or move away.

Unlike the foundation story, which traces an inevitable working out of "manifest destiny" and the free market for middle-class white Americans, the counter-narrative is often a tragic tale of struggle and defeat that begins with treaty violations or other illegalities. Some farmers opposed the construction of mill dams because they flooded their hay fields, and some fishermen opposed the construction of dams because the prevented fish from migrating upstream. Native Americans along the Columbia River told of how dams prevented salmon from spawning and so undercut the material basis for their way of life.[14]

A classic example of a counter-narrative is the autobiographical book *Black Elk Speaks*.[15] It begins with a depiction of the traditional life of the Oglala Sioux, including a powerful evocation of Black Elk's visionary experiences and his initiation into manhood. It describes the whites' invasion of the Sioux lands, solemn treaties and their violation, the coming of the railroad, the willful destruction of the buffalo, the loss of territory and population, the Phyrric victory against Custer, the continual arrival of more white settlers, the slaughter at Wounded Knee, and the precipitous decline of the Sioux. Black Elk tells this story through a white interpreter, John Neihardt, who transforms it from an oral tale into a written text. Neihardt presents it not as the inversion of a technological foundation story but as a surviving record of "the world we have lost." The railroad, the telegraph, and other technologies are powerful, but the story marginalizes them while validating a Native American cosmology expressed through Black Elk's personal visions, healing ceremonies, and out-of-body experiences. Counter-narratives reconstruct familiar events, sometimes by emphasizing a different ideological orientation or a fundamentally different epistemology.

Two variants of the foundation narrative and the counter-narrative are the utopian story (which can emerge only at the start of the narrative cycle) and the nostalgic tale (which can appear only at its end).[16] Although neither will receive much attention in this study, a brief characterization follows.

The utopian narrative anticipates a sudden technological breakthrough that allows people to create a society of ease and abundance. In the later nineteenth century, this story was often presented at world's fairs, notably the one held in Chicago in 1893. Early in the century, imagining the future was largely confined to written texts, such as J. A. Etzler's *The Paradise within the Reach of All Men, without Labour, by Powers of Nature and Machinery* (1833).[17] Etzler argued that humans could harness the wind (with immense sails driving mills) and the sun (with mirrors boiling water to create steam) to produce perpetual power for every imaginable purpose.[18] He imagined huge earth-moving machines capable of digging canals or ripping out trees and flattening the earth into perfect fields. He proposed making an infinite supply of building materials much as bricks were produced, and he believed that a substitute for cloth could be manufactured with equipment modeled on paper-making machinery.[19] Etzler's story anticipated many of the extravagant claims later made for atomic power, plastics, alternative energies, and the Internet. Such visionary stories usually seem probable only to a minority who believe that machines can radically reconfigure society. Whittier met this utopian speculator and reported: "He was possessed with the belief that the world was to be restored to its Paradisiacal state by the sole agency of mechanics; and that he had himself discovered the means of bringing about this very desirable consummation. His whole mental atmosphere was thronged with spectral enginery—wheel within wheel—plans of hugest mechanism—Brobding-nagian steam engines—Niagaras of water power—wind mills with 'sail-broad arms,' like those of Satan in chaos—by whose application every valley was to be exalted, and every hill made low—old forests seized by their shaggy tops and uprooted—old morasses drained—the tropics made cool—the eternal ices melted around the poles—the ocean itself covered with artificial islands—blossoming gardens of the Blessed, rocking gently on the bosom of the deep."[20] That Whittier had an ironic view of Etzler's visions (as did most of his contemporaries) is evident from his references to *Gulliver's Travels* and to the coming of the Messiah, who also would flatten the mountains, exalt every valley, and usher in a golden age.

Etzler's is an extreme case of a utopian narrative that describes future perfection based on the control of new technologies. Such true utopias shade off into the speculative designs of real estate agents, town planners, stockbrokers, and promoters of new inventions, all of whom traffic in visions of transformation. The foundation story proper is distinct from the utopian narrative in that it purports to describe changes that have already taken place rather than to project possibilities, and also in that these changes are gradual even if they break decisively with the past. The foundation narrative describes an evolutionary process in which new technologies increase the wealth and abundance available to the average person. It describes what seem to be actual recent events, and it is widely accepted by most members of the middle class as a factual representation.

A speech by Daniel Webster illustrates this evolutionary view embedded within the foundation narrative. One of the most famous orators of the nineteenth century, Webster was invited to speak to the Society for the Diffusion of Useful Knowledge in Boston in 1836, three years after Etzler's book appeared. Rather than paint fanciful visions of what might be done using the force of the sun and wind, Webster based his argument on the known productive power of the steam engine. He observed: "There has been in the course of half a century an unprecedented augmentation of general wealth. Even within a shorter period, and under the actual observation of most of us . . . vastly increased comforts have come to be enjoyed by the industrious classes." It seemed self-evident that "the present exceeds the past, in regard to the shelter, food, clothing, and fuel enjoyed by laboring families."[21] He then asked what were the causes of this progress, and concluded: "The successful application of science to art increases the productive power and agency of the human race. It multiplies laborers without multiplying consumers, and the world is precisely as much benefited as if Providence had provided for our use millions of men, like ourselves in external appearance, who would work and labor and toil, and who yet required for their own subsistence neither shelter, nor food, nor clothing."[22] For Webster, "this mighty agency, this automatic labor whose ability cannot be limited nor bounded" produced general prosperity. Etzler had imagined sudden transformation; Webster described a more gradual development. His narrative is the characteristic American story: change is rapid but piecemeal, ameliorative, and beneficial to all.

Like Webster, the influential newspaper editor Horace Greeley argued that mechanization was the root cause of progress and of the dispersal of wealth: "In our discoveries in science, by our applications of these discoveries to practical art, by the enormous increase of mechanical power consequent upon mechanical invention, industry and skill, we have made them a common possession of the people; and given to Society at large— to almost the meanest member of it—the enjoyments, the luxury, the elegance, which in former times were the exclusive privilege of kings and nobles."[23] Webster and Greeley articulated the master narrative of technological amelioration, in which the second creation emerges seamlessly out of the first.

In contrast, the narrative of technological nostalgia rewrites the second-creation story after its central tools or machines have become obsolete or outmoded. The nostalgic narrative is about an irrecoverable and static yesterday, not a dynamic present. For example, once Americans adopted automobiles and trucks, they rewrote the foundation story of railroading in nostalgic terms, just as they had reconceived the steamboat as an idyllic representation of the past after the railroad superseded it. A technology once celebrated as the source of American prosperity is now described as quaint. A mill that once called a town into being is now a restaurant and a reminder of the simpler life of the past. A canal, once an artery of commerce that caused a city to grow, is now a tourist attraction. A steam railroad is maintained by buffs and made available to the public on holidays. When new, each technology represented a sudden increase in power. In retrospect, each is diminished. Indeed, they eventually seem to be almost "natural." The axe, the saddle, the rifle, and metal tools were products of centuries of development, but in nostalgic stories they are often decontextualized and thereby naturalized. Thus, nostalgic stories do not reply to counter-narratives; they simply restate the major elements of a second-creation story, emphasizing the automatic unfolding of inevitable events.

The foundation narrative appears to be a transparent description of events, but it is not. Technologies are elements of the dialogue about how the world is structured. This dialogue takes the form of stories people tell one another to make sense of the transformations that accompany the adoption of a new tool or machine. They may be foundation narratives that seem to explain the origins of the present; they may be counter-narratives that dispute that story. They may project utopian visions of ease and abundance; they may focus on a way of life that is fading into the past. (In other stories, especially during the twentieth century, apocalyptic machines run amok; that subject requires another book.[24]) People seldom understand machines as purely abstract things in themselves. For Americans settling a new continent, technologies became central to stories explaining how they had developed their New World. By the 1830s, when Tocqueville took his tour, mechanical progress had, paradoxically, become an unalterable part of nature.

Let us now return to Whittier's refusal to protest the industrial use of Pawtucket Falls. As a sophisticated writer aware of the several narratives that might be employed to explain how Lowell's factories had come into existence, Whittier does more than simply present a series of facts. He dramatizes his version of the facts by contrasting it with Wordsworth's response to industrialization. Whittier recognizes that one might see the construction of dams as an aesthetic loss, and even evokes the possibility that white men have desecrated the lands of Native Americans, but he does so to highlight the practical advantages of the mills, to invite us to see them as masons and engineers do rather than as tourists do, and to accept nature's manifest intention that the Merrimack become a "cheerful, hard-working Yankee river." Whittier's narrative is defined by the evocation of counter-narratives, and it gains meaning through what it opposes. Thus, the narrative of second creation did not emerge separately from the counter-narrative. Rather, competing interpretations emerged in an interdependent process. If second-creation stories organized events into a description of the inevitable and benign assimilation of nature, counter-narratives marshaled the same facts into accounts of destruction and loss. During his journey through the United States, Tocqueville heard examples of both.

2

Surveying the Ground

Stories for granted take certain things about the structure of their world. Usually the envelope of assumptions that supports a foundation narrative is unvoiced and scarcely visible. As Pierre Macherey put it in *A Theory of Literary Production,* "the [literary] work is articulated in relation to the reality from the ground of which it emerges: not a 'natural' empirical reality, but that intricate reality in which men—both writers and readers— live, that reality which is their ideology. The work is made on the ground of this ideology, that tacit and original language: not to speak, reveal, translate or make explicit this language, but to make possible that absence of words without which there would be nothing to say. We should question the work as to what it does not and cannot say, in those silences for which it has been made. . . . The order which it professes is merely an imagined order, projected on to disorder, the fictive resolution of ideological conflicts."[1]

At the time of the American Revolution, a shifting ideological ground was transforming the possibilities of narrative as a new structure for society was being embodied in the laws and social practices of the new nation. Though the transformations wrought by the Revolution were many, the technological creation story rested particularly on four shifts in perception. Most important, traditional ways of dividing up land were abandoned in favor of an abstract grid. As Americans redefined space, they gradually shifted from regulated prices to a free-market ideology, from a psychology of scarcity to a belief in natural abundance, and from belief in a world of mysterious, spiritual forces to a Newtonian universe of clear, quantifiable causes and effects. Together these four changes constituted a fundamental shift in consciousness. From the pre-Revolution perspective, the world had few regularities and many peculiarities. Supplies of goods were uncertain. Scarcities were recurrent. Space was idiosyncratic, with no piece of land identical to any other. The state was necessary to keep order. From the post-Revolution perspective, spaces, prices, markets, and forces all obeyed natural laws and largely regulated themselves. The state's proper role was reduced to maintaining the "normal" conditions of resource abundance, price competition in a free market, and accessible land divided impartially on the basis of the grid. This change in perceptions was by no means instantaneous; it occurred gradually, late in the eighteenth century and early in the nineteenth, just as the first technological creation stories were emerging. Those stories took the post-Revolution perspective for granted, and they were based on this ideology.

The anthropologist Edward Hall once observed a striking difference between France and the United States that was visible in the two countries' road systems.[2] French roads radiate out from towns, forming clear centers and peripheries. The cathedral stands in the middle of town, often as the literal focal point of most highways, and its spire is the first object visible from a distance. A "star"-like pattern of lines flows out from the heart of older European cities and towns. The origin of the pattern is ancient, and the location of each road is attributable to a combination of topography, history, and convenience. In contrast, most of the United States west of the Alleghenies is laid out in a vast grid that imposes a design on the contours of the land without regard for topography and without reference to any history or custom of land use.

This new sense of space was largely invented at the time of the American Revolution. There were antecedents to such thinking in the layout of at least one ancient Greek colony, in the Roman army camps that evolved into towns and cities, and in the planning of some French towns during the late Middle Ages, but Americans imposed a far more comprehensive design on the entire landscape. In the 1780s and later, Americans formally embraced a new sense of space that found expression in a vast rectilinear grid projected thousands of miles in all directions. The idea of surveying federal land into perfect squares had been tried on a smaller scale in several states; New York, for example, had successfully introduced it in its western areas. But the national grid was a fundamental change that literally put a new frame around stories of migration and settlement.

The imposition of a strict geometrical pattern on much of North America was central to the imagined order that made stories of technological creation possible. In the colonial period, the community and not the individual had been central. The theocratic order of the first settlers was visible in the layout of the land: Americans reproduced the European village, with the church at the center and the roads radiating outward. The local governments of the first settlers did not conceive of land as generic; they evaluated it and divided it into woodlots, pasture, and farmland, distributing some of each to every household. A family's land was not all in one location but dispersed. The shape of each lot was by no means regular and was seldom a perfect square. The new grid system erased hierarchy and centrality from the landscape, substituting the values of individuality and equality.

In view of this contrast between pre-Revolution and post-Revolution America, it seems particularly significant that George Washington, before he commanded the American army, had been a surveyor, and that Thomas Jefferson introduced and promoted the new system of land division. Americans in general felt that imposing the new sense of space was among the most urgent matters of public business, and that it should be taken up as soon as peace was concluded with Britain. The national survey was adopted quickly and was one of the most important laws devised during the short period when the Articles of Confederation provided the framework of government. It is tempting to claim that the creation of the grid expressed an immediate and fundamental shift in consciousness, but a look at the discussions that preceded the adoption of the Ordinance of 1784 and the later revisions of that ordnance reveals that the legislation was a practical program as well as a reconception of space. It may have expressed the Enlightenment values of rationality, equality, and order, but the survey was actually intended to facilitate the selling of western land to pay off the national debt. Not only was the far side of the Allegheny Mountains largely uncharted, but the nation lacked a corps of skilled surveyors large enough to provide maps in a reasonable time. Thus, "policy makers faced an immediate, practical problem in linking supply to demand," and "they had to create a market for a commodity, unimproved frontier land."[3] They needed a system to survey it and guarantee title, and the grid hastened the process. It was not an ideal solution for most localities, since the grid ran roughshod over topography.[4] But it could be put into effect quickly, because it used longitude and latitude as the basis for all land divisions. This made it possible to survey any location, no matter how remote, and to demarcate its boundaries in accordance with an external, verifiable standard.

Some geographers have defended the system. "It may be the best system of land division ever invented," John Fraser Hart recently argued. "It provides an excellent frame of reference for orientation, and it conveys a sense of neatness, order, and stability." Perhaps even more important, it "has obviated an enormous amount of litigation by facilitating a brief but precise description of the exact location of any tract of land."[5] But if the National Survey made land transactions easily comprehensible to all, it also encouraged farmers to ignore the contours of the land. The grid dictated more than the layout of farms. It became the basis for a system of roads, counties, town plans, and eventually for telephone and power lines, all of which followed north-south or east-west boundary lines and defied the actual landscape of hills and valleys. The roads fatigued horses, which had to work much harder than they would have in a well-planned road system based on topography. American roads also frustrated the traveler seeking to move diagonally rather than along the four cardinal points of the compass, and occasionally they resulted in steep city streets.

Nevertheless, during the nineteenth century Americans accepted and naturalized the geometrical ordering of the land as the grid expanded from the Appalachians all the way to California. Accepting the grid affected not only Americans' sense of space but also their sense of the past. The grid reflected a radical break with the religious story of New World settlement, in which colonists were represented as carrying European spatial patterns to their new home. Only when people self-consciously saw themselves not as colonists but as Americans were they ready to reinvent their sense of space. This reinvention found expression in the national survey.

The grid was not a scientific solution, however. Mathematically, it made no sense. As some legislators realized at the time, carving land into squares based on the lines of latitude and longitude was theoretically impossible. Such a plan assumed that the earth was flat, whereas "squares" surveyed on a globe are really trapezoids, slightly narrower at the top than at the bottom. Timothy Pickering explained to anyone who would listen that meridians converged at the poles, and that surveyors would not be able to lay out thousands of contiguous "squares." Instead, every lot would have to be shorter on its northern boundary, or the whole system would soon be out of alignment.[6] Pickering was right. As soon as surveyors began carving up the land in accord with Congress's wishes, they found that they had to introduce approximations, including "correction lines" at regular intervals. Just as they had learned to deal with iron ore deposits that deflected their compass needles and with metal measuring chains that shrank in winter and expanded in summer,[7] surveyors now learned to adapt a system that assumed a vast, flat earth to the reality of a finite, curved earth.

Surveying was practical work. It was the essential precondition to owning land, building a mill race, or constructing a canal, a railroad, or an irrigation ditch. Yet surveying the land into squares was also, in Macherey's terms, a part of the underlying and unarticulated ideology that was a necessary precondition for a technological narrative. Surveying was a kind of writing on the land, turning it into a free-market landscape. Before the surveyor measured land according to a repeatable and verifiable system, legal ownership was impossible. By dividing their land into units, Americans articulated an egalitarian sense of space that had no center and no past.

This world was not the tightly bounded and circumscribed space of Europe. Rather, most white Americans defined their West as essentially empty space, a "virgin land" waiting to be appropriated.[8] In 1804 David Humphreys declared: "An almost unlimited space of excellent territory remains to be settled. Freehold estates many of excellent territory may be purchased upon moderate terms."[9] This expansive sense of space is still visible from the air in the layout of fields and roads in the Midwest. Not only did the system call for the division of lands into a checkerboard of 640-acre squares, or square-mile sections; it also assembled perfectly square townships, each consisting of 36 sections, into quadrangular counties, and ultimately it projected perfectly square states.[10]

The grid expressed a new set of philosophical ideas that Americans literally inscribed on the land.[11] But at the same time it ignored and dispossessed those who already lived on those squares. The grid declared that the land was unused, empty, and waiting for settlers. Adopting the grid made

it easier to believe in Manifest Destiny; filling in its spaces became a naturalized historical process. It asserted human dominion. As Denis Cosgrove
notes, "confidence that nature had been nailed down by geometry was
shared by both railroad boosters in New York and Chicago and isolated
homesteaders on their quarter or half-quarter sections."[12] This conception of space assumed that in every unit a single social reality could be
replicated, creating a society at once homogeneous and made up of self-
reliant parts.[13] Such a social space had no limitations; it was conceptually
unbounded, and it was perpetually open to new people and new technological systems.[14] Owing to the grid's uniformity, Philip Fisher notes, this
"social space is transparent and intelligible," and it seems that the "lives of
others become spontaneously intelligible" as if there were no coding to
experience. Americans soon came to see their singular geographical system as not at all arbitrary. This leads Fisher to an extremely important
observation: "Such a Cartesian space provides for no observers, for no
oppositional positions. There are no outsiders." To be an observer is "a
symptom of a divided social space."[15] The grid was a totalizing system
that made it difficult to write a counter-narrative and even more difficult
to win an audience for one.

The new abstract sense of space surprised European visitors. In 1836 the
French traveler Michael Chevalier commented: "No one can look at a map
of the United States without being struck by the appearance of straight
lines constituting the frontiers of most of the States; this method of bound-
ing a territory by meridians and parallels of latitude is absurd since it
requires an infinite number of geodesic operations, which have not been
executed, and cannot be so for a long time. Meridians and parallels do
very well for the divisions of the heavens; but for the earth there are no suitable boundaries but the beds of rivers, or watersheds in the mountain
chains."[16] White Americans overlooked natural boundaries when visualiz-
ing and settling the land, imposing a new geography that made it into an
abstraction, a commodity, and an item of speculation. At the same time,
they paradoxically celebrated the moral values of the settled agrarian life.

The geometrical vision was further expressed in the new towns that
nineteenth-century entrepreneurs and railroad companies laid out.
Through swamps and forests and on open prairie, surveyors marked off
farms, streets, and lots, almost always in uniform rectilinear patterns. By

imposing lines on the landscape, farmers and town planners made geo-metrical space actual. As they did so, they transformed land into a com-modity that could be precisely delimited, purchased, and registered in a land office, even if no one had ever seen it. In 1827 the British traveler Basil Hall wrote: "A district of country, intended for the market, is first surveyed, and laid out into square portions, a mile each way. At the corner of each of these square miles a stake is driven in, with a proper number or letter carved upon it; and the trees between this post and the next, which is always fixed due north or south, or due east or west of it, are marked by means of what are called blazes [cut on the trees]. . . . The unoccupied country is covered with a net-work of divisions a mile square, each con-taining 640 acres."[17] Settlers could purchase these portions of land, or fractions of them, through land agents.

Similarly, whole towns were laid out by permission of state legislatures. Wyandotte, on the banks of the Kansas River, is an example. In the 1860s, when the town consisted largely of white tents, "shares of ten building lots were selling at eighteen hundred dollars." How had such a place come into existence? "In founding a city, a few speculators become [in]corporated, by special act of the legislature, as a town company. Then, if the land is already open for pre-emption, they survey and stake out three hundred and twenty acres—the quantity which Government allows set apart for a town site—at one dollar and a quarter per acre."[18] Few speculators were willing to work on such a small scale, though; typically, speculators paid others to pretend to be honest homesteaders, to lay claim to the adjoining lots, and to later sell them to the town incorporators. Thus, Wyandotte had to compete with two other nearby speculators' towns: Kansas City and Quindaro.

By the time settlers began to venture west of the Mississippi, the grid had become fully naturalized. Virtually every new town, every road, and every farm was laid out according to its rules. As J. B. Jackson notes, "whereas in the older states of Ohio and Indiana and Illinois (where the heritage of the colonial farm lingered) the straight lines of the grid were valued as an efficient and democratic way of organizing individual landholdings, west of the Missouri the grid played a much more decisive role: it was the *only* practical and speedy method of organizing space." In the western half of the country, the grid's "long-range effect was to eliminate, once and

Figure 2.1
"Bird's Eye View of Frederick, Dak., 1883." Courtesy Library of Congress.

for all, the impact of tradition and traditional spaces, in the forming of the new High Plains landscape. A composition of identical rectangular squares extending out of sight in every direction, ignoring all inherent differences, produced a landscape of empty, interchangeable divisions."[19] One of Willa Cather's narrators, when first confronted with the empty spaces of Nebraska, put it this way: "There was nothing but land: not a country at all, but the material out of which countries are made."[20]

The first white resident who took possession of a newly demarcated farm confronted not a geometrical space, however, but a changing, living biotic community, traversed by wind and water, that bore the marks of the Native Americans whose trails often served as the first roads. The settler confronted what John Rajchman has called the "other geometries" of living. Rajchman's examples are "the geometry of a young Japanese woman walking down a Parisian street or a Dutchman made to feel clumsy, elephantine, in a traditional Japanese house or inn."[21] For European settlers in North America, a similar unease was unavoidable when they visited Native American settlements, whose spatial geometries remained inscrutable. Even within a single culture, "each of us has such geometries, composed of lines of different kinds, coming to us in various ways, which make up the arrangements of dispositions of space—the assemblages—in which we move and relate to one another."[22] Since the Native American sense of space remained indecipherable to most colonists, the usual response was to erase the physical traces of that sensibility and impose geometrical order. Accordingly, most foundation narratives do not mention any previous inhabitants.

Earlier I noted the four fundamental assumptions that framed the second-creation story. The grid was the most important of these. The others were a new belief in the natural abundance of resources (including land), the free market, and the conservation of force. Foundation narratives projected an endless supply of land and raw materials. On the assumption that this abundance existed in a free market, its availability to citizens was taken to be almost an individual right, as if it were a part of the pursuit of happiness guaranteed by the Declaration of Independence. This assumption that resources were abundant was well suited to a laissez-faire ideology in which the self-reliant individual had only to make use of personal powers to rise in the world.

The assumption of abundance contrasted sharply with the theories of inevitable scarcity then widely accepted in Britain. Malthus had argued in his *Essay on the Principles of Population* (1798) that societies with growing populations could not avoid mass poverty and periodic starvation, insofar as human fertility always outpaced agricultural productivity. Likewise, Ricardo's economics sought to prove that wages tended to stabilize close to the subsistence level. Taken together, their work ruled out abundance as a demographic or economic possibility.

Americans, on the whole, refused to accept such conclusions. Emerson retorted: "There has been a nightmare bred in England of indigestion and spleen among landlords and loom-lords, namely, the dogma that men breed too fast for the powers of the soil; that men multiply in a geometrical ratio, whilst corn multiplies only in an arithmetical; and hence that, the more prosperous we are, the faster we approach these frightful limits."[23] Emerson agreed with Henry Carey and his school of American economists, who rejected Malthus. He refused to believe that "the plight of every generation is worse than of the foregoing."[24] Instead, in a classic presentation of the technological creation story, Emerson argued that at each successive stage of development mankind is more adept at farming: "The first planter, the savage, without helpers, without tools, looking chiefly to safety from his enemy—man or beast—takes poor land. The better lands are loaded with timber, which he cannot clear; they need drainage, which he cannot attempt. He cannot plough, or fell trees, or drain the rich swamp. He is a poor creature; he scratches with a sharp stick."[25] Malthus and Ricardo assumed that the best lands were taken first, but Emerson adopted Carey's argument and declared that, as man's technological skill improved, he was able to master more valuable terrain, such as heavily forested lowlands: "The last lands are the best lands. It needs science and great numbers to cultivate the best lands and in the best manner. Thus true political economy is not mean, but liberal."[26] Emerson pointed to the example of his own town, Concord, where farmers had recently learned how to reclaim marshy land by placing tiles under the earth to improve drainage. The work was time consuming and costly; however, it expanded the local acreage available for cultivation, and the newly recovered land was more productive than the old. Malthus and Ricardo were proved wrong by "more skill, and tools and roads." At the same time, Emerson's presentation suggested that Native Americans were savages with sharp sticks, that their population never grew, and that they remained "poor creatures." Thus, Emerson, like Webster, Carey, and many others of his generation, embraced a vision of a second creation based on white American technological mastery.

The technological foundation story led the American eye to see not the landscape that was there but what it might become. Zachariah Allen described a conversation concerning the future of Pottsdam, Pennsylvania, during the coal boom of the early 1830s, in the course of which a passenger in a stagecoach proudly pointed out "his little parcel of ground," which he had bought for $450. He emphasized that it was "the convenient size of 50 feet by 180 feet, suitable for both a store and a dwelling house; fronting on Jackson Street on one side, and on a noble avenue extending down to the Sharp Mountain on the other side." "For these noble streets and avenues," Allen wrote, "I looked in vain. They are, without doubt, all nicely draughted and colored on the plat, and are thus rendered palpable to the imagination of the purchaser; but to the optics of an ordinary observer, there was apparent here only one muddy road, winding amid stumps, and not a house within a considerable distance of this choice lot; and the noble avenue of the Sharp Mountain appeared to be a glen tangled with bushes—a fit retreat for the fox or the rabbit. Several of the passengers in the coach were discussing projects of the layout of new towns as familiarly as if the process were attended with no more difficulty than the planting of a potato patch."[27] In the 1830s the mania for town planning and investing in unsettled lands expressed a belief in the transformative effects of technology. Surveying a site into lots seemed to be the first stage in an inevitable story of growth. In this case, Pennsylvania's coal mines, linked to New York and Philadelphia by canals, were expected to be the engine of development. Allen's stage passengers were making the relatively new assumption that they were investing in a free market.

Today we take for granted things that were unheard of two centuries ago. We take for granted that energy can be transported anywhere by ship, pipe, or wire, and that it is inexpensive. We believe that markets are free because tolls and shipping costs have been greatly reduced, forgetting that for millennia markets were almost entirely local and that cheap and fast transportation are essential to the arrival of bananas from Central

America, cars from Japan, or oil from the Middle East. We enter a super-market and purchase wine from Chile, cheese from Holland, crackers from England, and grapes from California. This is a profoundly unnatural state of affairs. In contrast, a new homesteader in Illinois in the 1820s "made a trip some eighty miles down the river and succeeded in getting a few sacks of flour and some coffee and some sugar."[28] He had to canvass the neighborhood for miles around just to acquire enough food for his family. For him there was no free market in the present-day sense. One generation later, however, an Illinois farmer had no such difficulties. As Abraham Lincoln noted in a public lecture, the young American of 1859 had access to the world's markets: "Men and things, everywhere, are ministering to him. Look at his apparel, and you shall see cotton fabrics from Manchester and Lowell; flax-linen from Ireland; wool-cloth from Spain; silk from France; furs from the Arctic regions, with a buffalo-robe from the Rocky Mountains, as a general out-sider. At his table, besides plain bread and meat made at home, are sugar from Louisiana; coffee and fruits from the tropics; salt from Turk's Island; fish from Newfoundland; tea from China, and spices from the Indies. The whale of the Pacific furnishes his candlelight; he has a diamond ring from Brazil; a gold watch from California, and a Spanish cigar from Havana."[29] When Lincoln delivered his address in Jacksonville, in Decatur, and in Springfield, his listeners were aware that the ready availability of the world's goods was a fundamentally new situation, and that it was based on a new infrastructure of transportation and communication. Lincoln reminded them that the "iron horse" was "panting and impatient" to do their bidding, and that the telegraph was "ready . . . to take and bring . . . tidings in a trifle less than no time."[30]

In colonial America, transportation and communication were more difficult, and most people believed in a quite different idea when it came to buying and selling: the idea of the just price. This idea originated with Thomas Aquinas and was widely accepted in the medieval period. State regulation of prices for some commodities (e.g., flour) continued into the eighteenth century.[31] The just price was appropriate to a world of limited and uncertain production, a world where reliance on nothing more than supply and demand resulted in enormous price swings and inequities. In wartime, for example, when transportation became difficult or impossible, one or two merchants might monopolize most of the flour available to a community. On more than thirty occasions when this happened during the first three years of the American Revolution, "men and women gathered in crowds to confront hoarding merchants, intimidate 'unreasonable' shopkeepers, and seize scarce commodities."[32] They then sold the goods at what the crowd believed a reasonable or fair price, and gave the proceeds to the owner.[33] According to the theory of the just price, often enacted into law, merchants could not simply charge what the market would bear. The concept of the just price was based on the idea that the cost of something was based not on supply and demand but on the work it required, the materials used, and the cost of shipping and storage. In a world where muscle power and hand-operated machines were used to produce most goods, people developed a clear idea of what something had cost in time and effort.

The just price was a traditional form of the labor theory of value, and many colonial laws were based on it. During the shortages that occurred between 1776 and 1779, ordinary men and women advocated it through "food riots." Revolutionary leaders, however, tended to oppose price regulations and "publicly endorsed free trade in the patriot press beginning in 1777."[34] The merchants and businessmen active in the Revolution solidified their opposition to these crowds and their demands, and by 1780 "it had become clear that neither Congress nor the states would pursue price control policies."[35] Food riots largely disappeared after the Revolution, but the social attitudes behind them did not fade as quickly. For another generation, many persisted in viewing prices, wages, and employment as matters for public regulation.[36]

The social construction of the free market was not based on economic theory alone; it was inseparable from the availability of inexpensive transport. Only the low cost of shipping made it feasible for Lincoln's Illinois audience to buy Irish linen, Pacific Ocean whale oil, and Brazilian coffee. Today rapid and inexpensive transportation of such goods seems "natural," but in 1859 Lincoln realized that it relied on a technological infrastructure that, historically speaking, was new.[37]

This infrastructure was anticipated to an extent in *The Wealth of Nations,* but Adam Smith's classic was written primarily as a critique of the mercantile system. Only one chapter attacked the agrarian ideas of the Physiocrats.[38] Smith was acutely aware of the American crisis, which had been building since 1770, and his critique of mercantilism can be read as a gloss on British policy toward the American colonies. *The Wealth of Nations* is often mistakenly abbreviated to the notion that laissez-faire economics ensures the fairest and most efficient system. In the details of his argument, however, Smith was able to see situations in which government rather than private interests ought to provide a service. Although in many areas the enlightened self-interest of individuals engaged in private competition would provide the most efficient administration and would supply the least expensive goods and services, Smith added an important caveat to such generalizations when it came to maintaining transportation. He posed as examples a canal and a public road. If a canal is not maintained in good order, the water runs out of it, the locks do not work, and the business is lost. A road, however, can deteriorate quite badly and still be passable, even if the traveler must proceed slowly and with discomfort. A private toll road therefore has less incentive to keep up a high standard of maintenance. Smith concluded that canals function well as private monopolies, but that roads do not and must therefore be overseen by the state.[39]

Smith did not consider the role of transportation in the development of a place such as the emerging United States. He did acknowledge that the cost of transporting goods (for example, taking grain to a grist mill) was a legitimate and unavoidable part of the cost of any commodity. But what happened in the new United States was more complex. The introduction of transportation linking remote areas to the market suddenly increased property values. Americans learned that, when good transportation opened up a region, land prices might double in a few years, with further value accruing to property that was partially cleared of trees and broken by the plow. These increases could be far greater than the profits from farming itself. Transportation, though only one factor in determining the price of agricultural goods, became central in determining the value of land and in creating wealth.[40]

No parallel situation had presented itself to Smith in his appraisal of European history. He observed how, after the fall of the Roman Empire, cities lost their importance as social power became concentrated in the hands of feudal lords based in the countryside, and how cities gradually emancipated themselves from this situation. Smith's particular version of this story need not detain us here. What is important, indeed crucial, is that the new United States inverted these conditions. In America, cities were never in thrall to the countryside. Rather, the countryside itself seemed to be called into existence and settled through the enterprise of cities and their representatives. An entirely different form of development emerged. It was not anticipated by mercantilism's monopolies, tariffs, and duties; it was not based on French-inspired agrarianism, nor was it the full expression of Smith's free-market ideas. Rather, in practice, the citizens of the new United States invented a paradox: an entrepreneurial agrarianism and a free-market industrial capitalism defended by neo-mercantilist tariffs. Though in theory such a formation might seem riven with contradictions, it seemed to work in practice as the United States rapidly industrialized and expanded westward.

Smith had emphasized the importance of good roads and canals. Improvements in the arteries of transportation lowered the cost of shipping more than enough to cover the toll. Faster and cheaper delivery increased profit to the merchant and lowered prices for the consumer. But Smith was describing roads and canals that passed through France and Britain, which had been inhabited and farmed for millennia. The United States was moving into vast new regions. The National Road, which by 1818 ran from western Maryland through the Appalachians to Wheeling on the Ohio River, opened lands to participate in the market for the first time, as would steamboats, canals, and railroads. The unusual elements in the American equation were an abundance of land that the government could sell or give away at low prices and a suddenly acquired ability to move easily into these regions as a result of innovations in transportation. Technological mastery of movement accelerated the exploitation of land. Jefferson had expected the settlement of the continent to take centuries. The expansion of population and the development of resources might have paralleled the slow movement of Spain (and later Mexico) into the American Southwest; it was the Anglo-American mastery of movement that hurried expansion.

Daniel Webster and many other politicians of his era were keenly aware that the mastery of space was directly related to technology, both in the obvious sense that steamboats and railroads moved more rapidly than horses and sailing ships and in the less obvious sense that increased productive capacity transformed land values. In an 1836 speech, Webster asked: "What would be the comparative value of the soil of our Southern and Southwestern States if the spinning of cotton by machinery, the power loom, and the cotton gin, were struck out of existence?"[41] Webster linked this insight directly to a critique of *The Wealth of Nations*. He agreed with Smith that "labor is the true source, the only source of wealth; and it necessarily follows, that any augmentation of labor, augments, to the same degree, the production of wealth." "But," Webster continued, "when Adam Smith and his immediate followers laid down this maxim, it is evident that they had in view, chiefly, either the manual labors of agriculturists and artisans, or the active occupations of other productive classes. It was the toil of the human arm that they principally regarded. It was labor, as distinct from capital." In contrast, with the advantage of 60 years' hindsight, Webster argued that "the true philosophy of the thing is, that any labor, any active agency, which can be brought to act usefully on the earth, or its materials, is the source of wealth. That is to say, this labor, like man's labor, extracts from the earth the means of living. . . . Now it has been the purpose . . . of scientific art, to increase this active agency, which, in a philosophical point of view is, I think, to be regarded as labor, by bringing the powers of the elements into active and more efficient operation, and creating millions of automatic laborers, all diligently employed for the benefit of man. The powers are principally steam and the weight of water."[42]

Mastering new technologies of transportation and production allowed Americans to erase differences between one space and another. All places increasingly became parts of one economic system, and the erasure of spatial difference was expressed in the grid's inscription on the land. Here Michel de Certeau's distinction between strategies and tactics is useful. Strategies belong to the powerful; tactics are "the art of the weak."[43] A strategy is a "calculation of power relationships that becomes possible as soon as a subject with will and power (a business, an army, a city, a scientific institution) can be isolated. It postulates a *place* that can be delimited as its *own* and serve as a base from which relations [can be conducted] with an *exteriority* composed of targets or threats." The goal is "to distinguish its own place . . . from an 'environment.'" Certeau's definition sounds almost like a general description of migration into the interior of the United States, particularly when he concludes that "it is an effort to delimit one's own place in a world bewitched by the invisible powers of the Other."[44]

Foundation narratives were written by the powerful as a part of the strategy of using new technologies to lay claim to the new space delimited by the grid. The foundation narrative, as expressed in Webster's 1836 speech or in Emerson's essay "Farming," asserts that greater access to technology will empower every individual and democratize wealth by increasing the geographical extent of the market and increasing the average family's income. In this story, more technical skill and more mechanical power mean a higher level of civilization. In contrast, the counternarratives that survive reflect the tactics of the weak. Such stories often are fragmentary, often are transmitted orally, and often exist only as transcribed hearings or trials. They point to the losses that result from the use of technologies to "improve" the land, and often they dispute the primacy of the grid and the market.

The expression of a technological "Manifest Destiny" required one final factor that was central to Webster's 1836 speech: an increasing mastery of force. To master land required new forms of power in order to take control of new territories, improve them, put them into cultivation, and link them to the market. When nineteenth-century Americans spoke of water wheels and steam engines, they thought in terms of an interlocking set of ideas underlying a narrative in which more power continually became available, enabling them to move into the empty grid of the West and link it to the national economy. The narrative took for granted not only a free market and abundant, inexpensive land, but also access to as much force as was desired. Creating a continental nation required the mastery of force.

The free market was inseparable from what we now call energy, which in the late eighteenth and the early nineteenth century was called force. The words 'energy' and 'force' did not mean what they do today. Samuel Johnson, in his *Dictionary of the English Language,* defined 'force' as "strength; vigour; might; active power." His first definition of 'energy' began with "power not exerted in action." His second, "force; vigour; efficacy; influence," sounds more modern at first, yet the literary passages Johnson cited to exemplify the term all concern the Deity: God's energy is eternal, and God fills the earth with ethereal energy. In the middle of the eighteenth century, energy was a spiritual attribute, whereas physical matter was "senseless and stupid" and made "no approach to vital energy." Johnson's final citation emphasizes this contrast: "How can concussion of atoms beget self-consciousness, and other powers and energies that we feel in our minds?"

Moreover, the conception of physical power scarcely included electricity, much less gasoline motors or atomic reactors. A standard definition of the "sources of power" offered in 1840 by a Harvard professor named Jacob Bigelow stated: "Men and animals, water, wind, steam, and gunpowder, are the principal agents, employed as first movers in the [mechanical] arts. Their power may be ultimately resolved into those of muscular energy, gravity, heat, and chemical affinity."[45] For Americans before the Civil War, the only practical sources of power were the muscles of men and animals, the falling weight of water in mills, the pressure of the wind on sails or windmills, steam engines, and explosives. To be sure, Bigelow's definition included other possibilities: "Certain other agents are also capable of producing motion, upon a more limited scale; such as magnetism, electricity, capillary attraction, etc."[46] This extended definition did not use the word 'energy'; it always used 'force', and the quintessential forms of force were falling water and expanding steam.

For most people, force was closely tied to observable phenomena, such as muscles or the movement of visible parts of machines. Inventors and scientists were interested in measuring force and in determining how efficiently falling water or expanding steam might be converted into useful movement, such as rotation. But in natural philosophy the idea of force still involved more than material causes and effects. For 2,000 years, from Aristotle until Kepler, the definition of force was quite different from the one it received in classical mechanics after Newton. Force was long inseparable from the study of the stars, the planets, and other astral bodies, all of which were thought to be kept in motion by divine intelligence.[47] Even Kepler found it difficult to give up the idea that "force presupposes for its activity the existence of an animated being."[48] Descartes attempted to eliminate the idea of force from his system, since it seemed to interfere with his absolute division between matter and spirit.[49]

Until after the European settlement of North America began, then, the concept of force was imprecise, bordering on metaphor—even in the work of Galileo.[50] Only in Newton's laws was force stripped of psychological or spiritual connotations and presented as purely physical.[51] Newton's formulation of the laws of motion and his mathematical explanation of the movements of the planets made it clear that the same principles applied to the heavens as to the earth and made force a more precise concept. In contrast, the word 'energy' still retained a spiritual connotation, and it was not commonly used to describe motion, gravitation, or mechanical work. Force was inseparable from the secular worldviews of Benjamin Franklin and Thomas Jefferson, and from the worldviews of most Americans for several generations after the Revolution.[52]

Workers, farmers, and engineers thought in terms of how much a horse could pull, how much a man could carry, or how much water a steam-driven pump could raise. Their preoccupation with force was registered in a general fascination with competition between its various forms: steam vs. water power, railways vs. steamboats, and so forth.[53] The average person did not worry about questions such as whether or not the action of gravity required the existence of an ether that permeated the universe, or whether the concept of force was an analogy or a precise concept. Most people, to the extent that they thought about such matters, focused on practical applications of force. To this majority (which included politicians like Daniel Webster, the entrepreneurs who built mills, investors in factories, builders of railroads, mechanics, engineers, architects, and workers), two things appeared incontrovertible: the force available for use increased and the efficiency of the devices that supplied it improved. By the 1840s this was palpable not only in the spread of steam engines but also in the diffusion of the water turbine, which developed roughly twice the power of a water wheel.[54] Whatever theoretical problems force might pose for philosophers or scientists, its empirical abundance was not in doubt, and the methods of its appropriation were clear. Furthermore, Americans believed that force was indestructible and that it was conserved during its conversion from one state to another. Just as Newton proclaimed that every action had an equal and opposite reaction, it seemed obvious that the sum total of force in the universe must remain always the same. Improving nature therefore implied no losses, only gains in utility and efficiency. (The concept of entropy was unknown until after the Civil War.) Second-creation stories were conceived not as tales of exploiting nature but as accounts of becoming more efficient or less wasteful in utilizing resources that ultimately could not be destroyed.

Each second-creation story posited mastery of force as the basis for a particular kind of community. As the control of force increased, the scale of the possible community increased proportionally. Pioneers wielding the improved American axe could create an agricultural community of scattered farms. The mill could become the nucleus of a small manufacturing town. The railroad founded a string of small cities. An irrigation project made an entire region habitable. The sequence of foundation narratives transferred every increase in technological power onto an enlarged urban scale, culminating in the imaginative projection of a world metropolis, a central point from which radiated lines of power and influence: rivers, canals, and railroads. This fantasy, though realized in part in Chicago and Denver, usually proved chimerical. Such narrative assertions culminated in the projections of what irrigation could do in the arid West, where mastery of technology was expected to transform a desert into a vast and fertile garden supporting a huge population.

Together these four concepts—geometrical space, natural abundance, the free market, and increasing access to force—defined an abstract narrative situation. Owing to the universality of the grid, each narrative became more than the description of a particular technology used at one location; it became a general account of the operation of necessity. Time and again, as authors described how axes, mills, canals, railways, and irrigation ditches could transform the world, the story was presented as a representative tale. The sequence of events seemed inevitable. What else but progress could result from the human control of increasingly abundant resources in a space that was inexpensive, mathematically organized, and fostered by a free market? The presumed law of the conservation of force seemed to be nature's guarantee that the second creation was both latent in the landscape and free of cost. Within this framework, each foundation story was less about idiosyncratic events than about inevitable results that flowed from the regularities of economic law. There seemed to be no possibility of failure, and ecological or human losses were largely excluded from the story by the very definition of the ground from which it arose. By 1830, white Americans defined abundant land and inexpensive energy in a free market as "normal" conditions that ought to exist and must be preserved. Paradoxically, an interdependent economic system created the "self-reliant individual" who had only to exploit new lands and forces in order to become "self-made."

Most people rejected counter-narratives that limited, challenged, or threatened the foundation story. They rejected them not merely because they seemed unappealing but also because they violated what seemed natural. A counter-narrative had to subvert at least one of the four under-lying concepts assumed in the foundation story. Were land, power, and resources really abundant, or were there natural limits? Was there really a free market, or did monopolies such as railroads distort or control it? Was the space of America a neutral geometrical expanse waiting to be possessed and infused with power, or did Native Americans, squatters, or other early settlers have valid claims? Counter-narratives tended to shift from abstraction toward the particular, the individual, and the local. They often recorded destructive effects, unforeseen consequences, or losses. The technological foundation narrative could easily be projected into the largely unknown space of the American West; the counter-narratives of contingency and loss required a more detailed sense of one's place on earth.

As Macherey suggests, the ideology of technological creation stories remained implicit and unspoken. They implicitly rejected European meth-ods of land division in favor of the grid, and they took the National Sur-vey for granted. They were a fundamental part of American nationalism, yet they said nothing about the state. Technological foundation stories did not overtly mention laissez faire or Adam Smith; they simply described the frictionless workings of a free market that had rejected economic controls and the "just price." Without mentioning either Malthus or Ricardo, such stories assumed the availability of plentiful resources and increasing access to force. Whether wielding an axe, building a mill, constructing a canal or a railroad, or digging an irrigation system, individuals deployed abundant resources and forces in a vast geometrical space organized as a free mar-ket. After the American Revolution, all these assumptions were natural-ized in the span of one generation. Even at the beginning of the twenty-first century, narratives grounded in this ideology still appeal to many Americans not as stories but as descriptions of reality.

Hayden White rightly emphasizes that history is always erected as a contrast to an earlier narrative. But foundation stories only pretend to be histories. The pioneer alone in the forest with his axe is less a historical figure than a rhetorical invention based on "that absence of words without which there would be nothing to say."[55]

3

Axe, Clearing, Cabin

A settler, especially from the New England states, often begins the world in that country with no other fortune than a stout heart and a good axe. With these he has no fears and sets merrily forward in his attack upon the wilderness.
—Basil Hall[1]

The American axe! It has made more real and lasting conquests than the sword of any warlike people that ever lived, but they have been conquests that have left civilization in their train instead of havoc and desolation.
—James Fenimore Cooper[2]

When nineteenth-century Americans imagined the settlement of their nation, the axe, the clearing of the forest, and the log cabin seemed inseparable. The story is paradigmatic: A settler enters the vast primeval woods. Using a new technology, the American axe, he transforms the forest into field and meadow, allowing it to be farmed for the first time. Initial settlement draws others to the area. As the population increases, a community emerges. As the land is "improved," its value rises. The region prospers. Elimination of the forest is equated with progress, hard work being tangibly rewarded for each acre cleared. Using an axe, the pioneer creates a new landscape—a landscape inscribed within a technological narrative that moves from desolate forest to clearing, from trees to logs, from wilderness to home, from empty space to civilization.

The narrative of the axe compresses into a deceptively simple story a set of philosophical assumptions: that nature is essentially abundant raw material waiting to be improved; that land properly belongs to the person who improves it (an idea given formal expression by John Locke[3]); that this labor itself was a form of muscle power or mechanical force in a larger system where all movement, action, and conversion of forces took place with no loss (i.e., there was no entropy); and that these forces and natural resources were best developed not through mercantilist-inspired protectionism but in Adam Smith's free market, where the measure of progress was the degree to which mankind increased its ownership of land or multiplied its control of forces. Americans prepared a setting in which to act out this story by surveying the land into perfect squares, ready for purchase or homesteading. The act of clearing land to build a log cabin created its own entitlement to the property, and it was the paradigmatic act of advancing civilization. For more than a century this narrative seemed to be the unvarnished truth. This chapter will describe its basis in history, its emergence between c. 1780 and 1820, and its popularity thereafter. Like all foundation stories, this one is saturated with cultural values, but it is presented as fact.

The entire opening chapter of Emerson Hough's popular 1903 book *The Way to the West* was devoted to the American axe. Hough begins: "I ask you to look at this splendid tool, the American axe, not more an implement of labor than an instrument of civilization. If you can not use it, you are not American. If you do not understand it, you do not understand America."[4] The tool Hough praised was not the straight, short-handled British axe that the first colonists had brought with them. The American item had a longer, curved handle, which maximized the force of the chopping motion. It was made from hickory, a hardwood that Native Americans had learned to bend and shape into lacrosse sticks and tools.[5] The a handle was tough yet springy. The metal head of the American axe was better balanced than its British counterpart, so that it wobbled less when swung and required less muscle effort to be guided accurately. The combination of the long, curved handle and the new head produced a far more efficient tool, one well suited to the colonist's needs. Carroll Pursell summarizes: "The axe first brought over from Europe had a fairly short handle, a thick wedge and narrow bite, and the handle was set well back from the cutting edge. Slowly, and by whose agency no one knows, this axe developed into a new type: one with a long, curved and springy handle set nearer the center of the head, which was broader at the edge than the center, and much narrower throughout."[6]

Unlike western Europe, whose forests had been removed centuries before, much of the Atlantic coastal plane was covered with trees, which colonial farmers cut down not only to clear the land but also to obtain building materials and firewood.[7] In 1828, James Fenimore Cooper praised the American axe as being admirable "for form, for neatness, and precision of weight."[8] The American axe was so much more efficient than its British counterpart that Parliament investigated it in 1841. An English expert witness called it "the most mechanically perfect and the best-constructed little instrument I know" and declared "a man can fell three trees to one, compared with those which are ordinarily made in England."[9]

When did this tool become a symbol of the American nation? The axe had already entered popular poetry by 1692, when Richard Frame described the clearing of land in western Pennsylvania:

Then with the Ax, with Might and Strength,
The trees so thick and strong
Yet on each side such strikes at length,
We laid them all along.

So when the Trees, that grew so high
Were fallen to the ground,
Which we with Fire, most furiously
To ashes did Confound.[10]

Yet no evidence suggests that a distinctive American axe existed in 1692. Although it was once thought that the design had emerged by the early eighteenth century, "the best archaeological evidence suggests a later appearance for the American axe."[11] The earliest datable axe is from 1776; it was recovered from an American gunboat sunk by the British in Lake Champlain. The earliest appearance of an American axe in a painting may have been as late as 1805.[12] The axe Richard Frame described most likely had not evolved into the American form, although it was already a poetic subject. In the eighteenth century the axe became inseparable from the narratives of settlement, and by the early nineteenth century Americans understood that their distinctive axe was the icon of the American woodsman. Abraham Lincoln had a reputation as a first-class "rail-splitter," which meant that he was good at driving wedges into logs to split them into rails used to build fences. Lincoln, like Andrew "Old Hickory" Jackson, seemed to partake of the virtues of the axe itself: lean, hard, practical, efficient, resilient. When Lincoln ran for the presidency in 1860, his supporters marched in parades carrying enormous symbolic axes too heavy for actual use.[13] The final sentence in Frederick Jackson Turner's *Rise of the New West*, which celebrated the rise of the Jacksonians to power, was "And on the frontier of the northwest the young Lincoln sank his axe deep in the opposing forest."[14]

For Hough, whose book appeared shortly before Turner's, the axe was "so simple and so perfect that it has scarcely seen change in the course of a hundred years," and its unadorned, curved, hickory shaft had "the beauty of utility."[15] "With the axe," Hough observed, "one can do many things. With it the early American blazed his way through the trackless forests. With it he felled the wood whereby was fed the home fire, or the blaze by which he kept his distant and solitary bivouac. With it he built his home, framing a fortress capable of withstanding all the weaponry of his time. With it he not only made the walls, but fabricated the floors and roof for his little castle. He built chairs, tables, beds, therewith. By its means he hewed out his homestead from the heart of the primeval forest and fenced it round about. Without it he had been lost."[16] Hough was restating what had become the standard nineteenth-century narrative of the axe, but this story would have been news to the settlers of Jamestown or Plymouth. Ask most Americans how the first settlers lived and they will talk about log cabins. It was not so, but later generations superimposed this vision on all of the American past.

Popular accounts also suggest that the settlement of new areas was individualistic, but during the seventeenth century such was seldom the case. As John Frederick Martin observed in his magisterial book on entrepreneurship and early town founding, "people did not simply move to the wilderness and build log cabins." In colonial Massachusetts, the General Court had to approve new settlements, after which the site had to be surveyed and the survey had to be approved by the General Court before the lots could be laid out. These lots were not geometric squares on a map. New Englanders made clear distinctions among "home lots, planting lots, wood lots, meadow, and swampland." And "lots of poor quality had to be augmented by quantity, river frontage had to be rationed, rocky land shared."[17] Those who took up grants of land had to improve it or risk forfeiture, and typically they had no more than 3 years to begin the improvements.[18] Yet in practice it was difficult to find enough settlers for the land, and entrepreneurs early became central to the process of town formation. Typically they paid for an initial survey and for the construction of bridges and other necessities, and they profited only over the long term as settlement increased the value of their holdings. Creating a town was a collective enterprise, regulated by the state and spurred along by the concerted plans of investors. Even after a community existed, strangers often could not freely purchase land and gain the rights of an inhabitant. Sales of improved lands to new people were generally supervised by the town, both to keep out undesirables and to limit the number of those with a right to future divisions of the town's unimproved lands. The depiction of a solitary woodsman carving a farm out of virgin forest, when attributed to colonial New England, is historical fiction.

In the early colonies, log cabins, either isolated or in groups, were almost unknown. The English colonists built houses as they had back home: square-hewn timbers were mortised and tenoned together to make a sturdy framework, which was then filled in. According to one study of the origin of the log cabin, "within ten years of their foundation the English settlements consisted almost entirely of framed wooden houses covered with clapboard, roofed with thatch or cedar shingle, and using various forms of filling such as wattle-and-daub, nogging, or plain clay daubing, between the outer wall and the inner sheathing."[19] In 1679 a Dutch traveler in what is now Delaware complained: ". . . the dwellings are so wretchedly built, that if you were not so close to the fire as almost to burn yourself, you cannot keep warm, for the wind blows through them everywhere. Most all the English, and many others have their houses made not otherwise than of clapboards."[20] Swedish log houses built before the arrival of William Penn and his Quaker colonists stood nearby, some within the present city limits of Philadelphia.[21] However, it was several generations before English colonists adopted the log building.

The techniques of Swedish settlers were better suited to a heavily forested area than English methods. Not only was a Swedish-style log house warmer in winter, but building one required less labor and fewer materials. The aforementioned Dutch traveler visited several such houses and found them "of hewn logs, being nothing else than entire trees, split through the middle or somewhat squared out of the rough." These were "laid in the form of a square upon each other as high as they wish to have the house." The entire house could stand "without a nail or spike," yet it was "very tight and warm."[22] German settlers later arrived with similar traditions. Far from being an American invention in response to the frontier, log structures were common in Russia and in northern Europe centuries before the English settlement of North America.[23] A "skilled German or Swedish cabin builder" was "capable of hewing logs with two or four flat surfaces, and interlocking them with carefully executed notches to produce tight, square, even corners."[24] Scotch-Irish imitators at first used rounded logs with gaps between them that needed to be caulked. By the end of the eighteenth century they had become more skillful, and most houses in western Pennsylvania and western Virginia were built of logs. As builders adapted the log cabin to local conditions, they added elements not characteristic in Scandinavia: a shingle roof, a front porch, an outdoor kitchen, a larger fireplace. By the end of the eighteenth century, the log cabin had become an American synthesis.[25]

In view of the origin of log cabins, American stories about them could not emerge until the eighteenth century. Indeed, it is hard to find seventeenth-century tales of one family's moving alone into an uncut forest. Early colonial writing describes the founding of towns and forts. Most of the first farmers were dependent on their immediate neighbors. Few lived alone in the forest. They lived together in communities that they delimited as their own space by clearing the land of trees and erecting forts, palisades, and fences. They were kept together not only by their colonial governments but also by fear of the unknown land (which they perceived as wilderness), fear of Native Americans, and inexperience with settling new territories. Poor transportation, which limited access to skilled artisans who could grind grain or repair tools and weapons, hampered settlement of the interior. Colonists needed goods that were available only in towns: bullets, horseshoes, glass, nails, wagon wheels, and imported foodstuffs such as tea.

Until the end of the French and Indian War (1756–1763), New Englanders were generally unwilling to live alone on the frontier in areas such as western Massachusetts, northern New Hampshire, and Vermont. New Englanders typically moved into new areas in groups; they continued this practice later when they settled the Ohio Valley.[26] Further from the threat of French-led attacks from Canada, Pennsylvanians and Virginians ventured into unsettled areas in small groups somewhat earlier. William Byrd, who traveled with a surveying party commissioned to delineate the boundary between Virginia and North Carolina, found isolated families in the forests and mountains of the interior. Many of them were squatters, and some lived far from neighbors.[27] In 1728, Byrd observed, "most of the Houses in this Part of the Country [were] Log houses, covered with Pine or Cypress Shingles, 3 feet long and one broad."[28] However, few could live long cut off from blacksmiths, millers, and other skilled workmen. Only as fears diminished and transportation improved did it become more common to break away from the community and live in the woods.

The log cabin stories familiar today suggest that clearing a farm took one man only a season or two, but in the seventeenth and eighteenth centuries this was hardly the case. Here is what one Peter Gibbon recalled from clearing a wooded area on a "rocky mountain" near the Connecticut River Valley of Massachusetts in the decades before the American Revolution:

When I got here I was as poor as poverty itself. . . . However, I had got into the woods and a howling wilderness it was. No roads in any direction to lead anywhere but by marked trees. The first summer I had to hire my team work when I could get it, and sometimes [the crop] came to nothing, and when it did well the vermin of the woods destroyed half of it. . . . But I kept clearing my land yearly and I gained slowly and in about eleven years I built a 26 foot barn and I cut hay and grain enough to fill it and had stock enough to eat it, and in fifteen years I built me a house so that I lived in it and kept up a finishing it and adding to it as I was able, and I got to live tallerable comfortable.[29]

Gibbon, a young man in his twenties, did not attempt this work alone; he hired a team to help clear the land and prepare it for farming. The process took more than 10 years, and only after another year did Gibbon build a modest barn. Much is missing from Gibbon's account. Perhaps assuming that any man of his time would know what was involved, he did not explain the work. How did he clear the land, and what happened to the trees he cut down? Did he build the barn out of logs, or lumber? Where did he live for the first 14 years? He does not say, but simple log structures were common enough. A few years later, in another town in the same region, almost half of the 65 houses were "round log huts."[30] There was far more wood than was needed for building; like many others, Gibbon burned much of it to cook and to keep warm. He may also have used part of the winter to haul logs to the nearest sawmill for barter, receiving lumber in exchange. Gibbon does not mention the blacksmith who must have shoed his horses, repaired his iron pots, and sharpened his iron tools. These omissions make it easy to imagine such a man as more independent of society than he in fact was. "No family could be completely self-sufficient." All had to buy "some finished products—tools, farm implements, household utensils, and the like." Furthermore, the more distant the merchant was, the more certainly he would want cash or a promissory note for cash, not barter, as payment.[31] The colonial farmer was tied to a community.

Eighteenth-century colonials could not imagine the movement west as a journey into virgin lands. This conceit was possible only for later generations. Rather, as maps from as late as the early nineteenth century clearly demonstrate, both the English and the French government recognized a complex pattern of Native American settlements.[32] "Indian towns," Byrd observed, "are remarkable for a fruitful situation; for being by nature not very industrious, they choose such a situation as will subsist them with the least labor."[33] The colonists who forced Native Americans out of such locations could not conceive of the woodsman's axe as the primary tool of conquest. The colonial foundation story emphasized conflict with Native Americans, often depicting them as the aggressors. The central technology in this narrative was the rifle. In contrast, the technological creation story, when fully developed, would virtually delete Native Americans, so that the central relationship was between the settler and the land.

During the transition between these two stories, observers described civilization's emergence as embodied successively in three different kinds of settlers: frontiersmen, then large landowners, and finally farmers. The first displaced the Native American and began clearing the forest; the second enlarged the arable land; the third completed the conversion of woods into field and meadow. Benjamin Rush wrote a typical account of this transition in 1786. In Rush's narrative, the first settler is "generally a man who has out-lived his credit or fortune in the cultivated parts" and has escaped to a new region:

His time for migrating is in the month of April. His first object is to build a small cabin of rough logs for himself and his family. The floor of this cabin is of earth, the roof of split logs; the light is received through the door and, in some instances, through a small window made of greased paper. A coarser building adjoining this cabin affords a shelter to a cow and a pair of poor horses. The labor of erecting these buildings is succeeded by killing the trees on a few acres of ground near his cabin; this is done by cutting a circle round these trees a few feet from the ground.[34]

The following May, the ground plowed around the dead trees is sowed with Indian corn; in October, it yields 40–50 bushels per acre. Until then, the family survives on a small quantity of grain they brought with them and on fish and game. After 2 or 3 years, neighbors begin to arrive; they thin out the game, both through hunting and by scaring it away. This forces the early settler to raise domestic animals for meat, and he begins to consider leaving. Furthermore, he "revolts against the operation of laws." He "cannot bear to surrender up a single natural right for all the benefits of government," and therefore he "abandons his little settlement and seeks a retreat in the woods,"[35] where he starts the process again.

In Rush's account, the land next passes to a "second species of settler"— "generally a man of some property" who owns 300–400 acres. A sawmill having been built in the neighborhood by this time, he expands the original log house, using sawn timber to create a second floor. He clears a meadow, plants fruit trees, puts up a wooden barn, and cultivates more crops on his expanded acreage. However, this wealthy settler, who in the South became the ideal figure in the rural landscape, is in Rush's account something of a failure. He "by no means extracts all from the earth that it is capable of giving." He does not plow well. He feeds his cows and horses poorly. He "delights chiefly in company," he "sometimes drinks spiritous liquors to excess," and he "will spend a day or two in attending political meetings." He contracts debts he cannot pay. Forced to sell, he leaves the land considerably improved for "the third and last species of settler,"[36] who will complete what the others began.

Rush describes the last kind of settler as "commonly a man of property and good character—sometimes . . . the son of a wealthy farmer" from a more settled region. He completes the transformation of the land into meadow. He farms more carefully than his predecessors did, using manure for the first time. He builds a stone barn, which keeps the livestock warmer in winter than the loosely built log structure already on the land; therefore, his animals require less feed. He "uses economy, likewise, in the consumption of his wood" by replacing fireplaces with stoves, which "save an immense deal of labor to himself and his horses in cutting and hauling wood."[37] He raises a greater variety of grains, starts a vegetable garden, builds a smokehouse, places a milkhouse over a spring, and plants more fruit trees. The "last object of his industry is to build a dwelling house," and this effort is often completed by his son. Many of these improvements are forms of energy conservation that demonstrate an understanding of how time, money, and muscle power are connected. By constructing a better barn, the farmer saves feed; therefore, he has more hay and grain to sell. By substituting stoves for fireplaces, he saves wood and the time required to cut it. The milkhouse and the smokehouse ensure that food will not spoil, further increasing the efficiency of his farming. Therefore, the settler has time for other tasks of improvement. By saving animal feed, wood, and food, he achieves tangible long-term advantages. Every hour released from supplying food and chopping wood is an hour that can be devoted to such tasks as planting and caring for an orchard. Such a settler appreciated Benjamin Franklin's frugal maxims in *Poor Richard's Almanack.*

Compared to the first two, Rush's third settler lives in luxury. "His table abounds with a variety of the best provisions. His very kitchen flows with milk and honey-beer, cider, and homemade wine are the usual drinks of his family. The greatest part of the clothing of his family is manufactured by his wife and daughters."[38] He does not run from legal institutions as the first settler did, nor does he fail to pay taxes as the second did. He pays his taxes punctually and gives to the school and the church. "Of this class of settlers are two-thirds of the farmers of Pennsylvania," says Rush, who even compares the first class to the Native Americans: "In the second, the Indian manners are more diluted. It is in the third species only that we behold civilization completed. It is to the third species of settlers only that it is proper to apply the term *farmers.*"[39]

Rush already was drawing on a tradition. Hector St. John de Crèvecoeur had described the development of farming in a similar way, though with less emphasis on the transitional second figure. The broad outline of Rush's story also resembles the account of settlement given by Benjamin Franklin in 1766.[40] Franklin likewise distinguished three successive species of settlers.[41] All three saw hunters on the frontier as a marginal group who opened up the land but who were not praiseworthy. The idealization of the frontiersman and the full creation of the narrative of the log cabin in a remote clearing came only in the closing years of the eighteenth century and later. The idea of a clear succession of stages had become standard by 1803, when Timothy Dwight toured western New York. Dwight emphasized the role of *"foresters* or *pioneers,"* whose "business . . . is no other than to cut down trees, build log houses, lay open the forested grounds to cultivation, and prepare the way for those who come after them."[42]

The marginalization of Native Americans in this narrative is striking, but there is another aspect of this story that can easily escape attention: this is a story about transforming the world through technologies. All three types of European settlers possess tools made of iron, which enable them to cut down trees, plow the land, harvest with sickles and scythes, cook in iron pots, split wood with wedges, saw boards into planks, and drill wooden beams with holes for mortise-and-tenon construction. In the foundation narrative, the axe stands for all these tools. Just as important, the settlers had horses and oxen, which gave them access to far more muscle power than the Native Americans had commanded. The settlers used these tools and this extra horsepower to literally displace Native Americans. With the axe, the plow, and European seeds (the product of more than 1,000 years of selection and development), they reconfigured the land itself. In a sense, they did not so much drive out Native Americans as they farmed them out. Rush, Crèvecoeur, and Franklin displace attention from this aspect of the story, however. All three conceive of the forested land as wilderness in which it is not possible to survive without introducing (European) tools and foodstuffs. When they acknowledge the presence of Native Americans, it is as if they have made no impression on the landscape, which seems to be untouched and wild. The story emphasizes the use of skills and technologies against nature, not against Native Americans.[43]

White Americans understood cutting down the forest as the first step in *improving* the land. When the work was done well, it was selective, leaving sugar maples and other useful trees standing. Often, however, they removed all the trees. An Englishman realized that the American perception of forest landscape was quite distinct from his own. Coming from a long-settled country, he associated forests with shade, shelter, and beauty, and with the protection of game; he felt that trees added "variety to park scenery" and made a "pleasant contrast with rich cultivation." Americans, in contrast, found cultivation beautiful and apparently believed that "every acre, reclaimed from the wilderness, is a conquest of 'civilized man over uncivilized nature'; an addition to those resources, which are to enable his country to stretch her moral empire to her geographical limits, and to diffuse over a vast continent the physical enjoyments, the social advantages, the political privileges, and the religious institutions, the extension of which is identified with all his visions of her future greatness."[44]

Because wholesale clearing of land was novel to most Europeans, Benjamin Rush advised them not to attempt it. He emphasized the distinction between "improving a farm" and "settling a new tract of land." The latter, he warned, "must be effected by the native American, who is accustomed to the use of the axe and the grubbing hoe and who possesses almost exclusively a knowledge of the peculiar and nameless arts of self-preservation in the woods." Citing Europeans who have "spent all their cash in unsuccessful attempts to force a settlement in the wilderness" and fallen into poverty, Rush advised new immigrants with "moderate capitals" to "purchase or rent improved farms in the old settlements of our states."[45] Almost invariably, one of the existing buildings on a newly cleared tract of land was a log cabin.

What then is a log cabin? "The log house," Wilbur Zelinsky notes, "can be defined by its material, method of construction, and function. The walls are made of round, undressed logs or of logs hewn with ax or adz (but never sawed) to rectangular shape. The logs are laid one above the other and are mortised at the corners of the building by some form of notching that eliminates the need for hardware. These buildings are lacking in affectation. . . ."[46] The size of such houses was determined by the length of the available logs, which usually made possible a square structure with each side 15–20 feet long. Often the inhabitants found this too small and added a second log square to the first, with a hallway (or "dog trot") in between. Log structures predominated in the Shenandoah Valley and in western Pennsylvania just before the Revolution. In 1772, Pittsburgh was a village of "about 40 dwelling houses made of hewed logs"[47]; 10 years later, a traveler still found only "paltry log houses."[48]

From an English immigrant named Morris Birkbeck we have a good description of a hunter's log cabin in 1817:

. . . the cabin in which he entertained us, is the third dwelling he has built within the last twelve months; and a very slender motive would place him in a fourth before the ensuing winter. In his general habits, the hunter ranges as freely as the beasts he pursues; labouring under no restraint, his activity is only bounded by his own physical powers. . . .

The cabin, which may serve as a specimen of these rudiments of houses, was formed of round logs, with apertures of three or four inches between: no chimney, but large intervals before the "clapboards," for the escape of the smoke. The roof was, however, a more effectual covering than we have generally experienced, as it protected us very tolerably from a drenching night. Two bedsteads of unhewn logs, and cleft boards laid across;—two chairs, one of them without a bottom, and a low stool, were all the furniture required by this numerous family. A string of buffalo hide stretched across the hovel, was a wardrobe for their rags; and their utensils, consisting of a large iron pot, some baskets, the effective rifle and two that were superannuated, stood about in corners, and the fiddle, which was only silent when we were asleep, hung by them.[49]

This Englishman's disdain should not blind us to the technologies the cabin represented. He does not mention an axe; however, the family clearly had one, or it could not have constructed three such houses in the previous year. Nor were these people cut off from civilization and manufacturing. The "superannuated" rifles were only temporarily useless. In an age when parts were not interchangeable, they required handcrafted replacement parts from a skilled smith. To fire a rifle required gunpowder, made in mills. The clothing, crude though it was, required a needle; the black iron kettle was assuredly not homemade. Birkbeck does not mention whether these hunters had horses, but it is almost inconceivable that they lacked them or that they rode them bareback. Saddles, bits, bridles, and horse-shoes are also old European technologies. From the perspective of the Atlantic coast or Britain, such a family seemed to live in a state of nature. But they had technologies unknown to Native Americans before Euro-peans arrived, and they relied on specialized labor to produce and repair their axes, needles, pots, guns, harnesses, and other equipment.

Birkbeck wrote just before the full emergence of the myth of the log cabin as the archetypal pioneer dwelling—the primordial site where American life began. In the first decades of the nineteenth century, Rush's three-part story of settlement evolved toward the story of the solitary fam-ily in the wilderness. A pivotal figure in that evolution was Daniel Boone, whom John Filson had epitomized as the ideal frontiersman in his 1784 book *The Discovery, Settlement, and Present State of Kentucke.* Filson's

Daniel Boone was not the rough and degenerate frontiersman that Rush and Crèvecoeur had described; he was a heroic pathfinder and bringer of civilization. Soon reprinted, Filson's book became a popular foundation story that built upon the captivity narrative.[50] Boone, primarily a hunter and warrior, had become expert in woodcraft. Though he was ostensibly also a settler, the stories of his life emphasize exploration and confrontation with "Indians." According to Richard Slotkin, "it was Filson's implicit vision of Boone the hunter as the central figure in a creation myth that later chroniclers and commentators emphasized."[51] In the repertoire of possible frontiersmen, "it was the hunter who proved the most appealing figure."[52] But other figures were also significant, including the solitary settler carving a home out of the wilderness with his axe.

Indeed, in the late eighteenth century, when the American axe appeared, "American agricultural practice changed in ways that put an emphasis on the wholesale clearing of land."[53] Girdling trees had been sufficient for the planting of corn or tobacco, but planting wheat and other European field crops require that trees be removed. By 1807, Timothy Dwight concluded, removal rather than girdling had "become the universal practice" in western New England and in New York, and this work was largely done with the new axe.[54] Thus, in the last decades of the eighteenth century, the widespread adoption of the log cabin, the innovation of the American axe, and the preference for clear-cut land occurred almost simultaneously. Axe, log cabin, and clearing merged into a symbolic cluster. Together they seemed characteristically American. In the 1824 presidential election, Andrew Jackson demonstrated the voter's attraction to birth in a log cabin, which after 1830 became a candidate's guarantee of pedigree as one of "nature's noblemen."[55] A man born in a log cabin was by definition a forceful pioneer who cleared his own land. Countless speeches and editorials asserted that such closeness to the soil made a man virtuous and freed his intuitive genius from the trammels of European tradition.[56] In the context of such popular beliefs, the foundation narrative of the axe had emerged fully.

To Rush, Franklin, and Crèvecoeur, the frontiersman was a precursor of the farmer. After 1820, the two figures increasingly became one. Dwight, while condemning frontiersmen as a class, conceded that "a considerable number even of these people become sober, industrious citizens, merely by the acquisition of property."[57] Ownership of land made legal institutions seem more necessary to them: "The secure possession of property demands, every moment, the hedge of law; and reconciles a man, originally lawless, to the restraints of government." Permanent residence also made personal reputation more important. For the first time, Dwight observed, "he sees that reputation also is within his reach. Ambition forces him to aim at it; and compels him to a life of sobriety, and decency. That his children may obtain this benefit, he is obliged to send them to school, and to unite with those around him in supporting a school-master." He likewise helps "to build a church, and settle a Minister."[58] The ownership of property had this effect on the formerly anarchic frontiersman, and it encouraged even better behavior among other settlers.

By 1832, when Zachariah Allen extolled the American "pioneer, now master of the soil which he has subdued *with this own hands*,"[59] most white Americans shared this evolutionary narrative. When Allen journeyed through Britain he was struck by the contrast between "the luxurious ease and elaborate enjoyments, resulting from the high state of cultivation and improvement in this anciently settled country" and "the struggles and privations encountered by those who first apply the axe to primeval forests, and open the way for the march of civilization and refinement."[60] Adopting the prevalent imagery of Jacksonian America, he told the foundation story of the axe:

It requires a firm heart for a solitary individual to enter the wilderness alone, for the purpose of encountering the arduous enterprise of clearing up new lands. With a wallet thrown over his shoulder, and a keen axe in his hand, the new settler penetrates many miles distant from any human habitation, beneath the twilight gloom of one unbroken canopy of foliage. Relying on the vigor of his own arm to regain the cheerful sunshine, he applies his axe with persevering strokes, and every hour witnesses the fall and crash of sturdy trees which have been . . . rooted in the spot for centuries. Acre after acre of fertile land becomes strewed with trunks and branches closely intermingled. For the first time, perhaps, since the creation of the world the light of day shines with unbroken ray upon that spot of earth. . . . Fire is applied to the fallen trees, and all the small branches are consumed, leaving the huge blackened trunks scattered about the field, smoking and smoldering, half consumed. These trunks are afterwards rolled together to form great heaps with fragments of the unconsumed branches. The . . . fire is again rekindled. . . . Naught remains of a noble tract of forest but scattered heaps of white ashes. . . . The smoke of a score of burnings . . . may sometimes be seen. . . .[61]

There only is a trace of ambivalence as the "noble tract of forest" is transformed into "heaps of white ashes." Allen regarded the opening of the land to settlement and agriculture as a part of a longer process from "the first clearing of the wilderness" to "the conversion of it into a sort of artificial paradise."[62] In such a narrative of technological transformation, the axe and the manipulation of fire are central; the Native American is absent. In contrast to the hunter narrative, the forest is presented as primeval, and the only actor on the stage is the "solitary individual" with his axe. The defining act, completing the conquest of the forest, comes as "the log house is next constructed from the straightest trunks of trees that have been selected and preserved from the fire." The pioneer is now "master of the soil which he has subdued *with this own hands*."[63]

Alexis de Tocqueville developed a similar view when he traveled in the western settlements. Finding the endless forests oppressive, he was relieved to hear "the echoes of an axe-stroke" and to see "in the center of the rather restricted clearing made by axe and fire around it . . . the rough dwelling of the precursor of European civilization . . . like an oasis in the desert."[64] Gaston Bachelard later meditated on the meaning of such isolated houses and concluded that "we are hypnotized by solitude, hypnotized by the gaze of the solitary house."[65]

The log cabin was far more than a rude shelter; it created a human center in the wilderness. This break in what white men otherwise saw as an insufferable, oppressive, repetitive emptiness seemed an oasis where the traveler was refreshed. In this narrative, the pioneer did not wield the axe merely to carve out a clearing in the trackless forests and build there a cabin. He may have opened the way for an abstract future civilization, but more immediately he created a safe haven, an opening, a release from the wild. The story of building the cabin is quite literally a foundation narrative, an account of how civilization confronts and overcomes solitude in a new place, and it is imagined on a personal level.

Log cabin stories were not common before the nineteenth century, and there is an ironic side to this. Advances in transportation and manufacturing made it easier for individuals to live in apparent retreat from civilization and still participate in a national market. The isolated farmhouse in the forest (or, later, on the windswept plains) was linked in many ways to the market economy, which made it possible to concentrate on lumbering, trapping, or a few crops. Barbed wire, pot-bellied stoves, iron cookware, pre-cut lumber, factory nails, canned food, boots, and clothing could be shipped, inexpensively, hundreds and even thousands of miles to a cabin.[66]

The axe, though a symbol of the frontier, usually was not produced there. Once made individually by blacksmiths, by the 1820s it was produced in mechanized factories. In Connecticut, Elisha Root of the Collins Company invented machinery that increased the automation of axe production so much that it was felt to merit an article in *Scientific American*. Root had eliminated the tedious and dangerous grinding of axes by inventing a shaving machine that cut the blade. Avoiding hand labor, his system passed the axeheads through furnaces to harden them, bathed them in brine, then tempered them for several hours in an oven regulated by a thermometer. Root's system produced uniform results at low cost, ensuring that pioneers could attack uniform tracts of geometrically laid out lots with an identical tool.[67]

Factory-produced axes were widely sold. In 1835, Edward Everett gave a speech in support of improving transportation between New England and the West. He recalled encountering "one of [his] neighbors from Charlestown, who had emigrated to the north-west corner of Arkansas." The neighbor told Everett that "in that remote region—the last foothold of civilization, where you have but one more step to make to reach the domain of the wild Indian and the buffalo—a settler did not think himself well accoutered without a *Leominster* axe. But give him that, give him, sir, that weapon which has brought a wilder realm into the pale of civilization than the sword of Caesar or the scepter of Justinian, give him a narrow Yankee axe. He'll hew his way with it to a living, in a season. . . ."[68] For Everett, the axe was an extension of Eastern culture, pushing back the line of forest, supplanting the "wild Indian and the buffalo," and continuing the spread of civilization. It was an instrument of permanent conquest. Everett did not see living on the frontier in terms of self-sufficiency; rather, he saw transportation and factory production as having made expansion sustainable.

Everett still retained the Jeffersonian sense that the land was the ultimate source of wealth:

The difference between the ultimate result of the same amount of labor applied in the manufactures in England and to clearing away forests in the newly settled territories of the United States, consists in the circumstance that manufacturing labor produces consumable articles, whilst the agricultural labor of clearing forests develops a fixed and permanent capital in lands, which must remain to enrich the children of the laborer. . . . Who can estimate the value of the lands watered by the tributary streams of the Mississippi? . . . This value to the civilized world has been created as it were by magic, within the few years since the toil of free men has been directed toward these new regions. . . . Aided by these natural advantages of unbounded capital in new and fertile lands, the wealth as well as the population of the United States must in the course of nature surpass those of the older [European] countries.[69]

The "new and fertile lands" were, however, more than capital. Many authors saw the West as a liberation from the confines of the genteel East.

Daniel Drake, a Cincinnati physician and man of letters, believed that the individual who moved to the West was released from a "limited circle of employments" characteristic of more settled regions and the narrowness of following a career that consisted of only "one path." Drake emphasized that "in a new country, the restraints employed by an old social organization do not exist."[70] The construction of a log cabin became a part of a process of national self-creation.[71] Drake wrote near the end of Andrew Jackson's two terms as president. Jackson's popularity marked the emergence of the mythology of the West as central to American politics. If Jackson had been the first to exploit an association between himself and a log cabin boyhood in his three campaigns for the presidency, the Whigs quickly adopted log cabin imagery for General William Henry Harrison's candidacy in 1840. Seizing on a slight from a Democratic journalist, the Whigs made hard cider and the log cabin their symbols, even though Harrison lived in a 14-room house and owned 3,000 acres of Ohio farmland.[72] Nevertheless, his supporters emphasized that Harrison was a farmer who had been born in a log cabin. One of the signature events of his campaign was a great parade through the streets of Baltimore with a log cabin as its central symbol: "The Fayette County delegation from the Allegheny Mountains of western Pennsylvania marched behind a log cabin on wheels drawn by six prancing horses. Deer and fox skins adorned the sides and roof of the cabin, and alongside it was a barrel full of hard cider, with a gourd suspended over it. From time to time a delegate would break ranks and refresh himself from the barrel. . . ."[73] Delaware also had a log cabin

Figure 3.1
Woodcut used in Harrison-Tyler campaign, 1840. Courtesy Library of Congress.

float.[74] The parade converged on the Canton Race Course, where "a full-sized log cabin had been erected, its logs well-plastered with clay."[75]

Emerson was aware of the celebration in Baltimore, and on July 4 in Concord he witnessed the Middlesex County Whig celebration.[76] The Lowell delegation had constructed an enormous log cabin float, "drawn by twenty-three horses," that "carried nearly one hundred and fifty persons."[77] The most industrial city in the nation thereby embraced log cabin imagery, which likewise dominated *The Log Cabin*, an election newspaper published by Horace Greeley. The response to the first issue was so great that Greeley upped the print run from 30,000 to 80,000 for subsequent issues; he estimated that he might have been able to sell 100,000, but he was not able to get that many printed.[78] The public embraced log cabin imagery and swept the Whigs to victory.

In part, the Democrats lost because the 1837 depression had depressed wages by 30 percent or more by the election of 1840.[79] Hard times help explain the shift in power, but they do not explain the powerful appeal of log cabin imagery.

Even the leading Whig orator Daniel Webster found a way to adopt the pretense of rusticity. Before a large crowd at Saratoga, he declared: "I have a feeling for log cabins and their inhabitants. I was not myself born in one, but my elder brothers and sisters were." Webster went on to claim that he made an annual pilgrimage to the site of that cabin, "on the extreme frontiers of New Hampshire," taking his children there "that they may learn to honor and to emulate the stern and simple virtues that there found their abode."[80]

Like Webster, the painter Thomas Cole portrayed the emergence of American civilization from the log cabin. Cole often conceived paintings (including his famous sequences "The Voyage of Life" and "The Course of Empire") as narratives. On inspection, even images that seem to depict only one moment in time often suggest the unfolding of a process. "Landscape" (1825), painted when Cole was 24 years old, depicts a rural scene with a man in the foreground, swinging an axe. His foot is on the trunk of a tree which he has felled and stripped of its branches and which he has begun to chop in two. Much of the land visible in this painting has been cleared and "improved" by the axe, opening pastures on the hillsides. The

Figure 3.2
Thomas Cole, "Landscape" (1825). Courtesy Minneapolis Institute of Art. Bequest of Mrs. Kate Dunwoody.

man with the axe is not merely cutting up one log; he is performing the representative act that creates the landscape. Before he arrived, the area presumably was completely covered with trees; now his work has transformed that space into a pastoral zone between wilderness and civilization. This painting emphasizes the labor required to produce the fields; it also suggests the relationship between the fields and the technology of the mill. A water-driven mill (likely a sawmill) stands just behind and below the man with the axe. The woodsman's spatial location in the foreground, up the hillside, thus has a temporal dimension, suggesting the process of settlement: his attack on the forest had to begin before the mill could be built, and his work continues to provide it with logs as raw material.

When an axe or a mill appears in a landscape of the Hudson River School, it cannot be considered an incidental detail. A contemporary who saw Cole's "Daniel Boone at the Door to his Cabin on the Great Osage Lake, Kentucky" (1826) identified this pioneer with the axe he used to "improve his land" and to build his cabin. The Boone painting depicts the initial stage of settlement, in which a solitary man constructs a log cabin in preparation for the arrival of his family. Boone, famous for having opened up areas beyond the Appalachians to settlement, was a paradigmatic figure in the axe narrative. In Cole's "Hunter's Return" (1845) and in his "Home in the Woods" (1847), a nuclear family has built a solitary cabin beside a lake in the wilderness. Little of the forest has been cut down, and the yard of the house in the foreground is littered with tree trunks and branches, suggesting recent settlement. More land has been cleared than in the Boone image, but farming has not yet begun. Hunting and fishing supplement the family's diet until the woods are cleared for the first planting. In these images, the wife and the children stand in the doorway or lean out windows to greet the man returning with fish or game. Near the center of each painting, between the women and children on the right and the man on the left, is an axe embedded in a log. These canvases visualize the foundation story of the axe, which in each case stands between the family and nature.[81]

Four 1849 engravings of "the pioneer settler and his progress" make the narrative of the axe explicit.[82] The sequence, which appeared in *Pioneer History of the Holland Purchase of Western New York*, begins with a solitary cabin in the midst of a new clearing during the winter. Outside, near

Figure 3.3
Thomas Cole, "Home in the Woods," c. 1846. Courtesy Reynolds House Museum of American Art, Winston-Salem.

a few stumps, a pioneer stands with his axe, preparing the trunk of a fallen tree to be hauled away by his ox. Legends explain that the pioneer built the house quickly in the fall, that the roof is of elm bark, and that the floor inside is of split logs. As to why the family had settled in the forest rather than on open land, experience showed that wooded land was always fertile, especially if it could produce oak, hickory, elm, and walnut trees.[83] The second panel represents the summer of the following year. The cabin now has a chimney. A fence, partly of split rails and partly of brush, surrounds the house and its garden. In the foreground, outside the fence, are cows, a new calf, a pig, and nursing piglets. The pioneer mother stands at the door of the cabin with a newborn child in her arms. The husband is busy with some neighbors clearing a wide swath of woods behind the cabin, where among the stumps the first crops—corn, potatoes, beans, pumpkins—are growing. The next two panels jump ahead 10 years and then 45 to show the opening up of the landscape, the additions to the house and outbuildings, and the increasing prosperity of the couple. By the

Figure 3.4
"The First Six Months." From Turner, *Pioneer History of the Holland Purchase of Western New York*.

Figure 3.5
"The Second Year." From Turner, *Pioneer History of the Holland Purchase of Western New York*.

Figure 3.6
"Ten Years Later." From Turner, *Pioneer History of the Holland Purchase of Western New York*.

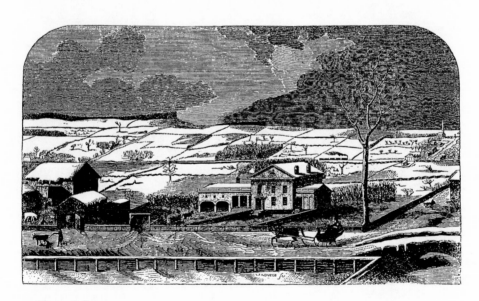

Figure 3.7
"The Work of a Lifetime." From Turner, *Pioneer History of the Holland Purchase of Western New York*.

third stage, they have added to the old log cabin a larger and more comfortable house made of sawn lumber. In the last image, the woods have been entirely removed, as has the log cabin. The original settlers now enjoy a comfortable old age. Their large farm is within sight of neighbors and a hillside village. A railroad crosses a rural checkerboard of fields almost devoid of trees. Farming has erased the forest and replaced it with geometric order. The once-isolated clearing has expanded to fill the landscape, and the railroad links it to the national market. Human and animal muscle power have cleared the land. Steam power has made the farm's agricultural products valuable in distant markets. And whereas Franklin, Rush, and Crèvecoeur had described three distinct species of settler corresponding to this sequence of landscapes, by the middle of the nineteenth century a single family effects the transformation. The notes explaining these images emphasize that the process of clearing the land, completed in western New York, continues further west, where the old couple's sons have migrated.

As they expanded westward, Americans also began to project the settlement process eastward and back in time. In 1857, when politicians converged on Jamestown to celebrate the 250th anniversary of its foundation, President John Tyler took it for granted in his speech that the first houses had been log cabins, and Governor Henry Wise of Virginia orated: "Here the White Man first wielded the axe to cut down the first tree for the first log cabin. Here the first log cabin was built for the first village. . . ."[84] By the eve of the Civil War, the building of the log cabin had become a pervasive story intertwined with the myth of American origins. The individual who built a log cabin, whether landowner, homesteader, or squatter, felt he was reenacting the foundation of society.

The year before the Jamestown ceremony, Walt Whitman published "Song of the Broadaxe,"[85] a poem that shows how axe, clearing, and log cabin had become elements of a foundation narrative.[86] The poem begins with a paradox:

Weapon shapely, naked, wan,
Head from the mother's bowels drawn,
Wooded flesh and metal bone, limb only one and lip only one,
Gray-blue leaf by red-heat grown, helve produced from a little seed sown,

Resting the grass amid and upon,
To be lean'd and to lean on.

Here the axe has almost human attributes. Shapely and naked, its head is drawn from the womb of Mother Earth. No mere object, it is an organic growth, "wooded flesh and metal bone." Within the larger structure of *Leaves of Grass*, the axehead is a "gray blue leaf" that has bloomed out of the heat of the forge; it is joined to the wooden helve, which has grown out of a single seed. Like the poet, the axe is "resting on the grass." The weapon has been naturalized. When it next appears in the poem, again it is part of a quiet scene:

The log at the wood-pile, the axe supported by it,
The sylvan hut, the vine over the doorway, the space clear'd for a garden.[87]

The destruction of the forest is only implicit. Whitman emphasizes the promise of the garden that is soon to come before turning to an evocation of emigration to America and "the disembarkation, the founding of a new city," a process he sees continually repeated in "the settlements of the Arkansas, Colorado, Ottawa, Willamette, / The slow progress, the scant fare, the axe, rifle, saddle-bags. . . ."[88] The axe becomes a central part of the pageant of settling and building the United States, and the poem evokes its many uses, from "the butcher in the slaughter-house" to "lumbermen in their winter camp." The wood cut down is for "the blazing fire at night" and for the building of houses and cities. A former carpenter, Whitman visualized the process of house construction, starting with

The preparatory jointing, squaring, sawing, mortising,
The hoist-up of beams, the push of them in their places, laying them regular,
Setting the studs by their tenons in the mortises. . . .[89]

After fifteen lines on house building, Whitman turns to the sparmakers:

The swing of their axes on the square-hew'd log shaping it toward the shape of a mast,
The brisk short crackle of the steel driven slantingly into the pine,
The butter-color'd chips flying off in great flakes and slivers
The limber motion of brawny young arms and hips in easy costumes. . . .[90]

The poet then weaves the axe into a succession of dramatic human activities, from firefighting to the axe's creation at the forge to its use in ancient civilizations and its symbolic incorporation into ancient Rome's fasces (rods bundled together with an axe, representing authority, that were carried by lictors on state occasions and in processions). Only at this point does Whitman focus on the axe's use as a weapon, announced in the first word of the poem. Yet it is Europeans who do battle with axes, and the executioner that the poem evokes is "the European headsman":

He stands mask'd, clothed in red, with huge legs and strong naked arms,
And leans on a ponderous axe.[91]

Whitman then dismisses the executioner as a figure of the past:

I see the blood wash'd entirely away from the axe,
Both blade and helve are clean,
They spirt no more the blood of European nobles, they clasp no more the necks of queens.

I see the headsman withdraw and become useless,
I see the scaffold untrodden and mouldy, I see no longer any axe upon it,
I see the mighty and friendly emblem of the power of my own race, the newest, largest race.[92]

The American axe is not a weapon of destruction; rather, it is an active agent in the construction of a new land:

The axe leaps!
The solid forest gives fluid utterances,
They tumble forth, they rise and form. . . .

The axe translates the forest into a succession of wooden objects: "Flail, plough, pick, crowbar, spade, / Shingle, rail, prop, wainscot, jamb, lath, panel, gable, / Citadel, ceiling, saloon, academy, organ, exhibition house, library, / Cornice, trellis pilaster, balcony, window, turret, porch,"[93] and many more items, leading up to "steamboats and clippers taking the measure of all seas."

The list continues through seven stanzas, each of which begins with "The shapes rise!" or a close variant. The American axe itself is no longer mentioned, but it is evoked by the work of the "cutters down of wood" in Maine. It is implicitly central to the lives of "dwellers in cabins among the Californian mountains or by the little lakes, or on the Columbia."[94] The "shapes" that "rise" include factories, railroads, bridges, ships, coffins, cradles, roofs, floors, courts, barrooms, and scaffolds. Through the agency of the axe arise the "Shapes of Democracy total, result of centuries, / Shapes ever projecting other shapes."[95] For Whitman, the axe transcends its immediate uses and becomes an organic symbol, linking nature and civilization and creating the United States.

In the 75 years between Rush and Whitman, the axe and the log cabin had become American icons closely associated with natural virtue. With the axe, the frontiersman cleared a space in the world and lived in primitive ease. The axe was the cutting edge of American society as it encroached on millions of acres of forest, yet Whitman made it both an organic expression of nature and the preeminent tool of nation building. The American axe made the grid real. It translated the surveyor's lines into a clearing in the forest, where it inscribed the nation's birthplace: a log cabin.

Not all Americans embraced this technological foundation story.

4

The Nurturing Forest

The small farm with its neat home, orchard and garden, its fields of yellow grain and tall corn—the home of the happy family, will become part of a cattle range, for these alone can retain a foothold, until that other more distant day shall come when the winds and droughts shall reduce the plains of Wisconsin to the condition of Asia Minor.
—Allen Lapham, 1865[1]

Behind the false front of standing trees, a fringe of virgin growth a quarter mile deep, was the real business of the national forest: timber farms, lumber plantations, field factories for the joist, board, pulp and plywood industry.
—Edward Abbey, 1975[2]

Throughout the nineteenth century a counter-narrative challenged the story of the axe. The two stories offered sharply different interpretations of the clearing of the land. By the 1860s, although a majority of white Americans still celebrated the foundation story, a vocal minority was convinced that deforestation undermined future prosperity. The challenge became even more pronounced during the twentieth century as environmentalists rewrote the narrative of the axe into a story of rapacious disregard for nature. Likewise, Native Americans criticized the government for failing to protect forests, which they perceived as a nurturing environment.

James Fenimore Cooper was one of the earliest critics of the master narrative. His description of a professional logger stands in striking contrast to Zachariah Allen's description of a solitary pioneer:

Commonly selecting one of the most noble [trees], for the first trial of his power, he would approach it with a listless air, whistling a low tune; and wielding his ax, with a certain flourish, not unlike the salutes of a fencing master, he would strike a light blow into the bark, and measure his distance. The pause that followed was ominous of the fall of the forest, which had flourished there for centuries. The heavy and brisk blows that he struck were soon succeeded by the thundering report of the tree, as it came, first crackling and threatening . . . finally meeting the ground with a shock but little inferior to an earthquake. From that moment the sounds of the ax were ceaseless, while the falling of the trees was like a distant cannonading . . . [finally] the logger would collect together the implements of labor, light the heaps of timber, and march away, under the blaze of the prostrate forest, like the conqueror of some city, who, having first prevailed over his adversary, applies the torch as the finishing blow to his conquest.[3]

This logger is quite unlike Allen's settler, who cuts trees to make way for a log cabin. He works for hire and, like a mercenary, will destroy any foe he meets in battle. What Allen saw as "improving" the land becomes here a war on nature. Cooper treats the burning of the woods as an act of arson committed against a fallen city. The logger wields his axe like a fencing master. Rather than the founder of civilization, he is a skillful destroyer. Allen's woodsman builds a home and cultivates the land; the logger slaughters and departs. The logger represents the specialization of labor that occurred as land speculators, notably Cooper's own father, purchased large tracts and developed them for prospective settlers.

Slavery offered an even more telling counter-narrative to the heroic story of the axe. The "peculiar institution' did not fit into the log cabin myth of self-sufficiency and individualism, either in the ideological or the practical sense. Apologists for the slave system depicted African-Americans as childlike and incapable of self-discipline and independence, and their log cabins as symbols of dependence and poverty. Frederick Law Olmsted, a Northerner touring the South, noted that slaves lived in "small and rude log cabins."[4] A minister in eighteenth-century Virginia noted that some slaves lived in a hut "formed of small pine trees, laid one upon another and fastened at the end by a notch; but . . . not plastered, either on the inside or the outside."[5] Slaves had little time and few resources to improve their dwellings. The walls of log cabins in the Deep South were not tightly caulked; insects easily found their way inside and were almost impossible to eradicate. Olmsted noted on the eve of the Civil War that Southern log cabins had "wide interstices between" the logs, and these grew wider as the logs dried out and shrank. "Very commonly," Olmsted noted, "as you pass along the road, you may often see all that is going on in the house; and at night, the light of the fire shines brightly out on all sides."[6]

As Olmsted toured what he called "the Cotton Kingdom," he became convinced that reliance on slave labor had impoverished the region. He determined that in Virginia the average cost of performing a given task of labor under slavery was "more than double" what it was in the neighboring free state of Pennsylvania, where comparable farmlands had been improved to the point that they were worth three times as much. Indeed, he found, "the citizen class of Virginia earn very little and are very poor— immeasurably poorer than the mass of the people of the adjoining Free States."[7] Reporting on his 3,000-mile tour of the South, Olmsted declared: "For every mile of roadside upon which I saw any evidence of cotton production, I am sure that I saw a hundred of forest or waste land, with only now and then an acre or two of poor corn half smothered in the weeds; for every rich man's house, I am sure that I passed a dozen shabby and half-furnished cottages, and at least a hundred cabins—mere hovels, such as none but a poor farmer would house his cattle in at the North."[8] In this landscape, the log cabin signified poverty. Olmsted readily admitted that the largest cotton plantations could turn a profit and could afford to pay high prices for slaves, but the smaller producers suffered. A farmer who owned two families of slaves could not sell more than $300 worth of cotton a year on average; once production costs were subtracted, he was dirt poor. Olmsted commented: "I have seen many a workman's lodging at the North, and in England too, where there was double the amount of luxury that I ever saw in a regular cotton-planter's house on plantations of three cabins."[9] Olmsted estimated that only about 7,000 slave-owning families had workforces large enough to amass wealth. He measured them by the pragmatic method of counting slave cabins, ten being the minimum for real wealth creation.[10] The slave's log cabin, whether viewed by Southerners as a part of a benevolent patriarchal system or by Northerners as an emblem of exploitation, symbolized poverty and dependence.

Such an abode would not do for the black hero of an anti-slavery novel. Ironically, the most famous log house in American literature—that in Harriet Beecher Stowe's novel *Uncle Tom's Cabin*—is historically inaccurate. With its clean "drawing room" and appointments, it resembles a middle-class home more than a slave cabin. Tom and Chloe's modest but comfortable home stands in contrast to the luxury of the master's house. Stowe had adapted the ideology of the log cabin to the anti-slavery cause.

Stowe was not the only author who wished to tidy up the log cabin. Other observers also associated them with sloth, filth, and barbarism. This was particularly true of many European travelers, Charles Dickens among them.[11] A German was surprised to find that Pittsburgh's early inhabitants mostly lived in log houses and were "extremely inactive and idle; so much so that they are recalcitrant when given work and opportunity to earn money, for which, however, they hanker."[12] Like these Europeans, eighteenth-century Americans often understood the backwoodsman as a barbarian. Notable in this context is William Byrd's account of his journey to fix the boundary between North Carolina and Virginia. Unpublished notes by Hector St. John de Crèvecoeur likewise suggest the moralizing attitude some early commentators took toward such settlers. Of a town where "several families dwelt, on the most fruitful soyl I have ever seen in my life," Crèvecoeur wrote:

... the warmest Imagination can't conceive any thing equal to it—these People raised what they Pleased—oats, Peas, wheat, Corn with Two days Labour in the Week—at their doors they had a fair River, on their backs high Mountains full of Game, yet with all these advantages Placed as they were on these shores of Eden, they Lived Poor as the poorest wretches of Europa who have nothing, their Houses were Miserable Hovels, their stalks of grain Rotted in their Fields, they were almost starved. . . . This people had not nor cou'd they have scituated as they were any Place where they might convey their Produce, they could neither Transport, sell, or Exchange any thing they had, except cattle, so singularly were they Placed, this annihilated all the richesses of their Grounds rendered all their Labours abortive rendered them Careless slothfull & Inactive.[13]

The backwoodsman's indolent opposition to the work ethic also suggested the possibility of a less exploitative relationship to the land, however. Thoreau's *Walden* bore the subtitle *Life in the Woods,* and this phrase could hardly have been added without an awareness of the preceding decades' political mania for log cabins. That Thoreau chose to construct his cabin primarily of previously used boards immediately suggests that he was building counter to the foundation story. Indeed, he seems not to have chopped down many trees even for firewood. The chapter in *Walden* titled "Economy" reduces the basic necessities of life to the need to preserve body heat. Although he farmed a small area to raise beans, Thoreau did not seek to "improve" the land by removing the trees. In *The Maine Woods* he remarked: "I know of one who deserves to be called the Treehater, and, perhaps, to leave this for a new patronymic to his children. You would think that he had been warned by an oracle that he would be killed by the fall of a tree, and so was resolved to anticipate them. The journalists think that they cannot say too much in favor of such 'improvements' in husbandry; it is a safe theme, like piety. . . ."[14] During his sojourn at Walden Pond, Thoreau created a personal landscape balanced between nature and civilization. His was not a heroic tale of conquering and transforming the land; it was a quite different story of learning to minimize his material demands on his surroundings and trying to understand the forest as an environment.

Painters of the Hudson River School also appreciated the value of the "unimproved" natural world. By the 1830s, stumps had become iconic in landscape paintings, representing at once the triumphant march of progress and the melancholy destruction of unspoiled woodland.[15] Stump-filled clearings in front of log cabins were common in their paintings, as they were in Currier and Ives prints. Many scholars have interpreted such landscapes as ambivalent statements about how Americans felt about their conquest of the natural world. Barbara Novak effectively used the work of Leo Marx to analyze Thomas Cole.[16] The destruction of wilderness, the attempt to preserve a middle landscape between nature and civilization, and the overarching concern that America would decline and fall as past civilizations had were recurrent themes in Cole's paintings.[17] Cole was acutely aware of both the heroic narrative of settlement and the dangers of accelerated deforestation.[18]

By the 1850s, when Thoreau completed *Walden,* the American public was beginning to understand that forests were a finite resource. As westward migration rapidly transformed the landscape, the Bureau of the Census became concerned: "It is lamentable, in view of present ruthlessness and the demands of posterity, to observe the utter disregard manifested by the American people, not merely for the preservation of extensive groves, but the indifference which they exhibit for valuable trees, the destruction of which is not necessary to good cultivation. . . . We have seen thousands of farms rendered less productive and of much less intrinsic value by the destruction of timber. . . ."[19]

The Census Bureau anticipated the seminal counter-narrative to the story of the axe: George Perkins Marsh's *Man and Nature* (1864), which reconstructed the history of forestry and agriculture in the Mediterranean Basin during classical times and read these events as a warning. Developing a point of view that anticipated the study of ecology, Marsh emphasized the complex interconnections of forests, temperature, humidity, and flooding and the "spontaneous arrangements of the organic and inorganic world."[20] He concluded that overcutting and land abuse had led to dramatic changes in the fertility of the soil and in the Mediterranean climate, devastating agricultural productivity. Modern abuses of the same kind, if not checked, would cause similar devastation, and America would decline "to such a condition of impoverished productiveness, of battered surface, of climatic excess, as to threaten the deprivation, barbarism, and perhaps even extinction of the [human] species."[21] Marsh's book became popular and influenced many—notably Frederick Starr, the author of "American Forests: Their Destruction and Preservation" for the U.S. Commissioner of Agriculture's Annual Report of 1865. Starr calculated that as much as 10,000 acres of land had been cleared of trees every day during the 1850s, and he took from Marsh the "warning of history" that other great nations had been turned into wastelands by overgrazing and wanton destruction of forests.[22]

Perhaps the most dystopian rewriting of the narrative of the axe was a report by Allen Lapham that was commissioned by the State of Wisconsin the year after Marsh's book appeared. Its cover bore a quotation from Marsh, and its contents summarized both Marsh's work and Starr's statistics. Lapham inverted the foundation narrative completely, foreseeing a sequence of landscapes that began like those in popular lithographs but ended in disaster. After the pioneers cut down the forest to create farmland, Lapham predicted, the farmland would gradually dry up: "The small farm with its neat home, orchard and garden, its fields of yellow grain and tall corn—the home of the happy family, will become part of a cattle range, for these alone can retain a foothold, until that other more distant day shall come when the winds and droughts shall reduce the plains of Wisconsin to the condition of Asia Minor." Wisconsin would gradually turn into desert. "Trees alone," wrote Lapham, "can save us from such a fate."[23]

Another early critic who spread such concerns to the rest of the nation was the newspaper editor and presidential candidate Horace Greeley, who warned the farmers of Texas in 1871 that they were cutting down too many trees. Echoing Marsh's argument, Greeley pointed to the destructive effects of deforestation: ". . . France and Spain, Italy and Portugal, have for the most part been denuded of forests, and suffer often not only in the scarcity of Timber and Fuel, but in the severity and duration of their droughts, the fierceness and devastations of their gales, the violence and aggravation of their floods."[24] Greeley pleaded for selective cutting, and for replanting in "the more naked and arid portions of Texas."[25] He realized how much the clearing of forests had accelerated during and after the Civil War.

In retrospect, we know that between 1860 and 1910 more land succumbed "to the ax for farming than during the whole of the 250 preceding years."[26] During the 1870s alone, more than 49 million acres were shorn of trees.[27] To clear so much forest required the equivalent of 400,000 men working full time over the entire decade. Farmers did much of the cutting as they created new farms in Michigan, Minnesota, Wisconsin, and the Pacific Northwest. A fully industrialized logging industry systematically cleared land not to "improve" it for farming but simply to get the wood. This had first been the case in Maine. Logging then became a full-time occupation for thousands of men in Michigan, Wisconsin, and Minnesota after c. 1850, in the South after c. 1880, and finally in the Pacific Northwest after c. 1900. The scale changed as water-driven mills were replaced by steam engines and as corporations acquired large tracts of land that cut timber systematically and used purpose-built railroads to haul out the logs. The corporations then sold the cut-over land to settlers.

A Boston clergyman who visited Maine in 1867 was appalled by the scale of lumbering. It seemed that the entire state was under attack. Wherever he went, he reported, "the lumbermen have been there before you . . . and lumbermen are the curse and scourge of the wilderness." He continued: "A lumbered district is the most dreary and dismal region the eye of man ever beheld. The mountains are not merely shown of trees, but from base to summit fires, kindled by accident or malicious purpose, have swept their sides, leaving blackened rocks exposed to the eye, and here and there a few unsightly trunks leaning in all directions, from which all the branches and foliage have been burnt away. The streams and trout pools are choked with sawdust, and filled with slabs and logs."[28]

In an 1885 article in *The Atlantic,* the historian Francis Parkman—recalling Marsh's litany of examples from Europe, and particularly his account of flooding along the Po River due to "the partial denudation of the mountainous country about the sources of streams"—warned that the same calamity was inevitable in the United States if its forests were not preserved. Parkman approvingly quoted the Census Bureau's conclusion that too much logging "means the ruin of great rivers for navigation and irrigation, the destruction of cities located along their banks, and the spoliation of broad areas of the richest agricultural land."[29] Like other civilizations, America might decline: "The arid and comparatively valueless condition of certain parts of Spain is due to similar causes."[30]

What had begun in 1830 as a heroic story of self-sufficiency based on an individual with an axe became, after the 1860s, a tale of thoughtless land exploitation based on the abuse of powerful technologies. Examples were easily found. Farmers had removed much of the forest from Virginia, Ohio, Indiana, and Wisconsin. Clear cutting to produce charcoal in New York's Adirondack Mountains had denuded large areas of trees. The expansion of railways into the treeless plains had taxed the forests of Minnesota and Michigan. Thoughtful observers, most of whom had read Marsh, began to worry that such practices would increase the severity of flooding in the Mississippi, Ohio, Hudson-Mohawk, and other river systems.[31] In 1885, such utilitarian concerns, combined with the desire to preserve the region for recreation, led New York's legislature to create an Adirondack Forest Preserve consisting of scattered parcels of land totaling 681,000 acres. These parcels became the basis for a larger park established in the 1890s.[32] Similar preservation efforts took place in New England, in Pennsylvania, and in the southern Appalachians.

Even so, much of the West remained unprotected, and federal lands were still logged. In 1897, John Muir lamented: "The axe and saw are insanely busy, chips are flying thick as snowflakes, and every summer thousands of acres of priceless forests, with their underbrush, soil, springs, climate, scenery, and religion, are vanishing away in clouds of smoke."[33] The federal Forest Reserve Act of 1891 and later legislation mitigated the worst abuses, as did new national forests and parks.[34] Particularly during Theodore Roosevelt's presidency, much land was protected. At the protected sites, cutting down trees was either prohibited or regulated. This appar-

Figure 4.1
Cut-over land in Mount Hood National Forest, Oregon. Photograph by Arthur
Rothstein. Courtesy Library of Congress.

ently amounted to an official revision of the narrative of the axe. Yet even
in National Forests large areas were cut over. As Michael Williams details
in *Americans and Their Forests,* since the late nineteenth century the question has been that of finding the proper balance of wilderness, parkland,
recreational areas, reforestation, and changing logging practices.

As logging became a specialized form of labor, the log cabin declined as
a form of housing. By 1900 the axe and the log cabin were sentimental
memories in the urban East and in much of the Midwest, and Americans
embraced a romanticized ideal of the pioneer's cabin. Frederick Jackson
Turner's frontier thesis drew on this popular imagery. Like Crèvecoeur,
Rush, and Franklin, Turner described a series of stages in the conquest of
western lands by successive waves of people, led by the frontiersman. As
early as 1890, inspired by Theodore Roosevelt's book *The Winning of the
West,* Turner composed a short essay on "The Hunter Type." It began with
these sentences:

On the western outskirts of the Atlantic colonies, pressing continually toward the west, dwelt the American backwoodsmen. (They found too little elbow room in town-life. They loved to hear the crack of their long rifles, and the blows of the ax in the forest. A little clearing, edged by the woods, and a log house—this was the type of home they loved.)[35]

Turner's frontiersman did not clear the land in order to farm; rather, he was another incarnation of Cooper's Leatherstocking, retreating ever westward until reaching the Pacific coast.

Many of Turner's contemporaries shared his nostalgia for life in the woods. They built log cabins not out of necessity but for recreation. William Wicks's book *Log Cabins and Cottages: How to Build and Furnish Them* went through four editions by 1900. The book begins as follows: "Some recent anthropologists regard the amusements of the chase, as cultivated by civilized men—hunting, fishing and the like—as 'traces in modern civilization of original barbarism.' If there is any truth in this theory, then the writer must confess that he is in a large measure a barbarian."[36] Wicks quite consciously recognized that, unlike the pioneers, his readers would "migrate to the woods, hunt and fish from choice; . . . for change, recuperation, pleasure, health." "We aim," he wrote, "to treasure up energies in order to better sustain the tension of civilization. Health is imperative, and demands a dwelling in the woods in many points resembling a civilized one."[37] Wicks explained different methods for notching the ends of logs, various kinds of roofing (including thatch), and how to make beds, chairs, and tables out of logs. The book concluded with sketches and floor plans of completed buildings, some of them grand structures—this was not really a do-it-yourself manual. Wicks advised hiring an architect, bringing him to the site, and working from detailed plans.

A few years later, Oliver Kemp published *Wilderness Homes,* a longer work intended for those who did not want or could not afford an architect but wished to build a summer cabin. Kemp's house plans were smaller and simpler than Wicks's. He listed the items needed and estimated that they would cost $215.25. And unlike Wicks, Kemp gave advice on what kind of axe to purchase, how to judge its fitness for the size of one's hands and fingers, and how to make sure the axe was "hung" correctly. His instructions on how to build a cabin explained in detail how to cut down and trim trees and even specified how many logs (119) one should buy in

Figure 4.2
Lobby of El Tovar Hotel at Grand Canyon. Courtesy Library of Congress.

order to avoid that labor. These differences aside, both Wicks and Kemp assumed that their readers wished to escape from the city into the restorative forests.

The log cabin, once celebrated as the emblem of the destruction of the forest, also was linked to the popular enthusiasm for the new national parks, where it served as the motif for much of the architecture. The train depot at Yellowstone National Park was a "modified log cabin, constructed of stone and wood." The Old Faithful Inn (1904) was made of unpeeled logs, local stone, and cedar shingles, and an early architectural critic wrote that this building seemed "a product of the forest, built with ax, saw, and hammer."[38] Rustic hotels at Yellowstone National Park and on the rim of the Grand Canyon also featured logs and local stone in their construction and Native American designs and artifacts in their decor. Other upscale western hotels were built along similar lines, and throughout the West log hotels, log gift shops, and log restaurants served tourists, most of them from the East. After 1920, with the growth of automobile touring, such structures spread to the rest of the country. They mimicked the appearance of log cabins, but they contained modern conveniences unknown to the pioneer, including electricity, indoor plumbing, and gas stoves.

A few log cabins were symbolically preserved in cities. In Salt Lake City, a one-room cabin moved from an early farm stands in Temple Square. In Dallas, the "John Neely Bryan Cabin," built in 1843 by the founder of the city, stands on the courthouse lawn.[39] The log cabins of pioneers, founders, and famous Americans (many of them reconstructed) have become tourist sites across the United States. Abraham Lincoln's alleged birthplace was exhibited at the Tennessee Centennial in 1897 and at the Buffalo Exposition of 1901, stored at various sites near New York City, then purchased and set up inside a modern building in Hodgenville, Kentucky. (In transit, the cabin somehow shrank by 4 feet.) One can also visit a replica of Lincoln's second log cabin, a restored log cabin in which Lincoln's parents are said to have been married in 1806, the girlhood cabin of Lincoln's mother, the foundations of the cabin Lincoln's father built when the family moved to Indiana, a replica of the log cabin law office where young Abraham apprenticed at the law, and the restored "Log Cabin Courthouse" where he first practiced law.[40] (Lincoln's assassin apparently was born in a log cabin, too.[41]) Biographers incorrectly declared that Presidents Andrew Johnson and Ulysses S. Grant were born in log cabins, and James Garfield also claimed that distinction.[42] Theodore Roosevelt, a well-born New Yorker, could not plausibly claim birth in a log cabin; however, a log cabin in the Dakota Territory in which he lived for a short time became a featured attraction of the Theodore Roosevelt National Memorial Park. Other log memorials attest to the pioneer origins of many Americans (including Brewster Higley, author of the poem that eventually became the song "Home On The Range").[43] As these examples suggest, the log cabin became the imaginary first dwelling of all Americans, the point of departure from which they could measure change from the country's settlement to the present.

In 1951 the Indiana Historical Society published *A Home in the Woods: Pioneer Life in Indiana,* written by an elderly man named Howard Johnson who claimed to be repeating stories his grandfather had told about settling the land in the 1820s. The first-person narrative begins: "It's hard to picture this part of the country as I first remember it. Here and there was a cabin home with a little spot of clearin' close by. The rest of the country was jist one great big woods and miles and miles in most very direction. From your cabin you could see no farther than the wall of trees surroundin' the clearin'; not another cabin in sight."[44] The book describes how a cabin was built and how the remaining trees were then systematically removed in stages, the smaller ones first. Large trees were girdled so that their leaves would wither and fall, opening up the land to the sun. Later came log rollings and large-scale burnings. The first-person narration makes it seem that one is hearing the primordial voice of the pioneer and implies that this was how it was always done from the time of the arrival of the first settlers in the seventeenth century. The reader receives no hint that it was otherwise.

In J. D. Salinger's novel *The Catcher in the Rye,* published shortly before Johnson's book, the protagonist imagines his escape from New York City. He wants to leave the "phony" world he has known and the prep schools he has attended, hitchhike west, get a job pumping gas, and live a simpler life in a cabin at the edge of the woods. Such modern clearings in the wilderness are less a new space inscribed in the world than a refuge from civilization. Another version of this same impulse is the high-tech business corporation that relocates to Idaho or Montana so its engineers and software developers can live adjacent to designated wilderness areas on large lots where the houses blend into the surrounding trees and underbrush. (Ironically, such houses are vulnerable to the fierce fires that are an unintended consequence of zealous fire-prevention efforts.[45])

By the 1970s, log houses had become a "lifestyle" choice. More than 10,000 were built each year, many of them assembled from pre-cut kits with numbered parts.[46] More than twenty firms sold such kits, though some aficionados preferred to construct cabins in the old way from nearby trees, a few even using hand tools. Most of the new log cabins had electricity, running water, and other amenities that would have been inconceivable to nineteenth-century pioneers. With the mass production of pre-cut logs for easy assembly, the narrative of the log cabin reached the point of logical self-contradiction. Not only had the machine displaced human muscle power, but the purpose of the structure itself was no longer shelter from the wilderness. Its purpose now was escape from civilization. Though log walls undoubtedly provide good insulation, the cultural meanings constructed out of logs are unstable. Not only is the imagination of such a structure as American culture's point of origin historically incorrect; it became a paradoxical, high-tech simulacrum for a simpler way of life. The praxis of log cabin building had been gradually transformed into its own negation.

The figure of the man with an axe was also transformed into a series of tall stories about Paul Bunyan and his enormous blue ox, Babe. Created by an advertising copywriter in 1914 (though conceivably based on loggers' stories that had not been widely disseminated), the legend of Paul Bunyan was the centerpiece of the Red River Lumber Company's advertising. In the following generation it was embraced, particularly in the upper Midwest. The folklorist Richard Dorson found that Paul Bunyan was kept in circulation by "tourist and resort promoters, writers and lecturers, professional artists, and feature-minded journalists," and that "various entrepreneurs attached it to their enterprises and made his name still better known."[47] There were festivals, contests, and pageants in his honor. The Minnesota town of Bemidji erected a concrete-and-steel statue. The California town of Fort Bragg put on a show titled "The Return of Paul Bunyan," and similar shows were staged in New Hampshire, Michigan, Washington, and Wisconsin.[48] At the 1939 New York World's Fair, the Health Building featured an enormous statue of Paul and Babe.[49] Dorson concluded that Paul Bunyan became a "pseudo folk-hero of twentieth century mass culture, a conveniently vague symbol pressed into service to exemplify 'the American spirit.'"[50] The massive muscular figure was the ultimate exaggeration of the nineteenth-century pioneer entering the woods with an axe.

Thus the foundation narrative of the axe and the log cabin invented in the late eighteenth century passed through a distinct series of stages. In its first incarnation, the frontiersman used the axe to prepare the ground for more sedentary and dedicated farmers, and the log cabin was a crude precursor of better housing. By the early nineteenth century, both the woodsman and the tool had risen in status, and presidential campaigns discovered the appeal of supposed birth in a log cabin. It had become the paradigmatic first house, from which all else began. Whitman gave this new narrative full expression; it could also be found in popular songs, in political parades, and in oratory. The counter-narrative, emphasizing the destruction of the forest and the loss of a primeval landscape, began to be articulated by Cooper and other writers, and the tension between the heroic story and the emerging counter-narrative was expressed in landscapes by the Hudson River painters. At the middle of the century, Thoreau and particularly Marsh inspired others to rewrite the narrative of the axe. At the end of the century, when the nation was rapidly urbanizing and far fewer log cabins were needed as primary residences, thousands of Americans built modern log cabins as vacation retreats, and the log cabin became the inspiration for a rough-hewn architectural style used in the hotels at the new national parks[51] and in luxurious second homes. In this nostalgic narrative, axe, clearing, and cabin were linked icons of the origins of America. Emerson Hough summarized this view of the American axe as an instrument of civilization: "If you can not use it, you are not American. If you do not understand it, you do not understand America."[52] In contrast, John Muir declared: "Any fool can destroy trees. They cannot run away; and if they could, they would still be destroyed—chased and hunted down as long as fun or a dollar could be got out of their bark hides, branching horns, or magnificent bole backbones."[53] The heroic figure of the man with an axe survived as the nostalgic figure of Paul Bunyan, invented by ad men and embraced a century after the foundation narrative first appeared.

Yet neither the nostalgic narrative of the log cabin nor the dystopian critique of logging was the final variation on the heroic story. Modern logging companies do not ravage the landscape. Rather, they replant systematically, imposing a new order on the land. Nineteenth-century German foresters first applied mathematical analysis to tree production so that they "could calculate the volume of wood in a given topography, project the growth rates of forests far into the future, and prescribe time frames for the felling of trees according to precise mathematical charts."[54] This systematic approach, adopted in the United States during the 1920s and the 1930s, was applied to cut-over regions such as northern Wisconsin, Minnesota, and Michigan.[55] This kind of forestry prescribed planting to replace the disorderly natural forest with rows of trees grown like wheat or corn. In 1919, Gifford Pinchot, who had helped establish the national forests under President Theodore Roosevelt, declared: "The forests which will be raised from now on will not be tangles of wilderness, left alone for a century or so and then ripped off so as to leave the country desolate and poor. Instead, they will be carefully tended and protected and, once established, will be permanently productive. Work in the forests will become a regular and permanent business."[56] Indeed, during the 1920s the Ford Motor Company used techniques of mass production to "harvest" trees from Ford-owned forests, saw them up, and replace them with seedlings.[57] Geometrical order was thus imposed on the woodland. Whereas Native Americans had seen the woods as a home environment, and late-nineteenth-century Americans had seen them as a sanctuary from the industrial order, modern forest management imposed a utilitarian order. The power saw replaced the axe; clear cutting replaced the clearing; replanting ensured an endlessly repeatable process. Adam Smith became the animating presence of the forest, which was transformed into another high-tech workplace. Forests had become renewable resources.

In 1981, President Ronald Reagan expressed this ethic in declaring National Forest Products Week:

The contribution forests must make to our Nation's welfare will remain just as great in the years ahead as in the past. To meet the needs of the future, our forests must benefit from effective timber management and from continuing research to find better ways to utilize forest products. Improved wood growth and usage will make more wood products available at affordable prices while helping to stimulate our entire economy. America has been greatly blessed with the resources of our forests. To allow them to waste away, when they could benefit so many, would be to ignore our responsibilities of stewardship. Our forests must be managed in ways that are environmentally safe and that ensure they will be available for the enjoyment and use of future generations. . . . The need and opportunity to commune with nature, to seek solitude, and to appreciate the beauty and grandeur of America's forests must be respected and preserved. With wise forest management, the demands of aesthetics and economics will remain compatible.[58]

In such rhetoric, the forests have an obligation to provide humans with their bounty; therefore, they *must* be managed, for otherwise, curiously, they would "waste away." Clearly, forests cannot take care of themselves. In this revised narrative of the axe, wisely managed forests can be cut down again and again forever. A perpetually renewable forest preserves the possibility of reenactment of the primal scene in which a solitary man with an axe enters the untouched virginal forest to make a clearing and build his first cabin.

But not all Americans celebrate "National Forest Products Week." Contemporary attacks on the logging industry can be found in the militant literature of the Earth First! movement. In Edward Abbey's novel *The Monkey Wrench Gang,* two environmental activists destroy logging equipment near the Grand Canyon's North Rim. Along the main road they see "what appeared to be, uncut and intact, a people's national forest." But "behind the false front of standing trees, a fringe of virgin growth a quarter mile deep, was the real business of the national forest: timber farms, lumber plantations, field factories for the joist, board, pulp and plywood industry."[59] In protest, the activists destroy a large bulldozer, a truck, and other expensive equipment. Such Luddism (not always confined to fiction) disputes a fundamental tenet of the foundation narrative: that nature is raw material waiting for development.

Ernest Callenbach's 1972 novel *Ecotopia* envisions a positive alternative to modern forest management. His environmental utopia rejects economic growth in favor of a steady-state economy that emphasizes mass transit, population limits, and recycling. Callenbach portrays humans living in harmony with the environment. Forests are cared for by communities of men and women who live permanently in the forest, which Ecotopians regard as a mystical center of awareness. They do not allow clear cutting, and they mix species of trees to minimize the danger of insect and fungi invasions and to preserve the biotic community. In Ecotopia, wood is widely used, since it is a renewable resource, yet trees are regarded as "being alive in an almost human sense."[60] Accordingly, Ecotopians reforest farmland and designate mountainous areas as wilderness. The idea of the clearing is rejected. One woman declares: "I feel best when I'm among trees. Open country always seems alien to me. . . . Among trees you're safe, you can be free."[61] *Ecotopia* idealizes neither the axe nor the clearing, but selective cutting and preservation.

The narrative of the axe had always ignored the original inhabitants of North America, and few Native Americans could regard the clear cutting of forests as "improving" the land. Their forest was not wilderness; it was home. Yet the narrative of the axe, as told by Zachariah Allen and other Jacksonians, did not mention Native Americans. The lands an axe-wielding pioneer entered had previously been taken from them, and yet the foundation narrative almost never depicted the axe's potential use in warfare. Fighting the savages was central to the narrative of conquest; it pitted the native tomahawk against the American rifle. Settling the land belonged to the entirely different story of completing the Creator's intentions. This division of the story of settlement into two narratives may seem illogical and hypocritical; however, the American axe usually belonged not to the story of combat but to that of settlement.

In denying the validity of forced land sales by Native Americans to the federal government, the nineteenth-century leader and warrior Tecumseh declared: "Sell a country! Why not sell the air, the clouds, and the great sea, as well as the earth?"[62] Tecumseh was eminently logical in his views, but most Americans had accepted the technological creation story. They believed that whites were manifestly intended to "improve" the land. Benjamin Harrison, the man who "negotiated" with Tecumseh, later asked a revealing rhetorical question while speaking to the Indiana legislature: "Is one of the fairest portions of the globe to remain in a state of nature, the haunt of a few wretched savages, when it seems destined, by the Creator, to give support to a large population, and to be the seat of civilization, of science, and true religion?"[63] Harrison saw providential intentions in the landscape, which seemed designed to "support a large population" and not to remain in an undeveloped "state of nature."

Perhaps the most effective answer to Harrison and to the story of the axe can be found in Louise Erdrich's 1988 novel *Tracks*, set between 1912 and 1924, a generation after the Ojibwa Nation lost sovereignty over its lands and was forced to live on reservations. The government, seeking to "Americanize" a tribe of Native Americans, no longer permits them to hold land collectively. Instead, the tribe is divided into family groups with separate holdings. The final chapter concerns the logging of the reservation by a white company, which has obtained a government contract to do so after land taxes are not paid on time. The reservation agent has exploited internal tribal divisions, pitted one family against another, and arranged the sale of the forest. After some resistance, only one patch of forest remains. It stands around the home of Fleur, a Native American woman. But Fleur's forest is an island in a cut-over sea of scrubland. Denuded by the axe, that land has become economically and spiritually valueless. Fleur's home neatly inverts the dominant narrative: an Indian home instead of a white cabin, a woman instead of a man, a nurturing forest instead of a clearing.

For Fleur, the axe does not open the landscape to production; it empties it of life and meaning.

As the logging company closes in, the narrator, an old man named Nanapush, goes to see Fleur in her isolation. For days the wind has been ominously still. Nanapush passes "among the twisted stumps of trees and scrub . . . through the ugliness, the scraped and raw places, the scattered bits of wood and dust and then the square mile of towering oaks, a circle around Fleur's cabin."[64] The geometrical figure of the circle is itself fraught with meaning, contrasting with the squares surveying parties carve out of the forest according to the national grid.[65] At the edge of the forest, the logging crew is ready to begin its destructive work. To Nanapush, this aboriginal forest is alive with memory: "The moment I entered, I heard the hum of a thousand conversations. Not only the birds and small animals, but the spirits in the western stands had been forced together. The shadows of the trees were crowded with their forms. The twigs spun independently of wind, vibrating like small voices. I stopped, stood among these trees whose flesh was so much older than ours, and it was then that my relatives and friends took final leave, abandoning me to the living."[66]

He finds Fleur at her front door. Her husband, who has joined the logging company, is trying to convince her to leave before the cutting begins. But Fleur refuses to go. She stands in her little clearing within the circle of oaks and waits for the wind to rise, knowing that with a stolen axe and saw she has cut the trees at their bases. When the first trees begin to fall before the wind, Nanapush realizes that around him "a forest was suspended, lightly held. The fingered lobes of leaves floated on nothing. The powerful throats, the columns of trunks and splayed twigs, all substance was illusion." In a startling reversal, the forest, which the loggers meant to destroy systematically and rationally, becomes their destroyer. In a gust of wind, the trees suddenly fall on men, wagons, and horses. "The limbs snapped steel saws and rammed through wagon boxes. Twigs formed webs of wood, canopies laced over groans."[67] When the wind dies down, the forest has fallen. Fleur leaves her home and her tribe, dragging a small cart. In subsequent years, Nanapush realizes that they have become "a tribe of file cabinets and triplicates, a tribe of single-spaced documents, directives, policy. A tribe of pressed trees."[68]

A foundation story in the first half of the nineteenth century, the narrative of the axe was almost simultaneously rewritten into a counter-narrative. It was then recomposed again as a nostalgic story. By 1900, the same axe that in the 1830s had been a new technology enabling the individual to carve a family farm out of the wilderness had become a representation of a simple life that had slipped away. All these variants continue to circulate in American culture.

Recent rewritings of the axe narrative imagine its reversal through reforestation or evoke a prelapsarian world of untouched woodlands. Callenbach's ecological utopia posits a future in which society rejects the principle of growth, maximizes recycling, reduces its population, and reforests much of the land. The citizens of his Ecotopia have recovered the kind of veneration for the wooded landscape once found in ancient Greece or in Native American societies. In contrast, Erdrich reimagines the past and evokes a pre-technological relationship between Native Americans and their lands. She asks her predominantly white readers to rethink the story of the axe. The heroic tale of a man transforming the forest from trees to logs, from wilderness to home, and from empty space to civilization has become a narrative that moves from a woman's forest home to a desolate clearing, from the triumph of the logger to his destruction, and from an ecosystem to a wasted land.

5

The Mill, or "Natural Power"

... an all-wise Providence has denied to the barren hills of New England the mines of coal, which would allow the inhabitants to congregate in manufacturing cities, by enabling them to have recourse to artificial power, instead of the natural water power so profusely furnished by the unnumerable streams. ...
—Zachariah Allen, 1832[1]

Men having discovered the power of falling water, which after all is comparatively slight, how eagerly do they seek out and improve these privileges. Let a difference of but a few feet in level be discovered on some stream near a populous town, some slight occasion for gravity to act, and the whole economy of the neighborhood is changed at once.
—Henry David Thoreau, 1843[2]

The second technological creation story to emerge was that of the water-driven mill. In the first half of the nineteenth century, many Americans viewed mills not merely as transformers of raw materials and local economies but as the seeds of new communities. In this second version of the technological creation story, the mill was a dynamic first cause; towns grew around it. Edward Kendall observed American conditions in 1808 and provided a typical narrative about how communities grew up in thickly wooded areas wherever there were good mill sites. This is a foundation story in the literal sense, explaining how society comes into existence. In it, the first building is not a fort, a church, or a house, but "a solitary saw-mill":

To this mill, the surrounding lumberers, or fellers of timber bring their logs, and either sell them, or procure them to be sawed into boards or into plank, paying for the work in logs. The owner of the saw-mill becomes a rich man; builds a large wooden house, opens a shop, denominated a store, erects a still, and exchanges rum, molasses, flower, and port, for logs. As the country has by this time begun to be cleared, a flower-mill is erected near the saw-mill. Sheep being brought upon the farms, a carding machine and fulling-mill follow. For some years, as we may imagine, the store answers all the purposes of a public-house. The neighbours meet there, and spend half the day, in drinking and in debating. But the mills becoming every day more and more a point of attraction, a blacksmith, a shoemaker, a taylor, and various other artisans and artificers, successively assemble. The village, however, has scarcely advanced thus far, before half its inhabitants are in debt at the store, and before the other half are in debt all round. What therefore, is next wanted is a collecting-attorney . . . whom the store or tavern-keeper receives as a boarder, and whom he employs in collecting his outstanding debts, generally secured by note of hand. The attorney is also employed by the neighbours.

But as the advantage of living near the mills is great, even where there is not (as in numerous instances there is) a navigable stream below the cataract . . . so a settlement, not only of artisans, but of farmers, is progressively formed in the vicinity; this settlement constitutes itself a society or parish; and, a church being erected, the village, larger or smaller, is complete.[3]

Harnessing water power calls forth an entire community, with the mill at its heart. (In later years the stimulus to development would be a canal, a railroad, or an irrigation dam, but the pattern to be abstracted from each story was similar.) An individual invests labor and resources to create a new source of power. The action that follows this foundational event seems almost automatic. As the water ceaselessly falls to drive the mill, laborers appear with timber, enriching the mill owner. A local market economy begins to emerge. Much of the action is described in the passive voice—for example, the countryside "has begun to be cleared." "Sheep being brought upon the farms," the machines needed to process wool "follow." Humans seem less agents than beneficiaries of this process. The mills "become" a point of attraction, and a settlement "is progressively formed in the vicinity" as artisans, shopkeepers, and a lawyer are drawn to it.

Yet, as the counter-narratives were later to make plain, this story also left out a good deal—notably the Native Americans who inhabited the land before the solitary mill owner appeared and the deleterious effects of a mill dam on migratory fish and fishing communities. It also is a secular story: the community is founded for economic reasons, and the church is the last institution to appear. The narrative of the village in the valley is not the colonial narrative of the city on a hill. The social center is the mill, not the Puritan meeting house, and a lawyer is needed before a clergyman. The narrative stops long before the countryside is stripped of trees and the original mill has closed down for lack of wood; nevertheless, it seems to reach a point of completion, as though the village were fully formed at the conclusion of the story.

This fable of origin, though a selective organization of fact, was rooted in historical experience. The evidence overwhelmingly supports Kendall's claim that the mill was central to community development.[4] If the first colonists of New England and the Atlantic seaboard founded communities and then looked for millers to serve them, the process was soon reversed. In Maine, the town of Berwick emerged in 1713 at the waterfall on the Salmon Falls River. Its water power attracted many mills and remained the basis for local industry until well after the Civil War. In colonial Connecticut, grist mills were built within 2 years of first settlement

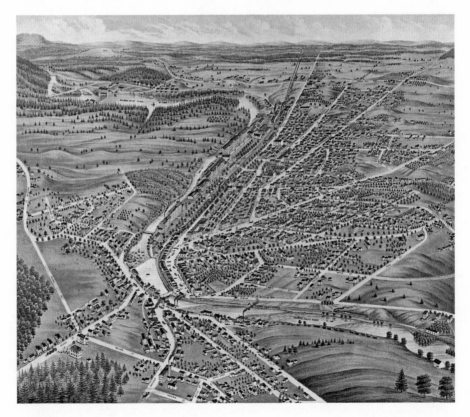

Figure 5.1
"Bird's Eye View of Great Falls & Berwick." Courtesy Library of Congress.

and "frequently preceded incorporation."[5] As settlement moved westward, a mill seat commonly was established first, as John Stilgoe found: "In western New York and Pennsylvania, and everywhere in the Old Southwest, in fact everywhere where New Englanders did not choose to replicate the clustered towns of home, mills served as economic and social centers."[6] Settlements formed around the nodal point of a mill, and Thomas Jefferson imagined a largely agrarian nation of farmers and mill seats of the sort that Kendall described. Even in New England, where the Pilgrim and Puritan fathers had focused on establishing religion, their descendants, as they pushed inland, did not start a settlement by building a church. Charles Clark found that settlers in northern New England changed their priorities: "The order in which the proprietors of New Marblehead attacked the various tasks of settlement after the distribution of town lots shows that they assigned priority first to roads, next to a sawmill, then to a cart bridge over the Presumpscut, and finally to a meeting house."[7] As in Kendall's account, the mill preceded the church.

Because a mill typically attracted farmers who lived within 10 miles in any direction, the miller often kept a store. While the mill did its work, the farmers talked and traded. Crèvecoeur observed: "There are but few people who are at any considerable distance from Grist Mills & that is a very great advantage considering the prodigious quantity of Flour which we & our Cattle consume annually."[8] The need for milling was frequent. The farmer did not grind all his corn at once, since it kept better as whole grain. Once ground, corn became moldy in a few weeks. In new settlements without good roads, a horse could carry no more than two or three bushels through forests and rough terrain. Frequent trips to the mill became a fixed routine, often delegated to an older child.[9] Cash payment was rare; the miller usually took between one-twelfth and one-eighth of the meal in trade, though he might lend the farmer money against a coming crop.

Grist mills were common in Britain, but sawmills were not. In the 1830s, Zachariah Allen found, to his surprise, that "a common saw mill, which is considered indispensable to the convenience of the inhabitants of every village of the United States, remains yet to be generally introduced throughout England as a species of labor-saving machinery."[10] The difference was due in part to the resources available in each country. There were fewer undeveloped mill sites available in Britain, where most forests had long since been cut down. Furthermore, sawyers had a vested interest in continuing their labor-intensive work. Until the late eighteenth century, "the English guilds of sawyers insisted that logs be sawed by hand." Using muscle power, two men could produce in a day "perhaps five or six 10- to 13-foot-long boards an inch or two in thickness."[11] With perfect health, a six-day week, and no holidays, they could produce no more than 2,000 boards in a year. But as early as 1652 one New Hampshire sawmill was cutting 60,000 boards and planks a year.[12] Using water-driven machinery, it could produce as much lumber as 60 English sawyers.

Despite such achievements, and despite the obvious importance of water power to early settlements, colonial historians seldom give mills a prominent place. In his otherwise excellent book on the settlement of the Connecticut River frontier in Massachusetts in the middle of the eighteenth century, Gregory Nobles almost ignores mills, focusing instead on such matters as patterns of migration, varieties of farming, soil quality, the Great Awakening, and domestic economy. Yet the locations of many towns were determined by access to water power. Crèvecoeur, who traveled widely in remote areas, often working as a surveyor, remarked: "You'd think by the Ingenuity displayed on Saw Mills erected on these Rough shores that the country has been Settled these Thousand years."[13] Crèvecoeur found one group of new settlers on the Susquehanna River who had almost no more resources than those that had been needed to transport themselves to the spot, but not a person was idle, and "they had already Erected good number of Saw Mills with which the settlement was supplied with all the boards & scantling they wanted. . . . They had already begun to float them down the River in Rafts very ingeniously fastened together."[14]

Sawmills not only led to the establishment of towns; they also encouraged the use of rivers to transport produce. Indeed, the mills were "built at a small distance from the Mouths of the Creeks on which they stand & Navigable for boats to their very Gates."[15] Such a location had important advantages. In his travels, Crèvecoeur noted areas where "the Roads were extremely bad, they had no bridges, no mills were to be found but at a considerable distance, no Mechanicks; they were sometimes gone 3 days before they co'd bring home 2 bushels of Flour."[16] Even in such remote locations, however, people did not grind grain by hand, since that took even longer. "To grind a bushel of wheat to flour," Eugene Ferguson notes, "requires about two horsepower-hours of work, the equivalent of two man-days of strenuous effort. In a horse-drawn mill, the process would require a third of a day, while a water mill of moderate size might produce fifty bushels of flour a day."[17] The farmers Crèvecoeur described, who spent three days getting two bushels ground to flour, were actually saving a day by not grinding it by hand. Thus, a town with a grist mill attracted anyone within two days' travel.

In Pennsylvania, Moravian settlers built a community around a good mill site. A French traveler noted:

The Moravians had built a sawmill the most beautiful and best contrived I ever saw. A single man only is necessary to direct the work, the same wheels which keep the saw in motion, serve also to convey the trunks of trees from the spot where they are deposited to the workhouse, a distance of twenty-five or thirty toises: they are placed on a sledge, which sliding on a groove, is drawn by a rope which rolls and unrolls on the axis of the wheel itself. . . . This mill is near the fall of a lake which furnishes it with water. A deep cut is made in a rock to form a canal for conducting the waters to the corn-mill, which is built within a musket shot of the former. . . .[18]

Many communities were created around mills. One example is Harrisville, New Hampshire, where a stream with three natural ponds falls 600 feet in 9 miles. Settlement near this promising power source, delayed by the French and Indian War, was not safe before the peace of 1763. The first settlers wanted not farmland but sites along the stream, where they immediately erected mills. In 1774, Abel Twitchell chose "a rough, rocky lot with [a] steep gorge and turbulent brook"; the lot could be of no agricultural use, but "it was undoubtedly these features that attracted Twitchell," who "soon had built a grist and a saw mill."[19] The next settler—Jason Harris, from Farmington, Massachusetts—built a blacksmith's shop with a trip-hammer a short distance downstream. In 1799 a third mill was built for fulling wool and finishing cloth. The mills attracted settlers, and Harrisville became an industrial town.

Mills also attracted settlers to western New York. After 1800, Dutch investors opened the huge Holland Purchase to settlement, but the absence of roads and mills made it difficult to sell undeveloped land.[20] Once a sawmill was built at Batavia, "newly arrived settlers took immediate advantage of [it] to procure boards for a house and business, preferring frame construction over log."[21] Other land companies followed suit, erecting grist and saw mills from Ohio to Kansas as a stimulus to land sales. In frontier Wisconsin in 1840, for example, 88 percent of all the capital invested in manufacturing was invested in mills.[22] As late as 1859, the arrival of a ready-to-erect mill in Manhattan, Kansas, pulled by twenty oxen, was "a greater event to the citizens . . . than the arrival of the Union Pacific Railway eight years later."[23]

Lumber and flour mills were given special privileges under the law. A Massachusetts law enacted in 1693 exempted one miller from each grist mill from military service.[24] The Mill Act of 1713 granted mill owners the right to build dams "regardless of the wishes of upstream riparian owners objecting to the flooding of and injury to their lands."[25] Any landowner who lost meadows to the rising waters behind a dam received damages, but these were assessed by a local jury, many of whom were likely to benefit from the mill. State law also encouraged the establishment of mills through the use of the power of condemnation. A mill owner who needed to gain control over the opposite bank of a steam from an unwilling landowner could reasonably expect a favorable outcome from a petition to the state. This was the case not only in New England but also in Virginia, Alabama, Kentucky, and Tennessee.[26] Nor was a mill owner restricted to flooding the area that was inundated by the first dam constructed. A 1796 Massachusetts law specified the right to overflow new areas in order to "continue the same head of water."[27] State and federal courts sustained such acts. They declared that mills, though private, served a public purpose and therefore could take private property from others under the laws of eminent domain. In *Head v. Amoskeag Manufacturing Company* the Supreme Court noted that "in most states the validity of the mill acts was assumed without dispute" until after the Civil War.[28]

State laws encouraged and supported the building of mills in part because the classic foundation narrative of the mill was widely accepted as a description of the formation and growth of communities. By the end of the eighteenth century, the narrative was underpinned by legal authority. Not only was the miller one of the first inhabitants in many a new community; he was also given stronger legal guarantees than other property owners. The special legal status granted to millers was rooted in colonial experience, but it would provide legal protections for industrialization too. During the western migration of the nineteenth century, not only did the foundation story seem to be accurate and true; it enjoyed judicial sanction. As presented by Edward Kendall and other early writers, it was the story of entering an unsettled country, taking control of its unused water resources, and attracting settlers to the site, where a town then quickly emerged. The natural landscape dictated the location of this second creation, and the power developed there was natural, coming from rain and melting snow. Ultimately, it seemed, little or nothing was removed from or added to the natural order. All the water was returned to the stream, flowing to the sea as before. The mill multiplied the force available for human use without depleting the water supply. Small mills seemed to be harmonious extensions of the original creation.

The large mills of the nineteenth century inherited the foundation story of colonial mills, their legal privileges, and the widespread perception that water power was natural. In comparison, steam power was not widespread. The 1840 census counted more than 66,000 mills driven by water[29] and fewer than 1,000 by steam.[30] Waterpower was by far the largest factor in American manufacturing. Water mills employed more than twice as many people as the iron industry and three times as many as machinery manufacturers, hardware manufacturers, cutlery manufacturers, and firearm manufacturers combined.[31] In 1836 an American traveler told what had become the characteristic tale of rapid second creation: "It is wonderful how all the western towns flourish which possess 'water privileges.' How extraordinary, for example, is the growth of Seneca Falls [New York]! Not long ago it was a mere hamlet, beside a little stream; now it is almost a city. . . ."[32] A British observer noted: "A retired valley and its stream of water become in a few months the seat of manufacturers; and the dam and water-wheel are the means of giving employment to busy thousands, where before nothing more than a solitary farmhouse was to be found. Such, in a few words, is the history of Waterbury, and all the Naugatuck settlements of Holyoke, Chicopee, Lowell, and Lawrence."[33] Thoreau commented: "Men having discovered the power of falling water, which after all is comparatively slight, how eagerly do they seek out and improve these privileges. Let a difference of but a few feet in level be discovered on some stream near a populous town, some slight occasion for gravity to act, and the whole economy of the neighborhood is changed at once."[34] As Thoreau's remarks suggest, the majority of these 66,000 mills were small and were located on tributaries. They greatly increased the force available to grind corn, to saw logs, to full wool, to turn spindles, and to perform many other tasks.

It would be strange had such radical changes in economy not found expression in a foundation story. In 1856 a journalist asked his readers:

What has made the city of Lowell? What is now making the city of Lawrence become a rival sister to her? What has cast the germ of a hundred cities, here and there, all about us in every direction, at present flourishing villages, where only a few years since was a dense forest, the stillness of which has given place to the multitudinous hum of business? The reader scarcely need be told. With the young the story has become a kind of instinct. The hammer and the file of the machine-shop, the dizzy whirl of the yarn-spindle, and the rattling of the weaver's shuttle, answer the question. Spinning by machinery has mainly done all this. For a moment imagine these germs never to have been thus spread broadcast over our country, and what should be now behold? The answer is obvious. Our wheels of improvement would be set backward half a century. So far as depending on this portion of our industry is involved, the geographies, the printed statistics, the newspapers printed sixty years ago, would tell you with startling accuracy what would now be our condition.[35]

Unlike the narrative of the settler entering the wilderness with an axe, the narrative of the mill was openly technological. The narrative of the axe became widespread in the early nineteenth century and then was gradually projected back to the beginnings of American settlement. It had such political appeal that both Whigs and Democrats used it to authenticate their candidates' virtues. The mill narrative did not immediately have such universal appeal. A clearing with a log cabin was egalitarian and agricultural; any man could obtain an axe. A mill was more hierarchical and proto-industrial; it required specialized knowledge and capital. The axe was the emblem of muscle power; the mill stood for the mechanical mastery of natural forces. Belief in the mill narrative often went hand in hand with support for tariffs on foreign-manufactured goods (to protect American mills) and with support for internal improvements (the better to knit together the American economy). Initially, it was the story told by Alexander Hamilton and by early industrialists. Later the Whig Party championed the mill narrative, which was not taken up in the South as readily as in the North. The narrative of the log cabin was more universal, being well suited to an expanding nation still primarily made up of farmers. The narratives of the axe and the mill seemed to be somewhat at odds, but by the 1850s many had harmonized them. In the North, Daniel Webster and Walt Whitman celebrated both stories. But in the South, where plantations were the nodal points of settlement, the mill story won only grudging acceptance until the eve of the Civil War.

The South lagged in industrial development.[36] Muscle power remained ascendant, and both the yeoman farmer and the plantation owner were idealized as self-sufficient.[37] In 1845, one Southern industrialist complained: "Even [John C.] Calhoun our great oracle . . . will tell you that no mechanical enterprise will succeed in South Carolina—that good mechanics will go where their talents are better rewarded—that to thrive in cotton spinning, one should go to Rhode Island. . . ."[38] But such comments from the middle of the nineteenth century can easily be misunderstood as suggesting that the North and the South were fundamentally at odds. In fact, many Northerners also initially thought industrialism was antithetical to agriculture. In the 1790s, when the United States had only a few modest factories, no steamboats, and no railroads, many Americans agreed with the French Physiocrats that agriculture was the only true source of wealth. They judged the clearing and settlement of the United States to be a permanent improvement in the national well-being, and they considered factories to be British forms of oppression. The mills of the English Midlands seemed a blight that Americans could and should avoid. Even Benjamin Franklin, whose extensive European travel and success as an inventor-scientist might have made him a proponent of factories, declared: "Manufactures are founded in poverty." Franklin based this conclusion on a widely shared belief that "it is the multitude of poor without land in a country, and who must work for others at low wages or starve, that enables undertakers to carry on manufacture."[39] In the 1790s, proposals that Americans pursue manufacturing were routinely attacked in the North and in the South. "In the manufacturing towns in England," George Logan wrote in 1792, "the poor appear to be in a state of the most abject servitude to their employers."[40] Logan argued that poverty and servitude forced workers into dependence and weakened democracy. When government encouraged manufacturing, many feared, it increased the power and wealth of an upper class at the expense of the common people. Many thought industry was incompatible with Republicanism.[41]

Even supporters of manufacturing often considered small-scale, independent production to be inherently more democratic than large enterprises and were suspicious of any legislative grant of rights that favored them. George Washington Parke Custis expressed the widespread view that "expensive machines can only be erected by companies, which soon form monopolies. This prevents competition, which is the life of all infant establishments."[42] Custis argued for reliance on labor power and felt that it was "advisable in the commencement of manufactures to lessen the labor of machinery and encrease the demand for workmen, since this will give the citizens a confidence in the utility of these establishments."[43] Most pernicious of all, it was thought, were manufacturers aided by the government. In the 1790s, when Alexander Hamilton, Tench Coxe, and others promoted a Society for Establishing Useful Manufactures, the aforementioned George Logan, a leading spokesmen for agriculture, attacked them: "The unjust and dangerous interference of government, in granting to a company of monied men, privileges, bounties, and favours, not enjoyed by individual citizens . . . will discourage citizens from acquiring a knowledge in productions and occupations, in which they may be at any time ruined by the arbitrary interference of government."[44] Coxe quickly replied that in fact no bounties or favors had been given them. Yet they had been granted permission to form a corporation, then an unusual privilege. This early investment scheme failed,[45] further suggesting that in the first decade of the nineteenth century by no means all Northerners embraced manufacturing.

Jefferson early proposed that the United States remain an agricultural nation. In a famous passage in *Notes on the State of Virginia* he declared: "Those who labor in the earth are the chosen people of God, if ever he had a chosen people, whose breasts He has made His peculiar deposit for substantial and genuine virtue. . . . Corruption of morals in the mass of cultivators is a phenomenon of which no age nor nation has furnished an example."[46] An agrarian nation would remain virtuous and would preserve its democratic traditions. In addition to this moral argument, Jefferson made a practical argument: In Europe land was scarce, and manufacturing was "resorted to of necessity not of choice to support the surplus of their people." In contrast, the United States had "an immensity of land courting the industry of the husbandman." "Is it best then," Jefferson asked, "that all our citizens should be employed in its improvement, or that one half should be called off from that to exercise manufacturers for the other?" To the young Physiocrat, this was a rhetorical question. Jefferson's arguments were long cited by opponents of manufacturing.

These were the views of the young Jefferson, however. His opinions changed during the remaining half-century of his life, as the United States created a federal government and endured the Napoleonic Wars. During the 1807–1809 embargo on trade with England and France, Jefferson became acutely aware that, however appealing an agricultural nation might be in theory, in practice Americans could not rely primarily on European manufacturing. Indeed, Jefferson manufactured nails on his own plantation.[47] By 1809, at the end of his presidency, he concluded that "we should encourage home manufactures to the extent of our own consumption of everything of which we raise the raw material."[48] Jefferson no longer felt that industry would lead to poverty and servility. Because future generations of American workers would still have access to abundant land, he believed, labor would be scarce, wages would remain high, and employers would have to treat their workers well. By 1810, opposition to manufacturing was waning but a more positive image of industry was still lacking. The "dark Satanic mills" of Britain remained undesirable, but an alternative vision of domestic production had not yet been articulated.

At the start of the nineteenth century, particularly in Massachusetts and Rhode Island, investors began to harness water power for larger-scale manufacturing. Samuel Slater pioneered this development. An English immigrant, Slater had memorized the Arkwright mill system and brought to the United States the technical knowledge that in 1790 made possible the first American factory for spinning thread. Many other small enterprises followed. It is easy to exaggerate the scope of this early industrialization. The Census Bureau found as late as 1840 that in the most industrialized state—Slater's Rhode Island—only one-fifth of the workers were engaged in manufacturing or the trades.[49] (The total number of workers in trades and manufacturing was 21,271.) Massachusetts was regarded as a leading industrial state at the time, with 85,176 of its citizens working in manufacturing or trades out of a total population of

737,699. That was roughly one person in nine. Agriculture claimed a slightly larger share. In Boston, then a city of 93,383, only 5,333 worked in manufacturing and the trades combined.[50] Boston, Providence, and other tidewater cities lacked falling water, and American manufacturing developed inland.

Southern investors were slower to develop their water power, but by the middle of the nineteenth century the South had begun to imitate the North. The English reformer Harriet Martineau visited a mill near Charleston and described it within the framework of the foundation narrative, albeit with some interesting additions that adapted the story to Southern conditions. Martineau prefaced her account of a new mill by remarking: "I could only wish that the slaves in the neighbourhood could see, as clearly as a stranger could, the good it portended to them." She then went on to praise "an enterprising gentleman" who had "set up a rice-mill":

. . . he avails himself to the utmost of its capabilities; but this is made much of in that land of small improvement; as it ought to be. The chaff is used to enrich the soil; the proprietor has made lot after lot of bad land very profitable for sale with it, and is thus growing rapidly rich. The sweet flour, which lies between the husk and the grain, is used for fattening cattle. The broken rice is sold cheap; and the rest finds a good market. There are nine persons employed in the mill, some white and some black; and many more are busy in preparing the lots of land, and in building on them. Clusters of houses have risen up around the mill.[51]

As in the Northern version of the story, the mill enriches not only the owner but also its neighbors, and an incipient community grows up nearby. Unique to the Southern version of the story is that the mill is taken to be an ameliorating influence on slavery.

In 1845, William Gregg, a leading Southern textile manufacturer, published a book calling for establishing mills in the South, rather than exporting raw cotton to Britain or to the North and importing finished cotton goods in return. Gregg estimated that there was local water power "sufficient to work up half the crop of South Carolina," yet noted that this abundant water power was considered to be "nearly valueless at the present time."[52] Because South Carolina's investors sent their money elsewhere, he complained, the state's roads were in poor repair, its farmers poorly provided with blacksmiths and agricultural implements, and its citizens underemployed. "What is to keep our business men and money capital in South Carolina?" Gregg asked. "If we listen much longer to the ultras in agriculture and croakers against mechanical enterprise, it is feared that they will be the only class left, to stir up the indolent sleepers. . . ."[53] Just as Crèvecoeur, Rush, and other critics of frontiersmen complained of their sloth and idleness, the advocates of mills often conjured up the image of an impoverished, inactive population.

By the 1850s a leading Virginia newspaper, the *Richmond Enquirer,* was repeating the mill narrative, not as an accomplished fact, but as a desirable end: "We have a limitless supply of water power—the cheapest of motors—in localities easy of access. So abundant is this supply of water power that no value is attached to it distinct from the adjacent lands, except in the vicinity of the larger towns. Numberless sites may now be found where the supply of water power is sufficient for the purposes of a Lawrence or a Lowell. Nor is there any want of material for building at these localities."[54] Such remarks were made in what had become the established tradition of boosterism, in which the press did not merely report events but called for particular developments and predicted the future greatness of its town or city. North and South, newspapers preceded and stimulated settlement,[55] encouraging Americans to translate the mill narrative into reality.

North or South, mills were fundamental to the foundation of two kinds of communities. Communities of the first kind were located on smaller streams in hilly or mountainous areas. These streams, seldom navigable, provided reliable power to many small mills along their banks. Communities of the second kind, larger and better known though less numerous, were founded at the sites of large dams or waterfalls along otherwise navigable streams. In such places, cities grew for two reasons: because the falls marked a terminus of navigation and because they supplied power for industry. In some cases, two separate communities evolved—for example, St. Paul (the last major port on the Mississippi) and Minneapolis (with its mills a few miles upstream at St. Anthony's Falls). More commonly, a single city served as both port and mill center—for example Rochester, on the Genesee; Louisville, on the rapids of the Ohio; and Richmond, at the falls on the James River. Richmond is a particularly clear example. Ships could pass up the James River from the sea to the city, where 5 miles of rapids and falls interrupted further upstream travel. In 1818 a visitor found at Richmond "several fine flour mills: some of them turn eight pairs of stones, and can grind and dress 1000 barrels of flour per week."[56]

In Massachusetts, Lowell was built on the Merrimack River just below Pawtucket Falls. The falls had been a barrier to navigation until 1796, when a canal was built around them. The canal soon lost most of its business to another artificial waterway that began further up river and led directly to Boston. However, navigation was not the only possible use for the water moving down the Pawtucket Canal and falling through its four locks. The Boston Manufacturing Company purchased the canal and adjoining lands in 1821. Five years later, its three mills produced 2 million yards of cloth per year and employed nearly 1,000 workers.[57] By 1840, Lowell, once the site of only a dozen houses, was the second-largest city in Massachusetts, employing 8,936 people.[58] An English traveler exaggerated only slightly: "A few years ago, the spot which we now saw covered with huge cotton mills, smiling villages, canals, roads, and bridges, was a mere wilderness, and, if not quite solitary, was inhabited only by painted savages."[59]

Lowell was soon emulated further upstream at Lawrence, Massachusetts and at Amoskeag, New Hampshire. By the 1840s, the Boston Associates (a group of investors) owned and controlled not only these towns but also the lakes and headwaters of the watershed, and they used a system of dams, canals, and locks to regulate its flow all the way to the sea.[60] Emerson reflects on this large enterprise in his journal. An entry from 1847 epitomizes the capitalist version of the foundation mill narrative and also suggests a critique:

An American in this ardent climate gets up early some morning and buys a river; and advertises for twelve or fifteen hundred Irishmen, digs a new channel for it, brings it to his mills, and has a head of twenty-four feet of water; then to give him an appetite for his breakfast he raises a house; then carves out within doors, a quarter township into streets and building lots, tavern, school, and the Methodist meeting-house—sends up an engineer into New Hampshire, to see where his water comes from, and after advising with him, sends a trusty man of business to buy of all the farmers such mill privileges as will serve him among their waste hill and pasture lots, and comes home with great glee announcing that he is now owner of the great Lake Winipeseogee, as reservoir for his Lowell mills at midsummer. They are an ardent race, and are as fully possessed of that hatred of labor, which is the principle of progress in the human race, as any other people. They must and will have the enjoyment without the sweat. So they buy slaves, where the women will permit it; where they will not, they make the wind, the tide, the waterfall, the stream, the cloud, the lightning, do the work, by every art and device their cunningest brain can achieve.[61]

In Emerson's retelling, the mill owner is far more prominent than in the versions that circulated a generation before. This capitalist buys rivers and watersheds, hires armies of laborers, commands expert advice, and carves the community out of empty space. The scale has changed. Emerson's American operates in several states, stimulates Irish immigration, and manufactures goods for a large market. With the larger scale, the meaning of mill construction becomes more abstract. Paradoxically, this intense activity and planning is prompted by a hatred of labor, and the ditch work of a thousand Irishmen promotes "progress in the human race." For Emerson, entrepreneurship and invention were liberating alternatives to either slavery for some or hard labor for all. Emerson visited Lowell and lectured to the mill girls.[62] Like Dickens and most other visitors, he was impressed by the thousands of orderly and attractive young women who saved much of their wages and, despite long hours, found time for cultural pursuits.

Rochester developed at almost the same time as Lowell. A mill built there in 1789 was a financial failure because the region was too thinly populated, communications were poor, and Native Americans had not relinquished their claims. The Genesee Falls went unused until 1810, when outside investors took an active role in developing them.[63] After the proprietors had drawn up town lots, the first settler was a miller from Connecticut.[64] Although the War of 1812 put the new town near the front line with Canada, it continued to grow. By 1827, Rochester's water power drove about 40 businesses, including seven flour mills, nine lumber mills, a nail factory, a window-sash factory, a cotton mill with 30 power looms and 1,400 spindles, and a shop that produced millstones for other mills. The grist and lumber mills together employed about 100 men, the cotton mill 80 children. Another 134 small establishments produced a wide range of items, including axes, rifles, scythes, pails, and 8 million bricks.[65] Much as Kendall had described, Rochester's location and growth depended on its water power.

Like Richmond, Lowell, and Rochester, the Georgia city of Columbus, on the Chattahoochee River, was founded on water power. Its development was almost explosive. Mrs. Basil Hall, an Englishwoman touring the United States, happened to be present when the former Indian lands around the falls were being sold. Her husband remarked that he had seen a town without inhabitants but had never before seen inhabitants without a town.[66] When the Halls arrived, in March of 1828 (the first year in which the site had steamboat service), they found that "as yet the town is a thick forest, with the exception of some temporary wooden buildings erected to shelter the numerous bidders from all parts of the Union who are waiting for the sale of lots which is to commence early in July." In the course of "an hour's walk from the water side along streets staked out amongst the trees" they encountered "little log and frame houses, most of them intended to be moved to some other situation a few months hence, but in the meantime exhibiting very imposing signs."[67] These bidders had come months before the land could be sold. They clearly had accepted the foundation story of water power, and they were active purveyors of that story to others.

The Halls were not the only curious travelers to visit the site and be told of its potential. Also in 1828, a Swedish traveler visited the unbuilt "town" and was taken through the area to admire the possibilities. He reported:

The first thing to which our attention was called was a long line cut through the coppice-wood of oaks. This our guide begged us to observe was to be the principal street; and the brushwood having been cut away so as to leave a lane four feet wide, with small stakes driven in at intervals, we could walk along it easily enough. On reaching the middle point, our friend, looking around him, exclaimed in raptures at the prospect of the future greatness of Columbus: "Here you are in the centre of the city!". . . After threading our way for some time amongst the trees, we came in sight, here and there, of huts made partly of planks, partly of bark. . . .[68]

The town was carved out of the woods, with lots demarcated by a surveyor according to the geometry of the grid. All the factors that led to the ideology of second creation were present: a geometrical definition of space, the belief in natural abundance, the operation of a free market in land, and access to new sources of power.

Columbus was built at the head of navigation on the Chattahoochee River, and one of its first buildings was a grist mill, with a small diversion dam. Four years later a water-driven cotton gin was constructed, and in 1838 the first textile mill was completed after being delayed by fighting

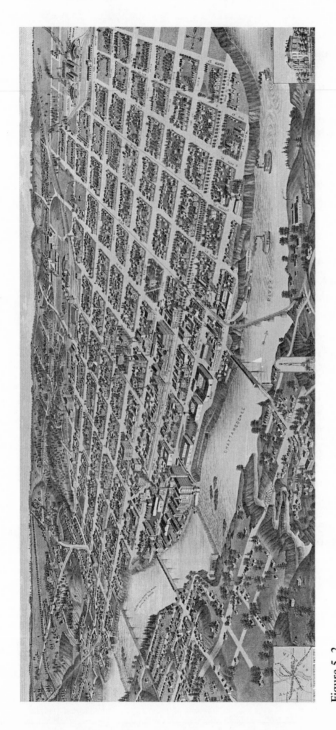

Figure 5. 2
"Perspective Map of Columbus, Ga., 1886." Courtesy Library of Congress.

with the Creek Indians. Yet the available power was by no means exhausted, and the local newspaper editors exhorted readers to look at the river and what "nature has done, and what art and the superior genius of man permits to lie idle."[69] By 1844, a 500-foot dam had been completed to siphon off water to a canal and a line of "water lots" where new mills could be erected.[70] Frederick Law Olmsted visited just before the Civil War and found that most of the river's power was still not being exploited at "the largest manufacturing town, south of Richmond, in the Slave States. . . . The water-power is sufficient to drive two hundred thousand spindles. . . . There are probably at present from fifteen to twenty thousand spindles running."[71] Olmsted noted a woolen mill, paper factory, a foundry, a cotton gin factory, and a machine shop. As in the North, most of the laborers were white, and in the spinning mills they were primarily young women, paid from $8 to $12 a month. After the Civil War, Columbus further developed its water power. As a panoramic map from 1886 shows, an extensive city laid out in a grid pattern had replaced the forests the Halls had seen in 1828.

Located on rapids or waterfalls of large rivers, Columbus, Lowell, Rochester, and Minneapolis all grew rapidly, and Americans became accustomed to the idea of the "boomtown" that grew from nothing or from a few houses into a sizable community in only a decade. Such places seemed dramatic demonstrations of the truth of the mill narrative. Many communities sought to imitate their meteoric rise, and any community that possessed water power was certain to proclaim this fact as the basis for its future greatness. "Froth and puffing is the order of the day," one early arrival in Rochester noted.[72]

But what counted as froth when a town's population grew by 25 percent in the single year of 1823, adding nine three-story brick buildings and 150 houses?[73] This was more than mere boosterism. All societies need to explain how they came into existence. The narrative of the axe was about the creation of an agricultural landscape; the narrative of the mill explained the origins of cities and towns, justifying them as natural outgrowths of unexploited rivers and streams. Such towns were a second creation, completing the work begun by the Almighty. In this view, rivers, streams, and waterfalls were providentially located at convenient locations, awaiting the hand of man. An Ohio newspaper editor put it this way in

1850: "Standing around scratching heads will never make Pomeroy the city nature intended it to be."[74] That "nature" could intend a site to be something man-made was a central conceit in the doctrine of second creation. Man's duty lay in completing God's original creation. By this logic, Pomeroy, Lowell, Rochester, Columbus, Minneapolis, and other instant cities were not impositions on nature or exploitations of the environment. For antebellum Americans, such towns expressed an intended design; their growth and prosperity measured providential intent.

Since these towns seemed natural outgrowths of their locations, they were not antithetical to humans. If in England some critics attacked industrialization, in America they were answered in the *North American Review* by one Timothy Walker, who in 1831 declared: "Where she [nature] denied us rivers, Mechanism has supplied them. Where she has left our planet uncomfortably rough, Mechanism has applied the roller. Where the mountains have been found in the way, Mechanism has boldly leveled or cut through them."[75] Mankind could use machines to improve the landscape and to reduce human labor, moving toward a more perfect state in which "machines are to perform all the drudgery of man, while he is to look on in self-complacent ease."[76] Mills thus not only created cities, they also improved mankind.

In an 1836 biography of Samuel Slater, George White wrote:

The effect of an extensive internal commerce, in as large a country as this, on morals and the arts, science and literature, as subservient to morals and religion, are too obvious and important long to escape the notice of an attentive observer. All experience proves that good morals never did, and never can exist, among an indolent people, and people who are poor in consequence of their indolence . . . in districts far from convenient markets, idleness is inevitable.[77]

Looking back on Slater's contribution to the creation of a factory system, White viewed the factory as a moral machine that transformed the laborer as well as the raw material. It produced not only cotton cloth but also good morals. A young man or woman who worked 12 hours a day 6 days a week had no time to misbehave. Such arguments were repeated by many mill owners in their responses to a questionnaire from White about the moral effects of manufacturing. They concluded unanimously that drunkenness and loose morals were scarcely found in the small manufacturing towns of New England. To them, Samuel Slater was not only the founder of an important industry that reduced the dependence of the nation on Britain; he was also a reformer.

Moral improvement was not limited to large factories. It could also be accomplished on a smaller scale in mill towns located on streams that could not be navigated. In contrast to agricultural villages, such towns scarcely affected the appearance of the landscape, being neither on plains nor on hilltops. James Fenimore Cooper remarked on a view in central New York: "I was told that a multitude of villages lay within the limits of the view; but as they were generally placed near some stream, for the advantage of its water-power, the uneven formation of the land hid them from our sight."[78] Instead of cities on hills, these were valley communities.

Mill towns had to be located where water could be drawn off into canals, but the mere presence of water was not sufficient. To found one of these communities required investment and artisanal talent. In 1850, one C. W. Blanchard wrote, in the manner of the day, a long historical letter to the newspapers. It described the creation of Clintonville, a Massachusetts town based almost entirely on the water-powered inventions of one man. First this inventor proved his skill by creating a mechanical loom to weave carpets that was installed in Lowell. Then, Blanchard continues,

he got up a gingham loom, which was equally successful. Well, he determined that his machines should be operated here [in Clintonville]. Capital was ready to pay the bills, for it saw a prospect of large returns; and this man and his brother, the [mill] operative, went to work. They built a machine shop . . . then they built machinery, and put up mills; and within the last six years, nearly two millions have been invested here under their direction. A village of three thousand inhabitants has sprung up in the midst of what was, eight years ago, woods and barren sheep pastures. But how?

As soon as it was known that work was to be done here, and money paid for it, Yankee enterprise pricks up its ears. . . . The carpenters, bricklayers, stone masons, iron founders, machinists, flock in. In their train follow tailors, shoemakers, butchers, bakers, and shopkeepers. . . . Those who have children of a suitable age put them at work, in the mills, as places occur for them, a portion of the year. Then the farmers all about the neighborhood—the citizens who do not come themselves— the doctors, lawyers, and ministers even—find their girls and boys have got the factory fever. The wages are so good, and paid every four or five weeks, too—sisters, cousins, acquaintances are going—if they shouldn't happen to like it, they can come home again—in short, they must go to the factory, and will! Occasionally, some of the parents, who are not getting along quite as well as they wish, come over to see what the prospect is. They find "its handy to meetin'" (we have three or four churches); "schoolin's so much better'n 't is up our way" (we have five or six schools, primary, grammar, and high) that they determine they will move to the factory village and take a boarding house, so as to be with their children, and enjoy the advantages of a larger community. The persons who came first, meantime, are getting rich, and men of note. They are being elected representatives, selectmen, or filling other responsible stations; but they see no impropriety in their children spending a portion of their time in the factory. . . . Their brothers, cousins, lovers, friends, or acquaintances are their overseers; and their fellow operatives are of the same character. And thus our factory village springs up, and thus its mills are filled, a large proportion of the operatives, however, always belonging to homes in the neighboring towns or states, whence they come to spend a year or two in the mill, and then return to marry some early acquaintance, or, marrying at the village, establish themselves in a new home.[79]

Blanchard's letter retells Edward Kendall's story and serves as a paradigmatic example of the technological foundation narrative. A manufacturing village "springs up" around mills built on "barren sheep pastures," emerging in a process of voluntary migration and settlement. The process seems to be virtually automatic, and the resulting village seems a harmonious whole where friendships and family ties link the factory to the community.

Edward Everett reprinted this letter in the collected edition of his speeches as an appendix to an address he had given in Lowell 20 years earlier. Everett was an ordained clergyman, the leading orator of his day, a former member of Congress, a former ambassador to Great Britain, and the president of Harvard College. The letter exemplified his belief in the connection between democracy and internal improvements. On July 4, 1830, appearing as the featured speaker at the new factory town of Lowell, he had praised the release of "popular energy" in that city as something "which cannot exist under any form of despotism." "It is usual," he noted, "to consider human labor as the measure of value." The great question was how to motivate people to produce more goods:

. . . there is as much mere physical capacity for labor dormant in a population of serfs and slaves, or of the subjects of an Oriental despotism, as in an equal population of the freest country on earth; as much in the same number of men in Asia Minor, or the Crimea, as in Yorkshire in England, or Middlesex in Massachusetts. But what a difference in the developments and applications of labor in the two classes of population respectively! On the one hand, energy, fire, and endurance; on the other, languor and tardiness: on the one hand, a bold application of capital in giving employment to labor; on the other, a furtive concealment of capital where it exists, and a universal want of it for any new enterprise: on the one hand, artistic skill and moral courage superadded to the mere animal power of labor; on the other, every thing done by hand, in ancient, unimproving routine: on the one hand, a constantly increasing amount of skilled and energetic labor, resulting from the increase of a well-educated population; on the other, stationary, often declining numbers, and one generation hardly able to fill the place of its predecessor.

For Everett the conclusion was inescapable: "It is the spirit of a free country which animates and gives energy to its labor; which puts the mass in action, gives it motive and intensity, makes it inventive, sends it off in new directions, subdues to its command all the powers of nature, and enlists in its service an army of machines, that do all but think and talk." The proof of this argument was not theoretical but practical: "Compare a hand loom with a power loom; a barge, poled up against the current of a river, with a steamer breasting its force. The difference is not greater between them than between the efficiency of labor under a free or despotic government; in an independent state or a colony."[80]

In Everett's formulation, the free institutions of a democratic nation ultimately produced the American mill town. Elsewhere in his speech Everett contrasted the vigor of industry in the United States with the lack of development during colonial times, when the British stifled manufacturing as a part of their mercantilist policy. He took pains to differentiate between American factories and those of Britain: "It would seem that the industrial system of Europe required for its administration an amount of suffering, depravity, and brutalism which formed one of the great scandals of the age. No form of serfdom or slavery could be worse. Reflecting persons, on this side of the ocean, contemplated with uneasiness the introduction, into this country, of a system which had disclosed such hideous features."[81] But Lowell's prosperity and the good order and comfort of its workers were proof that American industrialization would have a wholly different character. Blanchard's letter, which Everett came upon 20 years later, seemed but a further confirmation of the proposition that democratic labor in small water-powered cities produced prosperity for all.

Zachariah Allen observed in *The Practical Tourist* (1832) that "the manufacturing operations of the United States are carried on in little villages or hamlets, which often appear to spring up as if by magic in the bosom of some forest, around the water-fall which serves to turn the mill wheel," and that "these manufacturing villages are scattered over a vast extent of the country from Indiana to the Atlantic, and from Maine to Georgia." Allen warned the foreign visitor against seeing only a few of the larger mills, for such a tourist "commonly forms but an imperfect estimate of the extent of manufacturing operations carried on in the country."[82] In Britain and on the continent, most industries were urban and usually relied on steam power. As David Jeremy notes, Americans exploited water power much more vigorously than the English, who "preferred the capital intensive steam engine on city sites and neglected available water power."[83]

The United States industrialized in the countryside, and Allen did not fail to draw a positive moral from this fact. In cities such as Manchester in Britain, "where steam engines are in use instead of water power, the laboring classes are collected together, to form that crowded state of population, which is always favorable, in commercial as well as in manufacturing cities, to the bold practices of vice and immorality, by screening offenders from marked ignominy." In America, mills were "sprinkled along the glens and meadows of solitary watercourses." "God forbid," Allen exclaimed, "that there ever may arise a counterpart of Manchester in the New World." Allen speculated that it might "be intended as a blessing that an all-wise Providence has denied to the barren hills of New England the mines of coal, which would allow the inhabitants to congregate in manufacturing cities, by enabling them to have recourse to artificial power, instead of the natural water power so profusely furnished by the unnumerable streams."[84] The geography of New England was ideal: "Whilst a cold climate and an ungrateful soil render the inhabitants from necessity industrious, thus distributed in small communities around the waterfalls, their industry is not likely to be the means of rendering them licentious. . . ."[85] Allen had found the formula needed to answer agrarian critics. Industrialization could be divided into a smoky, urban, oppressive British variety based on "artificial power" and a pastoral and democratic American form based on "natural water power." The falls on the Merrimack, the James, the Genesee, and the Mississippi marked ideal locations for man's second creation.

Yet if water power was the predestined source for American industry, it was not sufficient. As Everett also emphasized, free institutions and widespread private ownership of land were also necessary to ensure that democratic industry would emerge. Americans understood technological foundation stories as accounts of what happened when free, enterprising men were not hindered in developing resources. A contrasting story told by the engineer Benjamin Wright to the readers of the *American Railroad Journal and Advocate of Internal Improvements* reported on Little Falls, New York, where development had been stymied. Located at a 45-foot waterfall near the Mohawk River and intersected by the Erie Canal, Little Falls seemed ideal for industrial development. Yet the village remained small because of the mistaken policies of an absentee English landlord. Wright explained:

It appears to have been his desire to retain the title of the property in himself, and to give nothing but leases, with ground rents, after the custom in general practice in England. This practice of leasing and having a village made up of tenantry, not being congenial to the minds of American-born citizens, had a tendency to retard the growth of the village; and this course of proceeding on the part of the proprietor, added to his refusal to permit the advantages of the water power to be used, has heretofore operated as a serious check upon its increase—enterprising business men being unwilling to settle themselves there, under all the disadvantages.[86]

Once again, American success was contrasted with British practices. Since before the American Revolution, this site had been owned by Alexander Ellice and then by his son Edward, a Member of Parliament. They instructed their agent to build a single grist mill in 1789, but otherwise they held back town development. The stagnation of Little Falls demonstrated what happened when Englishmen hindered American free enterprise. The post-Revolution ideology was compromised at Little Falls because absentee landlords kept land out of the free market, restrained the development of water power, nullified natural abundance, and suppressed latent energies. Only in democratic American hands was second creation inevitable.

The antebellum mill narrative was the story of a partnership with nature. It was at once a fable of origins, explaining how towns and cities emerged from the landscape, as Americans developed its potentialities, and a counter-narrative to British developments. The foundation narrative gained force through a series of contrasts: not the "artificial power" of steam but the "natural power" of water, not vast factories but small mills, not smoky cities but pastoral towns, not an anonymous crowd of workers but women and children from a tightly knit community working for a short time in factory, not European landlords but American freeholders.

As Hayden White suggests, "the primary meaning of a narrative would then consist of the destructuration of a set of events (real or imagined) originally encoded in one tropological mode and the progressive restructuration of the set in another tropological mode."[87] The primary meaning of this American foundational story lay in its inversion of the story of British industrial oppression.

Building up the mill narrative in terms of these contrasts also distracted attention from the complaints of environmentalists and the protests of workers.

6

Pollution and Class Conflict

All the men in the village worked in the mill or for it. It was cutting pine. It had been there seven years and in seven years more it would destroy all the timber within its reach. Then some of the machinery and most of the men who ran it and existed because of and for it would be loaded onto freight cars and moved away.
—William Faulkner[1]

At rows of blank-looking counters sat rows of blank-looking girls, with blank, white folders in their blank hands, all blankly folding blank paper.
—Herman Melville[2]

The mill narrative did not emerge and then spawn various counternarratives; rather, the narrative and the counter-narratives evolved simultaneously. From the American Revolution through the antebellum period, English industrialization remained a worrisome model of what the United States might become. Although most American mills were driven by water rather than steam (as would be the case until after the Civil War), and although American manufacturing was dispersed into the countryside,[3] some still objected to mills. By the middle of the nineteenth century, two critiques had emerged. The first attacked the mill because it despoiled the environment, the second because it exploited the worker.

The environmental critique applied even to simple grist mills and sawmills of the colonial period. Fish were an important and easily obtained food, and local residents fought to retain access to them. As late as the 1830s, annual migrations of shad, herring, and salmon still attracted thousands of local residents to fish the Susquehanna, the Connecticut, the Merrimack, and other eastern rivers.[4] In 1835, the citizens of industrial Lowell caught an estimated 65,000 shad.[5] According to a history of the

Pennsylvania township of Hanover, at the appropriate season "every family in Hanover had at least one of its members down at the river" in order to catch shad "in immense quantities."[6] Protection of fishing rights began early. In the 1640s, an ironworks at Braintree, Massachusetts had to shut down after fishermen destroyed its dam. In lawsuits during the eighteenth century, farmers contested the flooding of their lands by mill dams, and fishermen complained that dams prevented salmon, eels, and other migrating species from spawning in fresh water.[7] Erection of a dam on the Housatonic was delayed from 1839 until 1865 because local residents preferred to safeguard their access to shad.[8]

Early legislation mandated that mills provide fish ladders or fishways, but such legal safeguards tended to be forgotten over time. In the nineteenth century, salmon disappeared from Atlantic rivers and streams. Not only did dams often block their passage upstream to spawn, but industrial wastes poisoned the water. In 1857, George Perkins Marsh reported to the state of Vermont: "Almost all the processes of agriculture, and of mechanical and chemical industry are fatally destructive to aquatic animals within reach of their influence . . . [and] to fish which live or spawn in fresh water. . . . The sawdust from lumber mills clogs their gills, and the thousand deleterious mineral substances, discharged into rivers from metallurgical, chemical, and manufacturing establishments, poison them by shoals."[9] In the 1860s, Harvard Professor Louis Agassiz likewise blamed the decline of fish on sawdust, dyes, acids, and industrial waste.[10] In such accounts, technological intervention destroys or drives away the biological community. The second creation destroys the first rather than supplementing it. If the narrative of the axe had ignored the existence of Native Americans and the biotic community, the mill narrative overlooked the fish in the watershed. Each foundation story assumed the imposition of geometrical order on the land. Each assumed a conservation of energy and resources during any transformation; nothing of the first creation was thought to be lost in constructing the second. Each conceived of nature as an undeveloped space that could be made more fruitful by human intervention. The counter-narrative, by emphasizing pollution and the destruction of fish and wildlife, rewrote second creation as technological destruction. Native Americans were particularly disposed to see their experience in this way, but they were by no means alone.

In 1832, John James Audubon found that the Ohio Valley was "more or less covered with villages, farms, and towns, where the din of hammers and machinery is constantly heard." The woods were "fast disappearing," and "the greedy mills, told the sad tale, that in a century the noble forests . . . shall exist no more."[11] Henry David Thoreau was also concerned about the elimination of the forests. He talked to millers during his walks through the Concord countryside. After visiting a new sawmill that had been damaged by winter ice and flooding, he reflected: "How simple the machinery of the mill! Miles has dammed a stream, raised a pond or head of water, and placed an old horizontal mill-wheel in position to receive a jet of water on its buckets, transferred the motion to a horizontal shaft and saw by a few cog wheels and simple gearing, and, throwing a roof of slabs over all, at the outlet of the pond you have a mill."[12] Even such a small-scale affair was an instrument of tree "torture." After visiting another Concord mill that produced both lumber and flour, Thoreau noted: "The innocent forest trees, become dead logs, [and] are unceasingly and relentlessly, I know not for what crime, drawn and quartered and sawn asunder (after being torn limb from limb), with an agony of sound."[13] A few years later, Thoreau, "attracted by the music of the saw," spoke with the miller, who reported that about as many logs of all sorts were being brought to him as a decade earlier.[14] In another entry, made after hearing the saws working at night during high water in May, Thoreau noted: "It is a hollow, galloping sound; makes tearing work, taming timber, in a rude Orphean fashion preparing it for dwellings of men and musical instruments, perchance. I can imagine the sawyer, with his lantern and his bar in hand, standing by, amid the shadows cast by his light. There is a sonorous vibration and ring to it, as if from the nerves of the tortured log. Tearing its entrails."[15]

Thoreau observed Maine's logging industry in detail, canoeing through areas that had been cut over, stopping at logging camps, and viewing the extensive mills in the villages of Oldtown and Stillwater at the falls of the Penobscot River, where "the Maine woods are converted into lumber." There were 250 sawmills on the Penobscot and its tributaries, which together "sawed two hundred millions of feet of boards annually. . . . The mission of men there seems to be, like so many busy demons, to drive the forest all out of the country, from every solitary beaver-swamp and mountain-side, as soon as possible." The mills "relentlessly sifted" the logs.

One had sixteen sets of saws and used the water power to haul the logs up an inclined plane and to take away the cut lumber. Emphasizing the contrast between the particular tree and its reduction to identical wooden products, Thoreau wrote: "Think how stood the white-pine tree on the shore of Chesuncook, its branches soughing with the four winds, and every individual needle trembling in the sunlight—think how it stands with it now—sold, perchance, to the New England Friction-Match Company!"[16] The logging industry cut an entire region's trees to produce such ephemeral products as matches and firewood that would soon go up in smoke and fence posts that would soon rot in the ground.

During the second half of the nineteenth century, as had been done earlier in the Maine woods, logging companies cleared large areas in the South without calling many permanent communities into being. By the 1880s, every year 200 square miles of forest were being cut in the combined area of Mississippi, Alabama, and Florida (plus another 150 square miles for the production of turpentine). Laying rail lines into the forests, companies hauled wood to temporary mills that were "locationally footloose" and "as near as possible to supplies of lumber."[17] New counter-narratives attacked this industrialized logging, with its wholesale destruction of forests.

William Faulkner witnessed firsthand the logging of the Mississippi Delta region. Towns sprang up to house itinerant loggers and later disappeared. The lower Mississippi Delta had 150 mills in 1915 but only four in 1940.[18] Faulkner watched the transformation from forest to stumpage, and in "The Bear" he wrote that the plan to build a logging train had "brought with it into the doomed wilderness, even before the actual axe, the shadow and portent of the new mill not even finished yet and the rails and ties which were not even laid."[19] Faulkner expanded on this counter-narrative at the beginning of *Light in August*:

All the men in the village worked in the mill or for it. It was cutting pine. It had been there seven years and in seven years more it would destroy all the timber within its reach. Then some of the machinery and most of the men who ran it and existed because of and for it would be loaded onto freight cars and moved away. But some of the machinery would be left, since new pieces could always be bought on the installment plan—gaunt, staring motionless wheels rising from mounds of brick rubble and ragged weeds with a quality profoundly astonishing, and gutted boilers lifting their rusting and unsmoking stacks with an air stubborn, baffled and bemused upon a stumppocked scene of profound and peaceful desolation, unplowed, untilled, gutting slowly into red and choked ravines beneath the long quiet rains of autumn and the galloping fury of vernal equinoxes.[20]

The town briefly existed only to house those who worked in the mill, which produced not an agricultural landscape focused in a community, not a tidy quiltwork of fields in the Jeffersonian grid, but a space emptied of value. Its gutted remnants were desolate rubble presiding over the ravaged stumps and eroded land. Faulkner continued: "The hamlet, which at its best day had borne no name listed on Postoffice Department annals, would not now even be remembered by the hookwormridden heirs-at-large who pulled the buildings down and burned them in cookstoves and winter grates."[21] Such a mill annihilated even the possibility of a community. There was no settled place possible for Lena, who at the beginning of the novel walked away from its devastation. Even if she had not become pregnant by one of the itinerant mill hands, who then drifted on to other work and failed to leave a forwarding address, the town itself never had a name or a possible future. It would be too much to claim that the mill causes the action of the novel, but its disruptive effects lead directly to the rootlessness and aimless drifting of several of the main characters. In the foundation narrative a sawmill becomes the permanent core of a new community, but Faulkner's mill destroys the forest and destabilizes the rural world.

The counter-narratives discussed thus far—those of colonial farmers, Native Americans, Thoreau, Marsh, Agassiz, and Faulkner—all advanced environmental arguments. They attacked the first half of the mill narrative, and they rejected the desirability of constructing mills in the first place. They saw new technologies not as tools used to make a second creation but as weapons used to destroy the existing habitat. If Thoreau focused on the initial act of destroying the trees, and if Marsh focused on pollution, flooding, and other environmental effects, many later critics examined the rest of the mill narrative. How did the mill affect society? Did it bring prosperity? Was the landscape of this second creation desirable? Most important, what were the effects on workers? These concerns gradually became predominant.

As early as the 1840s, women struck the Lowell mills and agitated for a shorter working day. In the major manufacturing towns, mill owners had been known to ask women to work by candlelight, to stretch the work day by tampering with the factory clocks, to reduce wages arbitrarily, and to increase hours without warning.[22] Medical reports on mills found them damp, stifling in summer, and bone chilling in winter. Women in textile mills typically labored 70–75 hours during a six-day week. In the "Economy" chapter of *Walden,* Thoreau bluntly stated: "I cannot believe that our factory system is the best mode by which men may get clothing. The condition of the operatives is becoming every day more like that of the English; and it cannot be wondered at, since, as far as I have heard or observed, the principal object is, not that mankind may be well and honestly clad, but, unquestionably, that the corporations may be enriched."[23]

Even in the new communities of the West, the foundation story of the mill began to describe clear class divisions. In 1865 a woman traveling in the state of Washington noted:

Everywhere about Puget Sound and the adjoining waters are little arms of the sea running up into the land, like the fjords of Northern Europe. Many of them have large sawmills at the head. We have been travelling about, stopping here and there at the little settlements around the mills. We were everywhere most hospitably received. All strangers are welcomed as guests. Every thing seems so comfortable, and on such a liberal scale, that we never think of the people as poor, although the richest here have only bare wooden walls, and a few articles of furniture, often home-made. It seems, rather as if we had moved two or three generations back, when no one had any thing better; or, as if we might perhaps be living in feudal times, these great mill owners have such authority in the settlements. Some of them possess very large tracts of land, have hundreds of men in their employ, own steamboats and hotels, and have large stores of general merchandise, in connection with their mill business. They sometimes provide entertainment for the men—little dramatic entertainments, etc.—to keep them from resorting to drink; and to encourage them to send for their families, and to make gardens around their houses.[24]

This ingenuous passage is both another instance of the foundation narrative and evidence that undermines it. These sawmills do not create an ideal democratic community; they primarily enrich the mill owners. The logs are not cut from independent farms; they are a crop harvested by hired hands, who remain poor. These are not the yeomen farmers that Jefferson envisioned; they are an incipient proletariat.[25]

The larger a mill, the easier it seemed to find evidence of exploitation. In the 1840s, when Lowell epitomized the American system of water-powered manufacturing, John Greenleaf Whittier published *A Stranger in*

Lowell, a short volume of essays based on 6 months' residence in the city. He began by evoking the contemporary image of Lowell:

. . . a city sprung up, like the enchanted palaces of the Arabian tales, as it were in a single night, stretching wide its chaos of brick masonry and painted shingles, filling the angle of the confluence of the Concord and the Merrimac with sights and sounds of trade and industry. Marvelously here has Art wrought its modern miracles. I can scarcely realize the fact, that a few years ago these rivers, now tamed and subdued to the purposes of man, and charmed into slavish subjection to the wizard of mechanism, rolled unchecked towards the ocean the waters of the Winnipiseogee and the rock-rimmed springs of the White Mountains.[26]

Confronted with this transformation, the visitor "feels himself as it were thrust forward into a new century; he seems treading on the outer circle of the millennium of steam engines and cotton mills. Work is here the patron saint."[27] Nevertheless, after acknowledging this powerful first impression, Whittier suggested that the motto of "this New World Manchester" ought to be "Work, or Die"[28] and questioned whether mechanization really could provide more goods, higher wages, and more leisure time. He evoked the mill narrative only to debunk it.

Whittier was aware of J. A. Etzler's visions of an American utopia achieved by controlling more power sources for purposes of advanced manufacturing:

Looking down, as I do now, on these high, brick, work-shops, I have thought of poor Etzler, and wondered whether he would admit, were he with me, that his mechanical Forces have here found their proper employment in Millennium making. Grinding on, each in his iron harness, invisible, yet shaking, by his regulated and repressed power, this huge prison-house from basement to capstone, is it true that the Genii of Mechanism are really at work here, raising us, by wheel and pulley, steam and water power, slowly up that inclined plane, from whose top stretches the broad table-land of Paradise?[29]

By suggesting that the factory was a prison and questioning whether its grinding and shaking improved the human condition, Whittier showed that he was attuned to the growing complaints of the workers. No local resident or regular reader of the press could remain ignorant of their agitation for a 10-hour day and the conflicts between labor and management.[30] Whittier was also disturbed by the fact that the cotton processed at the mills had been grown and harvested by slave labor. "Every web which falls from these restless looms," he concluded, "has a history more or less connected with sin and suffering, beginning with slavery and ending with overwork and premature death."[31]

Slavery's defenders too attacked cotton mills. If free labor was more profitable to the capitalist than slave labor was to the planter, Henry Hughes argued in his *Treatise on Sociology,* it was because the capitalist took no social responsibility for the workers. Drawing on his reading of August Comte, Hughes presented slavery as a social system that was superior to industrialism because it guaranteed full employment, housing, medical care, and religious instruction for all workers. When hard times came in the capitalist North, Hughes argued, the workers starved. When a mill worker was sick, old, or injured, the capitalist simply hired someone else. He had no interest in keeping her healthy, and he lost nothing if she died young.[32] In *Sociology for the South: or The Failure of Free Society,* George Fitzhugh advanced similar arguments. He declared that to call free labor "wage slavery" was a libel on slavery.[33] Thus, both the abolitionist Whittier and Southern apologists for slavery criticized the mill narrative on the ground that it harmed the workers.

Herman Melville's allegorical short story "A Tartarus of Maids" takes the reader inside the mind of a businessman whose visit to a mill disconcerts him. Melville had lived in western Massachusetts, where mill development was visible virtually everywhere. Though Melville gives highly suggestive metaphorical names to the landscape (the hollow is "the Devil's Dungeon," the stream the "Blood River," and so on), he quite accurately begins his story by describing the location of a mill in a notch where "sounds of torrents fall on all sides upon the ear."[34] As was typical of such sites, the notch had first been exploited by a saw mill: "Conspicuously crowning a rocky bluff high to one side, at the cataract's verge, is the ruin of an old saw-mill, built in those primitive times when vast pines and hemlocks superabounded throughout the neighboring region."[35] Logging has opened up agricultural land (a precondition to the narrator's business, selling seeds which "were distributed through all the Eastern and Northern States").[36] The expanding business requires so many seed envelopes that the narrator has decided to buy them directly from the paper mill rather than through a middleman. In a story rich in symbolism this economic rationale can easily be overlooked, but it directly expresses the narrator's quest for the same efficiency and economies of scale that are evident in the paper factory.

The factory is almost invisible in the snow and murky gloom of the notch. Water power must be sought in mountain passes, and so the factory stands in a cramped location, where it looks like "an arrested avalanche." The buildings seem a jumble: "Various rude, irregular squares and courts resulted from the somewhat picturesque clustering of these buildings, owing to the broken, rocky nature of the ground. Several narrow lanes and alleys, too, partly blocked with snow fallen from the roof, cut up the hamlet in all directions."[37] This is by no means a fanciful description. Many factory villages had to be pressed into the narrow confines of a mountain stream, and their inhabitants were cut off from the outside world much of the time. Melville described a site that was more typical of American manufacturing than Lowell or Lawrence. The confusing arrangement of the buildings and the difficulty of finding the way become metaphors for the disorientation caused by industry.

The narrator's disorientation, the near invisibility of the mill, and its effects on the mill girls all suggest that "A Tartarus of Maids" is essentially about erasure. The sawmill erases the forests; the paper mill erases the sawmill, producing blank paper and whitening the faces of the pale mill girls who work inside all day. ("At rows of blank-looking counters sat rows of blank-looking girls, with blank, white folders in their blank hands, all blankly folding blank paper."[38]) The young mill girls, Melville and his readers know, are daughters of local farmers, many of whom purchase seeds from the narrator. The erasure of difference (or the production of blankness) in the paper mill is inseparable from the larger economic cycle, which contains forests, fields, seeds, mills, paper, and laborers. The mill drives this cycle faster; it also removes the women from the biological reproductive cycle. Like the narrator's seeds, they are enveloped in white paper. By telling the story from the businessman's point of view, Melville confronts capitalist assumptions with their human consequences.

The girls in "A Tartarus of Maids" remained nameless, but in sentimental women's fiction they soon achieved an identity. In 1871, *Bertha, the Sewing Machine Girl; or, Death at the Wheel* initiated a genre in dime novels and on the stage that persisted until the first decade of the twentieth century.[39] To the limited extent that earlier sentimental fiction represented working girls at all, it propagated the image of the educated Lowell factory girl who worked for a few years before leaving to get married. In the 1870s, however, factory girls were commonly depicted as victims of unjust employers and lewd supervisors, whom they fought to preserve their virtue. The plots of the new sensational fiction emphasized not work but abduction, escape, romance, and cross-class marriage. The once widespread ideas that the mill created new towns, reduced indolence, and increased the morality of workers had disappeared. In the 1870s, the factory was not deemed a decent environment for a respectable female character. The dime novel shifted attention from textile production to the private lives of women workers. The new heroines took pride in a good day's work, but paradoxically they were ladies. These stories almost invariably ended by freeing a young woman from factory work, either through marriage or by some other plot device that transformed her class status, such as the discovery that she was not an orphan but an heiress.

The mill foundation narrative had emphasized the creation of a town out of empty space, but by the end of the nineteenth century the predominant mill story was one of exploiting an existing community. One did not have to be a Marxist or a socialist to view the poor housing in many mill towns with alarm and deplore the long hours. Women continued to receive lower wages than men in textile mills, and in some cotton towns women made up three-fourths of the work force.[40] The strongest condemnations were reserved for child labor in the mills. From the inception of the textile industry, many workers had been children, including many who worked in Samuel Slater's first American textile mills. If initially the Lowell mills had employed young women, after the 1840s it became increasingly common to find whole families in mills. Because the children received little formal education, most remained unskilled as adults. They became the permanent working class feared by the earliest critics, who expected British-style industrial conditions to emerge in the United States.

As the photographer Lewis Hine noted, working children could not "escape the deadening effects of: long hours, monotonous toil, lack of proper recreation, loss of education, vicious surroundings." Above these words were two photographs, one of a well-dressed, healthy, smiling "normal child" and the other of a poorly clad, wan, listless "mill child."[41] Another Hine poster, titled "Making Human Junk," showed healthy children entering a factory and emerging as stunted and hardened teenagers with "no future and low wages."[42] Hine also presented powerful images of young girls working in cotton mills. "The Boss Teaches a Young Spinner in a North Carolina Mill" (1908) suggests the subordination of the child to industrial rigor. The floor is littered with wisps of fiber such as medical inspectors warned made mills unhealthy. The girl is diminutive compared to the machines she must tend, and she must stretch to reach the upper row of spindles. Hine put the camera at the child's eye level, reproducing her visual experience of the space. His selection of perspective also emphasizes the power of the supervisor and the worker's isolation.[43] Using such images, progressive organizations called for an end to child labor. The National Consumer's League (NCL) and the National Child Labor Committee encouraged middle-class Americans to boycott goods made with child labor and pressed for state laws to prohibit the practice. In 1901 the NCL permitted manufacturers who did not employ children to use its label

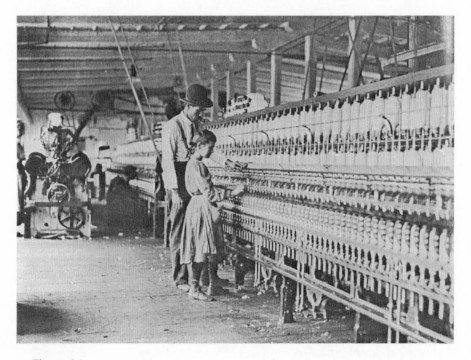

Figure 6.1
Lewis Hine, "The Boss Teaches a Young Spinner in a North Carolina Mill, 1908."

in advertisements, and by 1906 its 63 local chapters had found 60 manufacturers who qualified to do so.[44] Many clergymen preached against child
labor and assisted the NCL's campaign. Although some state legislatures
prohibited or restricted child labor, the textile industry remained identified
with long hours, drudgery, low wages, and worker exploitation.

As loggers and textile workers became an American proletariat, it was
difficult to see the mill as the center of a harmonious foundation narrative.
Instead, it stood at the center of a counter-narrative. The mill grew not as
the automatic effect of channeling natural forces but through the willful
exploitation of human labor. It created not prosperity but poverty. Its town
was not harmonious but divided. The physical landscape surrounding it
was no longer the desirable second creation suggested by the early Lowell;
it was a squalid town in a polluted valley. Such perceptions were reinforced
by the fact that the textile industry was no longer on the cutting edge of
industrialization in the late nineteenth century. Compared to the vast steel
mills in Pittsburgh, the new car factories of Detroit, or the burgeoning
electrical industry in Schenectady, cotton mills seemed old fashioned, even
if their machinery continued to be improved.

By the 1880s, textile factories had become classic settings for stories of
labor exploitation and strikes. At Cohoes, New York, investors built textile mills beside the waterfall where the Mohawk River cascaded down to
the Hudson. From 1837 until 1850 these mills remained small and generally lost money, but between 1853 and 1876 production increased ten-
fold and output per worker doubled without a corresponding increase
in wages. In the 1880s, after accepting paternalism for decades, workers
organized and struck to protest wage cuts, blacklists, and the arbitrary
power of management.[45]

Similar transformations of paternalistic textile towns into union communities recurred throughout the last decades of the nineteenth century.[46]
By 1912, strike narratives could be found not only in radical newspapers
or speeches but also in the Department of Labor's *Report on Strike of Textile Workers in Lawrence, Mass. in 1912*. The Department of Labor determined that the "general conditions of the industry in Lawrence are more
or less typical of the textile industry in all of the large distinctly textile
towns" and that "the strike in Lawrence and the conditions attending it
might just as easily have occurred in any other of these towns."[47] This was

a remarkable official conclusion, since at least 23,000 workers, from several mills, struck in Lawrence, most of them from January 11 until March 14. This came to be called the Bread and Roses Strike. The dispute grew out of a reduction in wages that the companies insisted must accompany a two-hour shortening of the work week mandated by the Commonwealth of Massachusetts. When the mills distributed the first envelopes containing reduced wages, the workers walked out, demonstrated, and broke a few mill windows. Half of the town's citizens were foreign born, including 7,700 French Canadians, 6,700 Italians, 5,950 Irishmen, 5,650 Englishmen, and 4,300 Russians.[48] The strike narrative emphasized that, although fewer than one-tenth of the workers were union members, the various ethnic groups were united in their desire for economic justice.[49] The lack of a radical past was characteristic in the strike narrative, which described not a carefully plotted and organized action but a spontaneous uprising by people who had not been politically self-conscious but who resisted reductions in pay and encroachments on their rights.

Against the workers the strike narrative posed not only the mill owners, with their tactics of blacklisting, spying, and intimidation, but also the police and the National Guard. In Lawrence, the Guard was called out to keep "public order." Working together with the police, members of the Guard killed two of the strikers, who on the whole remained orderly.[50] By emphasizing such facts, a strike narrative described nothing less than a social revolution in which the entire work force rejected the old paternalist vision of class relations and was radicalized by police violence. Furthermore, the strike narrative emphasized support from outside the community. The workers in Lawrence had no strike fund and relied on voluntary contributions from across the country to support themselves in the winter of 1912. Money poured in from all parts of the nation at the rate of $1,000 a day.[51] Orchestrating such facts, the strike narrative seemed to document the growth of a radical consciousness among American workers.

Artists, writers, and intellectuals found the story attractive. The outpouring of sympathy for the strikers at Lawrence and the widespread interest among writers and intellectuals continued after 1912 in other labor struggles—notably one in Paterson, New Jersey the following year. More than 25,000 workers walked out, uniting across ethnic and gender lines.

The radical intellectuals John Reed, Max Eastman, and Margaret Sanger supported the strike and lionized the radical labor leader "Big Bill" Hayward. The transformation of the Paterson strike into a narrative reached formal expression in a pageant held at Madison Square Garden. Thousands of workers came into New York by ferry on June 7, 1913, to serve as both performers and audience. The pageant was divided into dramatic scenes: going out on strike, police intervention, the death of a worker, the funeral, a strike meeting, the children sent away to safety. At the end came speeches and songs dedicated to solidarity. The event seemed to prove that workers and radical intellectuals shared a common cause. The connection proved ephemeral, and the strike failed,[52] but the strike narrative had become the radical alternative to the mill foundation story. The structure of both the Lawrence and the Paterson strike seemed to many intellectuals to prove that this narrative was the historical truth and not a selection and presentation of historical facts.

The strike narrative, emphasizing exploitation and struggle, became more widely known in the 1920s. When workers in Passaic, New Jersey stayed out on strike from January 25, 1926 until February 28, 1927, Mary Vorse wrote a short volume explaining the origins of the struggle, the lives of the workers, and even the strike activities of children. Vorse found that, whereas governments in Europe and New England generally did not permit women to work on the late shift, night work by women was "nearly universal in Passaic."[53] The National Consumers League learned that before the strike "the numbers of machine tenders had been reduced, increasing the work load on those who remained. Add to these the noise and shriek of the machinery, the oil-soaked floor, the close humid air, and the strain of the night work seems past belief. Some workers confessed to sleeping beside their machines."[54] Pregnant women also worked at night, moving constantly, chasing the spinning mule. Some had worked late into pregnancy, and several had given birth on the shop floor.

New Jersey textile companies commonly used spies and blacklists. In 1925, they paid stockholders dividends exceeding 15 percent, kept considerable amounts of cash on hand, and cut wages by 10 percent, provoking the strike. By the fourth week, 10,000 workers had gone out, from virtually all the mills. As at Lawrence and Passaic, strikers won public sympathy through with their orderly picketing and parading, which they kept up daily from 4 A.M. until nightfall. The companies tried to beat and intimidate the organizers, and they tried to defeat the strike by dragging it out for more than a year, but in the end the workers forced them to rescind the pay cut and to recognize the union.[55]

The mill was no longer a part of a process of settling a region or developing the economy. In place of the "inevitable" unfolding of events from the founding of a mill to the creation of a new town, the counter-narrative substituted a new inevitability: the conflict between labor and capital, leading inexorably to a strike. This conflict was then subsumed in the larger class struggle, leading to unionization, higher wages, better working conditions, and, in the most radical variant, the overthrow of capitalism. The mill foundation narrative had been individualistic and evolutionary; the strike counter-narrative was cooperative and revolutionary.

Labor unions told the story not only in print but also in films. *The Passaic Textile Strike,* a seven-reel epic, was completed and shown during the strike. It begins with the story of a Polish immigrant family who came to the "Land of Opportunity only to find industrial oppression and bitter struggle."[56] The father "dies of overwork, his daughter is raped by the callous mill manager, and his widow Katja is forced into the role of family breadwinner."[57] At this point, documentary footage shows orderly union meetings that include men, women, African-Americans, and immigrants, with representative figures from each group addressing the crowd. Violence comes from the state. Footage shot from rooftops shows how police charged "into lines of peaceful picketers . . . clubbing them until blood pours down their faces."[58] At the film's end, the strike's leaders join the United Textile Workers Union. Other union films made between 1913 and 1939 reached millions of working-class people, although it was difficult to get such films shown in commercial movie houses.

The shift from the foundation story to the strike counter-narrative was also registered in Sherwood Anderson's novel *Poor White.* Its central character is an inventor named Hugh McVey. A shy dreamer with little education, he invents a cabbage-planting machine and later a corn harvester. Manufacturing these labor-saving devices at first has precisely the effect on the small town of Bidwell that an Edward Everett would expect. Bidwell becomes a boomtown. In the process, it becomes physically ugly, and its citizens become strangers to one another. McVey himself has little interest in money, but his partners think of little else. By the end of the novel, the romance of invention and progress has been replaced by spontaneous industrial action.[59] The classes are divided against one another, and the first socialist speaker has appeared on the streets of Bidwell, denouncing capitalists.[60] Anderson undercut the foundation story and moved toward a tale of exploitation and class struggle. In contrast to the environmental counter-narrative, he attacked not industrialization itself but the way capitalists used machinery to exploit workers. Conflict suffuses McVey's thinking, which slowly moves toward a critique of the world his inventions have made. His partners are less imaginative and more confrontational. One declares: "I'd like to hang the men who are making trouble in the shops in town." He is certain that the way to get around labor problems is technological: "'We're inventing new machines pretty fast nowadays,' he cried. 'Pretty soon we'll do all the work by machines. Then what'll we do? We'll kick all the workers out and let 'em strike til they're sick that's what we'll do. They can talk their fool socialism all they want, but we'll show 'em, the fools.'"[61] McVey does not reply to this outburst, and the reader is left to ponder the justice of replacing men with machines. In 1927, Anderson put the matter more bluntly in a *Vanity Fair* article: "Man's inheritance—his primary inheritance—is being taken from him perhaps by mass production, by the great factory, by inventions, by the machine." He feared that "the age of the individual has passed."[62]

In 1920, when *Poor White* appeared, the textile industry was in the throes of rapid change. New mills in the South were displacing those in New England, Pennsylvania, New Jersey, and New York. Because the South had industrialized more slowly than the North, until after the Civil War it was commonly seen as backward, yet as in some ways idyllic.

In 1867, the young John Muir took a thousand-mile walking tour from the banks of the Ohio River to the Gulf of Mexico. In the Cumberland Mountains he found rural grist mills that were "remarkably simple."[63] Indeed, Muir found "all the machines of Kentucky and Tennessee" to be "far behind the age." "There is," he wrote, "scarce a trace of that restless spirit of speculation and invention so characteristic of the North."[64] Southerners also commonly depicted their region as pre-industrial, and some claimed that their manufacturing system had evolved independently of the North's.[65] Broadus Mitchell and other historians of the 1920s developed a regional narrative that attacked Northern industrialism while enshrining their own mills, particularly in the Piedmont region of North Carolina. They argued that these mills embodied a "planter mentality" that rejected the liberal capitalism of New England. Within a generation such ideas had become the orthodoxy expressed in W. J. Cash's book *The Mind of the South*. "The plantation," Cash declared, "remained the single great basic social and economic pattern of the South—as much in industry as on the land."[66] Indeed, "on their own private property the mills, typically speaking, built their own private villages. They not only provided houses for the workers, just as the plantation had always provided them for slave and tenant, but like the plantation, they provided, and owned, the streets on which those houses stood. The provided commissaries where the workman might get advances of food and clothing against his future earnings, exactly as on a plantation."[67]

But the claim that plantation and mill were nearly identical was fallacious. Whites, not blacks, filled most mill jobs. Whites had regular earnings; slaves did not. A slave had no freedom of movement; a white mill hand could leave at any time. If the mills paid for churches, workers did not have to attend. And if they sponsored schools, children could be sent to live with relatives elsewhere. The work rhythms and the technical demands on the factory also made it impossible to run one as a new form of plantation. Nevertheless, a Southern version of the mill foundation story flourished during much of the twentieth century, denouncing heartless Yankee capitalism and claiming that "planter capitalism" had been more benign. This new avatar of the mill narrative had no more basis in fact than the paternalistic vision of early Lowell. Southerners did not create their mills unaided; in a process of technology transfer that began before the Civil War, they drew heavily on Northern expertise, Northern machine shops, and Northern management methods.[68] The growth of Columbus, Georgia after 1828 (described in the previous chapter) showed that antebellum Southerners also invested in mill towns. If the North and the Midwest embraced the narrative of the mill earlier, making it a self-fulfilling prophecy of community development, Southerners soon followed suit. They lagged behind, but they did not go in a different direction. The pretense that Southern mills were modeled on plantations was an invention after the fact. The ideological use of such a fantasy was to sharpen the sense of sectional differences between North and South, inventing a paternalistic alternative to the industrial culture of the North. In this historical fantasy, factories only moved below the Mason-Dixon line after the Civil War. This version of the mill foundation story was widely believed until late in the twentieth century. Indeed, it became a part of a Marxist argument that presented the South as a pre-industrial region that exported raw materials and imported finished goods, becoming the North's exploited colony until long after 1865.[69] Fitzhugh had made a similar argument in the 1850s, and generations later Cash revived it in *The Mind of the South*.[70]

Textile mills expanded in the South during Reconstruction, partly because North Carolina and Georgia were closer to where the cotton grew than Massachusetts or Britain but chiefly because the South had cheaper labor. By the middle of the 1920s, North Carolina had surpassed Massachusetts as the principal textile state.[71] "The South," one New England cotton manufacturer explained, "has the advantage of longer hours of labor, lower wage scales, lower taxes, and legislation which gives a manufacturing plant a wider latitude than is usually possible in the North in the way of running overtime and at night. . . . The South is . . . fortunate in having a supply of native American labor which is still satisfied to work at a low wage."[72] Yet Southern textile workers did not have a lower cost of living. Government surveys established the reverse: ". . . for a family living in Greenville and Charlotte, even considering the low rates paid for shelter, fuel, and light, the cost of living was higher than in Fall River and Lawrence."[73]

Despite the South's "advantages," New England's productive capacity continued to grow in the early twentieth century, although slowly. At first the finer cloth was mostly made in the North, while the new Southern mills focused on coarser products, but by the 1920s the South had taken the lead in this area too.[74] Southern mills had easier access to cotton, lower labor costs, fewer legal restraints, better locations, and newer equipment.[75] They also had fewer labor problems than the Northern mills until the late 1920s, when severe labor unrest rocked the Piedmont region. In 1929, South Carolina alone had 81 strikes involving almost 80,000 workers, few of them union members. Their protests were spontaneous rebellions, like those at Lawrence and Paterson, and descriptions of these events followed the strike narrative. Sinclair Lewis visited the mill town of Marion, called it "barbarous," and concluded that there were "scores of mill towns quite as bad." Lewis's newspaper accounts were gathered in a pamphlet,[76] and Lewis urged his readers to send money to the striking workers. Many other writers also became interested in Southern textile strikes in that period. "Most of the proletarian novels," Malcolm Cowley later recalled, "were cast in one mold, and the fact is that many of them deal with the same events, usually a Communist-led strike like the one in the cotton mills of Gastonia, North Carolina."[77]

Gastonia's first mill was opened in 1887; by 1900 there were six more.[78] Gastonia, then with a population of 5,000, grew rapidly to become the center of the densest concentration of mills in the United States by the 1930s. Its native white work force was local and seemed docile. The United Textile Workers tried organizing in 1919 but lost a strike at the Loray Mill and seemed to make little headway.[79] The 1929 Gastonia Strike (not organized by a union) attracted wide national attention and became the subject of at least seven novels. The best known of these was Sherwood Anderson's *Beyond Desire* (1932). Grace Lumpkin's *To Make My Bread* was published in the same year. Mary Vorse's *Strike!* (1930) was told from the viewpoint of a Northern newspaper reporter. Dorothy Myra Page's *Gathering Storm: A Story of the Black Belt* focused on Communist Party efforts to organize both black and white workers. Olive Dargan's *Call Home the Heart* (1932) and *A Stone Came Rolling* (1935) focused on the experiences of a North Carolina girl who went to work in the Gastonia mills and got caught up in the labor struggle. All these novels were based on the same 1929 events, which themselves mirrored the pre-existing strike narrative in many ways. First, the strike occurred without outside leadership or local union organization. In the Southern mills at this time, "spontaneous outbreaks were far more in evidence than well-financed, organized efforts to establish recognized trade unions." Second, the strike revealed deep-seated anger: ". . . the bitterness, endurance, and solidarity displayed by the workers surprised not only the manufacturers, but the union officials, who had all but abandoned Southern textile workers. . . ."[80] Third, socialists and communists soon came from outside the region to join the struggle. Fourth, the state government backed the mill owners by sending in the National Guard, and that led to violent confrontations. Finally, the strike provoked a murder and a trial. This broad outline of events fit the pre-existing strike narrative so well that it seemed a representative moment in the larger crisis of American capitalism.

Page's *Gathering Storm* presents the Gastonia strike not as a defeat but as a part of the formation of class consciousness that would ensure the eventual triumph of the proletariat. Poor whites move down from the mountains into the mill town after a well-dressed stranger promises them work. "When we come to the mill," the narrator recounts, "we found that stranger had lied—plenty. Thar warn't no paved streets, no [water] pump in each block, 'n no schools, 'n thar warn't no church, nuther. . . . It was plain as day, the mill men diden mean to have no school, 'cause all the chillen, some no mor'n knee-high to a grasshopper, was at the mill."[81] The work was long and hard. "Thar was no limits on the hours, then. We went in afore mornin' light, 'n we worked until the bossman said stop."[82] The first half of the novel depicts poverty and exploitation; the second half moves toward armed confrontation, ending in gunfire and "another civil war." One of the organizers declares: "Because the Russian masses had sense enough to drive out their Rockerfellers, Morgans, 'n Jenkins, now all the wealth they produce goes not to a small class of blood-suckers but for the common good."[83] If the foundation narrative presented the mill as the automatic source of prosperity and community development, the radical rewriting of that story presented it as the central site of inevitable class conflict.

Anderson developed a more nuanced but ultimately similar view, delving into the psychology of mill work, which varied by class and by gender. He first depicted the town and its mill through the eyes of Red Oliver, a young man with one year of college education, who responds enthusiastically to the machinery: "The long rows of spindles in the spinning-room of the mill flew at terrific speed. How clean and orderly it was in the big rooms! It was so all through the mill. . . . The mill was in sharp contrast to the life of the town of Langdon, to the life in the houses, on the streets and in stores. Everything was orderly. . . . Fingers of steel moved. There were in the mill hundreds of thousands of tiny steel fingers handling thread, handling cotton to make thread, handling thread to weave it into cloth. . . . Little steel fingers picked up just the right-colored thread to make the design of the cloth. Red felt a kind of exaltation in the rooms. He felt it in the spinning-rooms. There was thread dancing in the air in there."[84] For the first time in his life, Red "got a sense of the human mind doing something definite and in an orderly way" and "some days all the nerves in his body seemed to dance and run with the machines." He feels exultant as he sees "the path of American genius. For generations before his day, the best brains of America had gone into the making of such machines as he found in the mill." Indeed, there "were other marvelous, almost super-human machines in great automobile plants, in steel plants, in plants where food was put up in tins, in plants where steel was fabricated."[85] The machines also invade his dreams, leaving him restless, and when working at night "he was sunk utterly in the strange world of lights and movement. Little fingers seemed always playing on his nerves." He grew tired from "watching the never-ceasing speed of the machines and the movements of those who attended the machines."[86]

Molly, a mill girl who has worked longer that Red, has a harsher understanding of the work: "It was hot in the mills on the long summer nights in North Carolina. You sweated. Your clothes were wringing wet. Your hair was wet. Fine lint, floating in the air, clung to your hair. In town they called you 'lint-head.'"[87] She suffers from headaches caused by the incessant noise and the pressure to keep the looms running, and feels as though she were in another world, one "of flying lights, flying machines, flying thread, flying colors." The flying shuttles move too fast for her to follow them. "Everything in the room jerked. . . . Molly got the jerks sometimes when she was trying to sleep in the daytime. . . . The loom room in the mill was still in her head. It stayed there. She could see it. She felt it."[88] Such experiences were not represented in the foundation narrative. "Mill girls" were presented as transient figures who temporarily lived in a Lowell or a Lawrence, leaving after a few years. Anderson's workers were proletarians, immersed in the mill for a lifetime.

Yet *Beyond Desire* focused less on the mill hands than on two disaffected members of the middle class, one the grandson of a Confederate doctor and one the daughter of the local judge. Each sympathizes with the workers. Rather than emphasize class differences, Anderson suggested a hollow middle class whose younger generation was becoming interested in communism. Ned Sawyer, the soldier who kills Red, is very much like him. Both are sons of professional men with some college education. Ned's sister is doing graduate work at Columbia University. Her father is a Jeffersonian, but, she notes, Jefferson "might have been all right in his day only. . . . He didn't count on modern machinery."[89] She warns her brother, who has joined the National Guard: "We may be shooting down the workers on whom everything rests. Because they are workers who produce everything and who begin to want—out of all this richness that is America—a new, a stronger, perhaps even a dominant voice."[90] Her brother prepares to lead a company of the National Guard against striking mill workers, Red among them. Likewise, many in the Guard had worked in the cotton mills. Anderson presented not a simple case of class warfare, but a nation torn apart, confused, grasping for explanations, groping toward transformation.

The mill foundation story was a capitalist romance about the benefits that flowed automatically from building a sawmill, a grist mill, or a textile plant. Anderson evokes this story near the beginning of *Beyond Desire*. The local capitalists hire a revivalist preacher to shape public opinion in favor of the new mills. "Presently the revivalist dropped everything else, and, instead of talking about a life after death, talked only of the present . . . of a new glowing kind of life, already lived in many Eastern and Middle-Western towns and that, he said, might as well be lived in the South, in Langdon."[91] The preacher described how "a sleepy little place" was transformed when "suddenly the factories came. People who had been out of work, many people who had never had a cent to their names, suddenly began to get wages." Their lives improved as they acquired automobiles, phonographs, and new homes.[92] But Anderson quickly underscores that these are false promises. The new mills almost fail at first, and the houses of the mill workers are cheap and unpainted. The superintendent's goal is to introduce as much automatic machinery as possible, so that workers will feel lucky even to have a job. "The thing to do," he tells Red, "is to keep plenty of surplus labor about. Then you can cut wages when you choose."[93] *Beyond Desire* discredits the foundation story and substitutes one about Red's struggle to throw off his middle-class consciousness and join the workers in a struggle for justice in the workplace. Ultimately, the strike narrative is not about the mill, but about how class struggle changes the major characters. It seeks to inspire readers to identify with workers. In contrast, the mill foundation narrative implicitly asks the reader to identify with the capitalist and the free market. It is not ostensibly about consciousness, but about how technologies can guarantee prosperity and build community.

The belief that American mill owners would not create a proletariat, widespread in 1830, had become unsustainable in 1930. Had the mill improved the land when the forests were gone and the salmon no longer spawned in the streams? What heroic image of the textile mill was possible after people had seen the photographs of Lewis Hine? How could idealized mill girls be believable after people had learned of the pregnant women working the night shift in Passaic? And yet the strike narrative also had its limitations, as it organized some facts and deleted others. Focusing primarily on the experiences of industrial workers, it did not address the creation of the mill itself or reflect upon its environmental effects. These matters were left to other counter-narratives. Furthermore, one learned little of the actual work or the machines in the mill from the strike narrative. The strike narrative excluded the perspectives of consumers, managers, investors, or those in research and development. The mill foundation story was heard in the boardroom and reproduced in company histories; the strike counter-narratives were written for unions and their sympathizers; the environmental counter-narratives were composed by people outside the industry. In comparison with the narrative of the axe, the second-creation story of the mill met more persistent and convincing opposition from its counter-narratives.

Academic specialists similarly subdivided their work on mills: historians of technology reworked the foundation story; labor historians rewrote the strike counter-narrative; more recently, environmental historians revived the ecological critique.[94] The best research managed to integrate two or more of these narratives into a single story. For example, in *Work, Culture, and Society in Industrializing America* Herbert Gutman began to move beyond institutional histories of unions and the strike narrative. He argued that for a generation after the mills were built in a major industrial city, such as Paterson, they seemed alien elements in the community. "They disrupted tradition, competed against an established social structure and status hierarchy, and challenged traditional modes of thought." The factory owner, far from being a founding figure, was an outsider whose "power was not yet legitimized." He faced "unexpected opposition from non-industrial property owners" and "did not dominate the local political structure."[95] Just as important, the emerging working class had many close ties to the local community and could often count on its support when conflict arose. Gutman's narrative radically reworked the foundation story. The mill owner is no longer the founder; he is an outsider invading a pre-existing community. The new technology is no longer the means used to create the town; it is a disruptive force that upsets class relations, attracts immigrants, exploits workers, and disrupts community life. When strikes come, small tradesmen and local government side with the workers. When the editor of a socialist labor newspaper is put on trial for libeling local industries, grocers and saloonkeepers raise money for his defense and join 15,000 of his supporters in a parade. Gutman's account wove the strike narrative into the community's history and pushed the technological creation story to the margins.

Anthony Wallace's *Rockdale* masterfully combines anthropology and history in an extremely detailed account of a small town called into existence by the mills in antebellum Pennsylvania. Wallace interweaves the foundation story (and the owners' perspective) with the laborers' experiences and their reasons for periodically striking. Wallace also takes account of environmental change. In 1843 a great rainstorm and flood pushed water levels 23 feet above normal. Because mills were located at the bottom of narrow valleys, they were always vulnerable, but the threat became worse with each passing year "as a result of human action. The progressive denuding of once forest-covered hills, the building of drains and culverts, and the construction of roads and covered areas all helped rain and melt-water to flow more rapidly. . . ."[96] Because dams diverting water to mills necessarily stood upstream, when one broke it released a wall of water that destroyed everything in its path. The torrents swept away in moments a new mill filled with new machinery, along with a stone warehouse, a dye house, and stone tenements for workers. Confronting these and other hardships, capital and labor at Rockdale found a common ground in evangelical religion, which subsumed human suffering within the larger apocalyptic tale of Revelation.

By the late nineteenth century, however, such religious "solutions" to the conflicting claims of narrative and counter-narrative were no longer convincing. In *Figured Tapestry*, which deals with the Philadelphia textile industry from 1885 until World War II, Phillip Scranton shows how larger market forces squeezed both management and labor. They faced not only competition from Southern mills but also new cost-accounting and inventory practices among buyers and wholesalers, who shifted "the risk and expense of stockholding back on to producers of seasonal goods."[97] In the 1920s textiles had become a buyer's market, and the Great Depression only exacerbated an existing crisis. In 1931, factories in Philadelphia were running at only 40 percent of capacity; their Southern competitors were running at 90 percent.[98] The problem was not only the competition from non-union workers in the Carolinas and Georgia but also the failure of the textile manufacturers to adopt modern management methods. Scranton's argument replaces both the foundation story and the strike narrative with a more inclusive analysis of economic and technological change. Few Philadelphia textile managers or workers were able to see their situation clearly, and even those who could were unable to act effectively in an industry that was regionally divided, suffering from excess capacity, only partially unionized, and old-fashioned in its accounting and inventory. Many went bankrupt in the 1930s.

New England textile workers often were on strike during the Depression. In 1936, New Hampshire's Amoskeag Manufacturing Company— once the largest mill in the world, with 17,000 employees—went out of business. After a 1922 strike, the paternalistic company had lost the support of many workers, and its directors had moved liquid assets to a holding company. This protected their investments, but it took away the money needed to modernize the mill, dooming it to eventual economic collapse. In a book titled *Amoskeag*, Tamara Hareven and Randolph Langenbach used these facts to frame the oral histories of those who worked in the company's mills, exploring the views of workers and managers. Even 40 years after it occurred, the closure remained a defining moment in the lives of many of these individuals. Some workers were convinced that industrial action had destroyed the company. A man who had worked there for 26 years, until the last day, declared: "They had to close because of the strikes."[99] But others did not feel that the mills had been victims of worker militancy. One said: "The people around here, like me, who did something

Figure 6.2
Mill pickets being arrested. Courtesy Library of Congress.

with their lives, they didn't do it because of the mills; they did it in spite of them." [100]

A decade after the collapse of this Northern mill, Charles Sheeler painted the Amoskeag factory buildings along the disused canal that had once delivered its water power. Sheeler had earlier painted "Primitive Power," the image of an overshot waterwheel in Alabama. Karen Lucic suggests that this old mill projects "a historical lineage for the country's contemporary dominance in technology." [101] In the context of the 1930s, with its search for a usable past and its celebration of the colonial heritage, "Primitive Power" is by no means dismissive of the water wheel. Rather, it depicts the beginning of the mill foundation story; Sheeler's "Manchester," the image of the Amoskeag mills, represents its end. Neither painting contains any hint of the strike narrative. In 1949, when Sheeler painted "Manchester," the Amoskeag mills had been closed for 13 years. That painting's perspective leads nowhere, and the absence of workers is prophetic not

Figure 6.3
Charles Sheeler, "Manchester" (1949). Courtesy Edward Joseph Gallagher III Memorial Collection, Baltimore Museum of Art.

of a high-tech future but of a melancholy closure and unemployment. The vista is stagnant and closed off. If "Primitive Power" visualizes the mill foundation story, suggesting a harmonious use of natural forces, "Manchester" quietly protests the story's denouement. The canvas is at once a flattened abstraction and a realistic depiction of a canal and a railway track passing between two mills. The impulse toward abstraction moves the painting toward atemporality, sidestepping historical specificity. Yet the site remains identifiable, and the lack of workers, the stagnant water, and the empty railway track all suggest temporal closure—the death of a very particular form of industry. Sheeler thus uses the stasis of an atemporal modernism to essentialize the exhausted political economy of the water-driven textile mill. "Manchester" is about an industry that has succumbed to entropy and become a capitalist dead end.

The City, a film shown at the 1939 New York World's Fair, suggests that Sheeler's paintings might be viewed as a part of a larger sense of history. In the film, colonial America stands for the virtuous life of the small town focused on a mill. The mill community lives in harmony with nature and community. In contrast, the middle portion of the film emphasizes the frantic pace of modern factories and skyscraper cities. It shows ramshackle houses where children play on slag heaps, where wet laundry hangs in filthy air, and where families must rely on polluted water that they hand pump out of the ground. This powerful presentation of the mill counter-narrative does not lead to a wholesale rejection of industrialization, however, nor does it lead to the strike narrative. Instead, the final section of *The City* suggests that Americans have the technology to move into greenbelt suburbs, which integrate work, leisure, and domestic life. These planned communities provide good housing, a clean environment, and a setting for harmonious family life. The film as a whole thus uses both the mill foundation story and the environmental counter-narrative to suggest the outlines of a future society.

Taken as a group, mill narratives naturalized the story of the axe. The clearing carved out of the woods looked natural when compared to a mill town. The cabin constructed of logs seemed far less artificial than the frame house built from sawn lumber that replaced it. Later, the mill village based on the waterwheel would seem a relic of a pastoral age when compared to twentieth century suburbs. This process of naturalization, in which each new technology makes its predecessor appear quaint, obscures the foundation narrative's meaning during the period of its emergence. Just as the first white settler's axe was a transformative technology in the wilderness, the mill, when new, was considered a transformative mechanical system that greatly expanded the power at a community's command. Those who built mills did not think them quaint.

Yet during early industrialization the mill had been considered pastoral, and it was not difficult to reuse many mills in picturesque ways. As mills fell into disuse, they became symbols of a vanishing era celebrated in such songs as "By The Dear Old Village Mill Down In The Valley"[102] and "Down By The Old Mill Stream." Mills also came to represent a pastoral birthplace of industrialization. When Henry Ford paid for the restoration of the Wayside Inn in Sudbury, Massachusetts, he had a mill built alongside it. The mill ground its first flour on Thanksgiving Day, 1929. The first building on the site, 190 years earlier, had also been a grist mill, later supplanted by a horseshoe nail factory. Ford removed its remains and hired a hydraulic engineer to design a new stone building with an 18-foot waterwheel. It drove two sets of millstones and powered a hydraulic grain handling system modeled on Oliver Evans's famous 1787 mill, the first automated production site. Thus Ford inscribed in the New England landscape the predecessor of his assembly line, suggesting a smooth and direct lineage from Evans to himself. To ensure that traffic did not disturb the pastoral scene, the world's richest automobile manufacturer paid the Commonwealth of Massachusetts to relocate a highway.

Figure 6.4
Wayside Inn, Sudbury, Massachusetts, c. 1930. Courtesy Library of Congress.

Today the Wayside Grist Mill continues to hold weekend milling demonstrations and to sell the flour.

Many abandoned mills were recycled as restaurants. Most of them emphasized the beauty of the mill pond, not the utility of the waterwheel. The gears and the millstones became decorative elements. Larger mills were retrofitted as commercial sites, most of them emphasizing tradition rather than change. One near Sturbridge, Massachusetts sells arts and crafts. The Nora Mill in Georgia advertises itself as a "granary" that "was built in 1876 on the banks of the Chattahoochee River in the North Georgia mountains. The mill itself is a large four-story building that has a 100-ft. wooden raceway powered with a turbine. On the same site, there is evidence that a mill has stood for almost 200 years. . . ." The mill sells stone ground natural flour and "a nice selection of kitchen ware." The Old Mill Inn in Vermont emphasizes that it was "built in 1786, tucked away on five acres of woods, meadows, and farmland, with chickens, barn cats, dogs, and miniature horses." The Old Mill Winery in Ohio sells antiques and its own wines and attracts people by sponsoring rock and folk concerts. In these and countless other instances, mills seem connection points to a past shorn of tensions or problems. Mills at Rochester, Minneapolis, Richmond, and other major sites have become offices, restaurants, and boutiques. It is difficult to learn much about the workers or their struggles at such sites. A few photographs may recall the time when the mill was still in production, but the long work hours, the din of machinery, the low wages, the strikes, and the hardships endured are effaced. The lived reality of the working textile factory disappears into the picturesque, and the foundation story reemerges.

7

"Let Us Conquer Space"

Let us . . . bind the Republic together with a perfect system of roads and canals.
Let us conquer space.
—John C. Calhoun, 1817[1]

The age has an engine, but no engineer.
—Ralph Waldo Emerson, 1853[2]

These remarks suggest the beginning of a third foundation story and its critique. Calhoun was speaking in Congress as an advocate of internal improvements. Calling upon his fellow legislators to act, he saw himself as a leader who could shape history. Half a lifetime later, the changes Calhoun called for were rapidly being accomplished. The United States had built an extensive system of roads and canals, supplemented by a growing system of railroads. The conquest of space and time had also been furthered by the telegraph. The republic was more closely linked than ever before. But the "us," the sense of collective agency, was missing. Sectional animosity had increased, not disappeared. Calhoun himself had sharply curtailed his support of federal funding for internal improvements by the late 1820s, for these were Whig Party policies. The canals and railroads were central parts of the infrastructure of industrialization that had increased tension between classes and begun to make cities less appealing places to live for the well-to-do. Urban elites used the railroad to move their families into suburban enclaves, and by the middle of the century the coincidence of political and economic power in one urban location was disappearing.[3] With such unexpected consequences in mind, Emerson wrote the sentence quoted above in his journal. Machinery proliferated more rapidly than the political means to govern it. Indeed, as early as 1839 Emerson wrote in his journal: "This invasion of Nature by Trade with its Money, its Credit, its Steam, its Railroad, threatens to upset the balance of man, & establish a new Universal Monarchy more tyrannical than Babylon or Rome."[4]

The contrast between Calhoun and Emerson suggests the difference between a foundation narrative about improved transportation and its critique, between the confident story of expansion of the first half of the nineteenth century and the widespread acceptance of counter-narratives thereafter. Whereas the narrative of the axe concerned solitary individuals, and the narrative of the mill described the emergence of towns, the foundation narratives of the canal and railroad projected new cities and massive growth for entire regions.

In the eighteenth century, roads were poor; the most efficient way to travel in the colonies was by ship, up and down the Atlantic coast. After independence, however, inland communication became important. In 1787, in *The Federalist*, James Madison discussed the need for better transportation, both to bind the nation together politically and economically and to ensure that armies could move swiftly to defend it.[5] In the 1790s, Tench Coxe and Alexander Hamilton argued for such internal improvements. Little was accomplished, however. In 1808, when the embargo on trade with France and England gave a sense of urgency to the problem of internal communications, Albert Gallatin presented his *Report on Roads and Canals* to the Senate. Gallatin proposed building four coastal canals to speed navigation along the Atlantic coastline, plus roads, canals, and river improvements to facilitate east-west trade. If little of Gallatin's vision was realized immediately, he did map much of the system that eventually would be built. Before c. 1820, however, Americans balked at the expense and still did not think in national terms, identifying instead with states and communities.[6] While the national survey was carving the West into geometric figures, ordinary Americans had not yet realized to what an extent improved transportation might stimulate the economy.

When James Monroe became president, it seemed likely that the federal government would at last take an active role. It was then that Calhoun championed the cause in Congress. Like other early advocates of roads, canals, and railroads, Calhoun had embraced a new foundation narrative. In it, the construction of transportation links became the equivalent of constructing the nation itself, in a process that could be seen epitomized in the growth of commerce and the establishment of new towns. "If we look into the nature of wealth," Calhoun argued, "we shall find that nothing can be more favorable to its growth than good roads and canals." He further noted that improved communications would knit the nation together politically and militarily. But Calhoun failed to win a majority over to his point of view.

The resistance was partly a matter of scale. The sawmill and the grist mill had founded towns; the water-driven factory served as nucleus for a small city. But the canal and railroad narratives reached beyond the local and regional to the national level. They described the founding not of one community but many, not increased wealth for a neighborhood but for the country as a whole, not the settlement of a locality but continental unification. The "conquest of space" could be seen in the work of clearing the land, but canals and railroads dramatically accelerated the process and linked the individual's actions to a national market.

Despite the nascent appeal of such a story, Monroe vetoed several internal improvement bills during his tenure as president. He was worried that the federal government's right to build roads and canals was not explicitly authorized in the Constitution. Nevertheless, Monroe was prepared to sign a General Survey Bill authorizing preparations for a program of internal improvements, and one of his last acts as president was to sign a bill that provided $300,000 toward the construction of the Chesapeake and Delaware Canal. During Andrew Jackson's two administrations, such cautious support of internal improvements continued. Jackson was a strict constructionist, yet the Constitution did speak of providing for the common defense, of post roads, and of regulating commerce between the states. Jackson vetoed funding for a road that lay entirely within one state (the Maysville Road), but he was willing to support some interstate ventures. Indeed, the average annual appropriation for internal improvements under the supposedly hostile Jackson—$1.3 million—was almost twice what it had been under John Quincy Adams, who had long championed such improvements but had met resistance from Congress.[7] The economic advantage of individual projects, in practice, often overcame objections based on a strict construction of the Constitution.

Yet many projects that clearly could benefit more than one state and that could speed the movement of the mails were long denied federal support.

For example, by 1819 the need for a canal to bypass the rapids on the Ohio near Louisville was obvious to any traveler. "In navigating the Ohio," one observer noted, "the saving of time, expense, and waste of property, by means of a canal . . . above the falls, is incalculable. It has been estimated, that Cincinnati alone, for several years past, has paid an extraordinary expense for transporting goods around the falls, exceeding $50,000."[8] Yet a 2½-mile canal to remedy this was not completed until 1830, and it was paid for by the state of Kentucky. The long delay in initiating such a useful improvement was due in part to its difficulty. Much of the excavation was made through solid limestone, the canal had to be wide enough to admit steamboats, and boats had to be raised as much as 60 feet. When completed, the canal—which made it possible for boats to pass all the way from New Orleans to Pittsburgh—was profitable right away.[9]

Opposition to federal funding of internal improvements was so strong that even works that clearly assisted the interstate movement of produce, the military, and the mails often had to be funded locally. The expansive narrative of internal improvements was fundamentally at odds with the strict construction of the Constitution or a jealous regard for states' rights. As a national narrative that envisioned population growth and movement to the west, it was most strongly embraced by the western states (whose representatives voted overwhelmingly in favor) and most strongly opposed by the older states, notably the slave states. The latter rejected liberal constructions of the Constitution, high tariffs on imported goods, and plans to spend funds collected from such tariffs on internal improvements. In addition, the states that had already invested their own money in canals (notably New York and Pennsylvania) often resisted federal funding for competing states, and even some parts of New England resisted the Whig vision of internal improvements at first.

For all these reasons, constitutional, sectional, and competitive, there was an absence of concerted federal action before 1820 and only sporadic investment in canals thereafter. The initiative passed to the states, particularly New York, where the Erie Canal from Albany to Buffalo was planned. Although it would link the Great Lakes with the port of New York, opponents of the Erie Canal claimed the costs would never justify the benefits. But Governor DeWitt Clinton won over the legislature in 1817, and 8 years later the canal was finished, stretching 363 miles, twice as long as any canal in Europe. The 84 locks were not distributed evenly; they were clustered in a few places, particularly at Lockport in the western part of the state and near Albany in the east. For much of its length the canal was level, a feat attained in some places by means of high embankments and aqueducts.

Contemporary stories about the Erie Canal emphasized second creation. Such stories began with a remote area outside the stimulus of the market. A new transportation technology—the canal—was introduced. Towns arose along its banks and mushroomed into cities. These technological creations were made possible by developing the latent potential of the landscape. The Mohawk River indicated where to build the canal. As Carol Sheriff notes in *The Artificial River*, those who "spoke of how Nature had left a gap in the Appalachians" meant "that God had created that break."[10] They believed that building the Erie Canal was "following God's plan to build a water highway to the West."[11] The "artificial river" flowed through a channel that seemed implicit in the form of the land itself.

The Erie Canal generated so much traffic that tolls paid for its maintenance and expansion, and it had dramatic economic and social effects, stimulating rapid growth in existing towns and creating new ones. Shortly after its completion in 1825, Captain Basil Hall passed through Rochester, a "bustling place" where "everything . . . appeared to be in motion. The very streets seemed to be starting up of their own accord, ready-made, and looking as fresh and new, as if they had been turned out of the workmen's hands but an hour before-or that a great boxful of new houses had been sent by steam from New York, and tumbled out on the half-cleared land."[12] Rochester's location at the major waterfall on the Genesee made the city's location ideal for shipping and manufacturing. Hall saw "great warehouses, without window sashes, but half-filled with goods" and new mills already in production even though the "carpenters were busy nailing on the planks of the roof."[13] He found some unnamed streets "nearly finished" and other streets already named but not yet begun. Everywhere were "churches, court-houses, jails, and hotels . . . all in motion, creeping upwards."[14] Construction had proceeded so quickly that in some streets the tree stumps had not yet been removed, so that in areas "where shops were opened and all sorts of business actually going on [it was necessary] to drive first on one side, and then on the other, to avoid the stumps of an oak, or a hemlock, or a pine-tree."[15] Hall even suggests the sound of this scene: "These half-finished, whole-finished, and embryo streets were crowded with people, carts, stages, cattle, pigs, far beyond the reach of numbers—and as all these were lifting up their voices together, in keeping with the clatters of hammers, the ringing of axes, and the creaking of machinery, there was a fine concert, I assure you!"[16] Other cities grew almost as quickly as Rochester, including Syracuse, Utica, and Lockport. Virtually overnight, the Erie Canal stimulated explosive growth and prosperity. "Canals created towns. Canals were the dominant feature of towns such as Utica, where civic life focused on the docks."[17] Wherever locks interrupted canal traffic, businesses were established and small communities emerged. The canal's social and economic effect was far greater than the sawmill's. Nathaniel Hawthorne recorded the new foundation narrative in an account of a trip on the Erie Canal: "It causes towns with their masses of brick and stone, their churches and theaters, their business and hubbub, their luxury and refinement, their gay dames and polished citizens, to spring up, till in time the wondrous stream may flow between two continuous lines of buildings, through one thronged street, from Buffalo to Albany."[18] In Hawthorne's language, the canal was cause and civilization an automatic effect. Another traveler noted: "It seemed to my eye as if more than half the city of Buffalo had been but yesterday redeemed from the wilderness."[19]

Even as the first canals were built, however, they were being compared to railroads. As the Erie Canal was completed, a detailed work appeared whose long title summarized its contents: *A Practical Treatise on Rail-Roads and Carriages, showing the Principles of Estimating their Strength, Proportions, Expense and Annual Produce, and the Conditions which render them Effective, Economical, and Durable; with the Theory, Effect, and Expense of Steam Carriages, Stationary Engines, and Gas Machines.* This book, by Thomas Tredgold, compared canals and railroads in detail and concluded that railroads were preferable for several reasons: they cost half as much per mile to build and therefore could charge lower rates for haulage, they were faster, and they were little affected by freezing and could be operated all year. An extensive review in *The New England Magazine* the following year did not fault Tredgold's calculations; indeed, it argued that New England's cold weather and rugged topography made the argument for railroads even more convincing. And yet, the reviewer could not suppress doubts about the practicality of a railroad from Boston to the Hudson River. He wondered "whether the merchants of Boston would ever be able to do business upon such a fascinating plan as to induce the manufacturers and agriculturists beyond the Hudson, to clamber over or plough through the intervening mountains, with their merchandise and produce, when it is so easy to drop them upon the deck of a sloop or a boat, and let the current waft them to a much nearer and a much larger market." "We believe," he continued, "that the sober people of Massachusetts will hesitate long before they offer violence to their purses in any such cause."[20] The reviewer had grasped the economics of constructing railroads, but not their foundation story. As his remarks suggest, the first railroads, like the early canals, were built with the object of increasing local prosperity by creating links to a national market. For this reason, Boston did eventually build a line west to the Hudson at Albany. Likewise, Baltimore sought a railway link from its port to the Ohio, and Charleston moved to reach beyond its immediate hinterland.

Before arguments of economic self-interest could convince these cities to invest in railroads, however, their merchants and bankers had to be convinced that river valleys were not the only "natural" channels of trade. To many in Boston, as the above reviewer suggested, it seemed too expensive to "clamber over or plough through" the Berkshire Mountains to reach Albany, in order to compete with ships that glided down to New York on the cost-free wind and tide. In 1835, Edward Everett addressed such skeptics at a public meeting where the idea of building such a railroad was being debated. He attacked "the general vague objection, that it is impossible, by artificial works, to divert commerce from its *great natural channels.*" "Abstractions prove nothing," Everett continued. "There are two kinds of natural channels—one sort made directly by the hand which made the world; the other, constructed by man, in the intelligent exercise of the powers which his Creator has given him."[21] A railroad or a canal thus became an extension of divine will exercised through man. The "natural" changed as a people became technologically more sophisticated. Everett hammered home the point: "It is as natural for a civilized man to make a railway or canal, as for a savage to descend a river in a bark canoe, or to cross from one fishing place to another, by a path through the woods."[22] By this logic, the improvement of nature was "natural." In closing, Everett cited George Washington, whose travels in upstate New York had convinced him of "the goodness of that Providence which has dealt his favors to us with so profuse a hand. *Would to God we may have wisdom enough to improve them!*"[23] The argument that the landscape contained latent within it the rudiments of a grand design, which it was man's destiny to carry out, had earlier been used to justify water power and steamboat navigation of western rivers. It would reappear in the 1890s to explain why irrigation was the destiny of the arid West. Technology, in such narratives, is a divinely ordained means to complete the construction of the world. Each foundation story could be justified as more than the mere assertion of human will for profit, as more than a patriotic necessity to expand the nation, and as more than a means to hold it together. Technological expansion became teleological. The axe, the mill, the canal, the railroad, and the irrigation project all provided new ways for Americans to make use of "God's favors." This was the teleology of second creation. The natural world as God had made it was the first creation; man's constructions were supplementary completions of the order that lay dormant within it.

Many saw engineering works such as railroads as elements of a larger, divine plan for a second creation. The idea was stated in the opening pages of a volume on the completion of the Baltimore and Ohio Railroad, commemorating the celebration held all along the line in 1857. William Prescott Smith began this book not with the railroad, but with philosophical remarks about nature as a divine creation that was ruled by "that principle of eternal progression—planted alike in humanity and the earth—which, from the first hour to this, has continually developed both in glorious harmony, so that just as the advancing capacity of man required new and increased fruits of nature's labor, his needs have been met and profusely supplied."[24] Smith had in mind the Ohio and Mississippi Valleys, which a benevolent creator had prepared for the arrival of white settlers but had wisely placed beyond the mountain barrier of the Appalachians: "Mountains and oceans constitute the chief barriers dividing nations from each other, and confining them to particular districts, thus provoking the concentration of effort and active exercise of reasoning and inventive faculties, necessary to advancement, and indispensable to a people's preparation for its new duties, when it shall take possession of another and larger sphere of action." The mountain barrier, which plausibly might have seemed to be a divinely ordained limit to the United States, was thereby converted to a part of God's plan for national expansion. The Alleghenies helped Americans to concentrate their energies until "we became so strong that the waves of population could not be restrained. These rose higher and higher, until a little stream of emigration began to flow down into the beautiful valleys beyond; and then, how wonderful the change! In a few brief years—a space of time almost incredible when we look at the results—the great West teemed with towns, cities, factories, steamboats, and industrious, busy men."[25] The settlement of the West and the building of the railroad system could thus be understood as a natural process in which the American people, like a mighty river, overflowed their banks and surged into a new region. Expansion seemed a partnership between man and nature, ordained by the creator.

By such logic, even the arbitrary geometry of the states' boundaries could be seen as providential. Everett was well aware that the individual states had "charters granted with little reference to geographical features; so that, in many cases, the state lines run east and west, while the rivers and hills tend north and south." The result was not only geographically anomalous but also perhaps economically injurious:

Owing to this cause, no small part of the business connections of Massachusetts are without the limits of the state; I mean those connections on which she depends for the necessary supplies of life; and precisely what is wanted is an artificial communication, which shall run across river and country, and unite her distant portions. That, sir, is the problem which Providence has committed to us. . . . It is a law of our moral natures, that the great boons of life are to be obtained by a strenuous contest with natural difficulties.[26]

For both Everett and Smith, the conquest of space by breaking through the mountains became a providential problem designed to exercise and develop man's moral nature. Persuaded by such arguments and by more mundane economic concerns, during its 1836 session the Massachusetts legislature allocated $1 million to assist construction of the Western Railroad, and an additional $1.2 million in 1838. Maryland was even more generous with the Baltimore and Ohio line, awarding it $3 million in 1838 alone, which matched $3 million from the City of Baltimore. By the 1850s, when that railroad finally reached the Ohio River, the cost of the whole line had risen to more than $31 million, with little support from the federal government and much of the funding from the City of Baltimore.[27]

As the new railway lines went into operation, Americans began to realize that, just as commerce could be made to flow through man-made channels, a town could be invented on any spot where a railroad chose to plant one. Laying out towns on speculation pre-existed the railroad, of course, and had become common by the early nineteenth century as steamboats opened many areas to commerce. But compared to steamboats and canals, the railroad was far less constrained by weather, topography, or gravity, and its directors could build lines into areas that water transport could never reach. It followed logically that these new artificial lines of commerce could concentrate the industry, wealth, and economic life of newly settled regions at a few nodal points.

Edward Everett, called upon to speak at the celebration of the completion of the railroad from Boston to Springfield in 1839, discussed how the continuation of the line would link Massachusetts to the rapid growth of the Midwest. He could not resist an aside on the potential growth of Chicago. In 1839, it was a small city, but one most favorably placed to grow into a metropolis:

Here, at last, you are brought into direct contact with the most extensive internal communication in the world. You are now on the dividing ridge of the waters, which severally seek the ocean through the St. Lawrence and the Mississippi. Here commences a system of travel and transportation by canal, railroad, and river, and mainly the latter navigated by steam, unparalleled by any thing on the surface of the globe. Did we live in a poetic age, we have now reached the region where the genius of steam communication would be personified and embodied. Here we should be taught to behold him, a Titanic colossus of iron and of brass, instinct with elemental life and power, with a glowing furnace for his lungs, and streams of fire and smoke for the breath of his nostrils. With one hand he collects the furs of the Arctic Circle; with the other he smites the forests of Western Pennsylvania. He plants his right foot at the source of the Missouri—his left on the shores of the Gulf of Mexico; and gathers into his bosom the overflowing abundance of the fairest and richest valley on which the circling sun looks down.[28]

Such visionary rhetoric personified the commercial potential of steam transportation. The "Titanic colossus" had a continental reach.

The expansion of steam power into new regions seemed to be the same thing as the spread of civilization itself. The author of an 1836 biography of Samuel Slater wrote: "Let our legislators be assured, that while they are extending towards its completion that system of improvement planned and hitherto carried forward with so much wisdom, they are putting into operation a moral machine which, in so far as it facilitates a constant and rapid communication between all parts of our land, tends most effectually to perfect the civilization, and elevate the moral character, of the people."[29] The railroad was not only a "colossus of iron and of brass"; it also improved the citizenry. Other authors were certain that the railroad was sublime, and that it worked upon the sensibilities of those who saw it just as powerfully as the finest mountain scenery.[30] Man's "second creation" could have as profound a moral effect as nature.

In 1846 the geologist Charles Lyell visited the United States and heard "much characteristic conversation in the [railroad] cars, about constructing a railway 4,000 miles long from Washington to the Columbia river." "Some of the passengers," Lyell noted, "were speculating on the hope of seeing in their lifetime a population of 15,000 souls settled in Oregon and California."[31] These passengers had accepted the premises of the technological creation story, in which settlement and progress were railroad creations. As it happened, the discovery of gold 3 years later brought more than 15,000 people to California a generation before such a railroad was built. But the desire for the transcontinental line long pre-existed it. Most Americans would have agreed with Emerson, who wrote in "The Young American": "An unlooked for consequence of the railroad is the increased acquaintance it has given the American people with the boundless resources of their own soil. . . . It has given a new celerity to time, or anticipated by fifty years the planting of tracts of land, the choice of water privileges, the working of mines, and other natural advantages. Railroad iron is a magician's rod, in its power to evoke sleeping energies of land and water."[32] Nature was not being invaded or destroyed but awakened, and the metaphor of "sleeping energies" that could be aroused was characteristic. "Railroads in Europe," Horace Greeley wrote, "are built to connect centers of population; but in the West the railroad itself builds cities. Pushing boldly out into the wilderness, along its iron track villages, towns, and cities spring into existence, and are strung together into a consistent whole by its lines of rails, as beads are upon a silken thread."[33]

In 1868, Greeley's conceit was visualized in a popular Currier and Ives lithograph, titled "Across the Continent," showing a train pushing out

Figure 7.1
Currier and Ives, "Across the Continent" (1868). Courtesy Royal Library, Copenhagen.

across the Great Plains.[34] The relationship between the railroad and the landscape is unambiguous. The railroad is rapidly developing the American West. The track ahead is still being laid, but the area already served by the railroad contains a prosperous new community with a prominent public school. The power to move across the land is translated into expansion and settlement. Human creations transform the landscape, which in the distance remains vague and unformed, whereas in the foreground a new town has been impelled into life by the railroad. The legend is a quotation from Bishop Berkeley's often-repeated declaration "Westward the course of empires takes its way."[35] This notion that human history records a continual westward shift in power, from the Middle East, to Greece, to Rome, to Northern Europe, and then to the New World, had become a staple of the American self-conception by the time this image was crafted and sold to a large public in 1868. By that date the first transcontinental railroad to the Pacific was nearing completion. The Currier and Ives image was both a vision of the recent past and a prediction of the future. It used space to represent time: the new community in the foreground was the present; the empty land ahead of the train was the future, which extended to the vanishing point of perspective. The image was also rather un-self-consciously imperialistic. The land ahead was presented as empty space awaiting the coming of white civilization. A few Native Americans on horseback were literally on the margins, watching, with the smoke from the train blowing over them and obscuring their view of the land and thus of the future.

This conception of the railroad's role in westward expansion was shared North and South. In 1860 the *New Orleans Picayune* articulated this foundation narrative in detail, beginning with what had become the traditional contrast between Europe and America. First the paper noted that American lines had been built far more cheaply than those in England: "The total cost of all the railroads in England is £320,000,000; that of railroads in the United States only £240,000,000. Or, in other words, 30,000 miles here have cost but little more than two-thirds of the expenditure in England for 10,000 miles." American dividends were 50 percent higher. "But the difference between the receipts of the roads of the two countries is not fully presented by this test of value," the *Picayune* continued. English trains served "thickly populated districts under the most perfect cultivation," and these regions were "already developed to their full power of production." Furthermore, "English railways do not create new industry, and fail to scatter abroad the seeds of new productive energy, bringing into existence along every line a business before unknown."[36] The technological foundation story seemed unique to America:

Nine-tenths of our roads when first traversed by steam pass through long ranges of woodlands in which the axe has never resounded, cross prairies whose flowery sod has never been turned by the plow, and penetrate the valleys as wild as when the first pioneers followed upon the trail of the savage. They connect distant centers of population, and open markets for an agricultural population in the very heart of the continent. But no sooner is the great work achieved, than population pours into the rich wilderness, is scattered over the wide sweeping prairies, and settles in the fertile vales, and away business springs up destined to increase each year for centuries to come. Villages soon dot the margin of the road. Factories turn to service the water power of the little streams it crosses . . . and golden grain covers the lands that since the primeval time have been overshadowed by the dark forest.[37]

Logically, the railroad could not pass through a woodland where the axe had never resounded, because the railroad right of way had to be cleared before it could be built. Nor was the land empty wilderness before the railroad arrived; both Native Americans and early settlers preceded it. But this newspaper story is faithful to the logic of the technological foundation story. A new technology makes society possible. The railroad opens the way for the axe, the plow, villages, factories, and towns. As settlers "pour in," the new region becomes a part of the national economy.

The ultimate expression of this story was the transcontinental railroad, much discussed before the Civil War (Jefferson Davis favored the idea and advocated a Southern route) and built immediately after it. A line from the Atlantic to the Pacific was not merely an engineering project or a business proposition; it was the fulfillment of Manifest Destiny. The rapidity with which track could be laid was widely celebrated as a measure of progress. In 1867 a travel writer described how the track advanced at the rate of 2½ miles or more per day, with an army of 12,000 men to do the job. That writer made explicit the latent content of "Across the Continent": "We found the workmen, with the regularity of machinery, dropping each rail in its place, spiking it down and seizing another. Behind them, the locomotive, before the tie layers; beyond these, the graders; and still further, in the mountain recesses, the engineers. It was Civilization pressing forward—the Conquest of Nature moving toward the Pacific."[38]

Many saw the new railroad as a passage to the Orient, and Whitman's "Passage to India" extrapolated the notion. A British geographer who had surveyed the new line concluded that even English commerce and travel would be affected. A journey from London to Japan would take 10 days less via New York and San Francisco than by sea around Africa. Travel to all parts of China east of Shanghai and to New Zealand would be affected too.[39] The painter Thomas Prichard Rossiter likewise suggested the centrality of the railroad in his 1858 canvas "Opening of the Wilderness,"[40] which depicts not a covered wagon, a lone scout, or a military patrol but five steam engines congregated at a small station beside a river at twilight. Their rising smoke provides the strongest vertical accent in an otherwise largely horizontal composition. The wilderness is suggested by the dark forested hills in the background. The only building fully visible is a small roundhouse, though there is the suggestion of a town beyond the depot.

Figure 7.2
Thomas Pritchard Rossiter, "Opening of the Wilderness," c. 1858. Courtesy M. and M. Karolik Collection, Museum of Fine Arts, Boston.

Rossiter emphasized the power and movement of the steam engines to transform a dark and static landscape.

The settlements along railway lines were, in almost every case, laid out on a grid pattern. During the two generations before the railroads, town planning had become so familiar as to be almost routine. The railroad carried the grid westward.[41] Americans imposed a rectilinear pattern of lines based on latitude and longitude throughout the Ohio Valley, across the Mississippi, and on to the flat topography of the Great Plains. In contrast to land speculation along rivers, however, canals and especially railroads transformed the speculator's situation. Any landowner with river frontage could claim importance for a site, when in fact the land might be swampy, unstable, easily flooded, or badly located in other ways. The construction of a canal or a railroad introduced new elements into city planning and growth, because they assured investors that a town would have regular, inexpensive transportation and a steady supply of coal and wood. With transport and energy guaranteed, all towns on canals and railroads became potential centers of manufacturing. Railroads could stimulate local development far more than a road and a good deal more than a canal, which could freeze in winter and which was a slower form of transport. Railroads also had it within their power to underwrite the creation of a town by guaranteeing to build a station there. Even in thinly populated regions and deserts, railroads established at least one town as a division point for every hundred miles of track, because this distance was the limit of a crew's day. Such towns had extra tracks, water towers, and repair shops. They also provided hotels, restaurants, warehouses, and other services. In addition, the board of directors could stimulate the economy of a town by locating a roundhouse and major repair shops there, or by designating it as a district headquarters. In a few cases, towns mushroomed into important cities based largely on the railroad, notably Indianapolis, Atlanta, and Denver, and to a considerable extent Los Angeles.

In view of these expected benefits, public lands were conferred on many railroads. The practice originated as early as 1803 in grants of land to assist in the building of roads. It was extended to canals in the 1820s, and then extended in the 1830s to several companies for railroads in Illinois, Florida, and Louisiana, none of which, however, were built.[42] From 1850 to 1870, railroads were given land along the right of way by the state legislatures or (more commonly) by the federal government. In all, government gave 131 million acres of land to the railroads, most of which they later sold for $7–$10 per acre.[43] A map of the lands given to the Little Rock and Fort Smith Railway in Arkansas shows a typical grant, a checkerboard of alternating sections, not only along the immediate right of way, but several miles in each direction. Land grants became even more valuable wherever towns were planned, generating quick profits for railroad investors. The relationship between railroads and land sales was understood from the beginning. The first railroad to be based on a land grant, the Illinois Central, "developed a highly sophisticated and successful technique for inducing immigrants to its lands,"[44] which other railroads imitated further west. Newspaper reporters received guided tours of new regions (typically in early summer, when the crops looked their best). Railroads paid lecturers to testify to the superb opportunities and flooded eastern states and foreign countries with glossy pamphlets.[45]

The Illinois Central had a charter that specifically prohibited it from laying out cities along its routes. This provision was intended to prevent land speculation.[46] Here, too the railroad's directors set the pattern others would follow, as they simply formed another company to create and sell new communities. They established a standardized design and used it for 33 towns founded at 10-mile intervals along the line.[47] The results were dramatic. In 1850 there were only 10 towns located in the vicinity of its tracks; a decade later there were 47. Subsequently, town planning was accepted as a part of railroad building, and virtually all western lines founded towns as well as selling land that had been granted to them by the state and federal government. The Chicago, Burlington & Quincy Railroad, for example, formed companies to acquire town sites along its lines. Typically, such companies donated lots for churches and schools, and shaped the dominant grid pattern by a rudimentary zoning into commercial and residential areas. For example, in 1863 the town of Lanark, Illinois stood where 6 months earlier there had been empty prairie.[48] Called into being by the construction of a railroad line from Chicago to the Mississippi River, this instant town already had a clean and comfortable hotel, twenty shops on Main Street, and 300 inhabitants living in wooden-framed houses. As was typical of the technological foundation story, at first Lanark had no church, and only established one later.

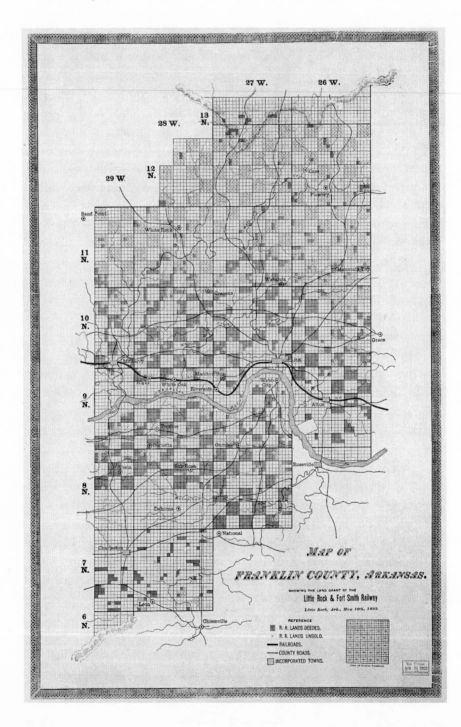

The two builders of the first transcontinental line, the Union Pacific and the Central Pacific, understood that city lots sold for more per acre than open land. As they built their lines, temporary communities were laid out on a grid system to house workers and suppliers. Some of these, such as Cheyenne and Laramie, survived to become important cities; others were taken down and moved to the front of the ever-advancing construction. Not only were ramshackle, temporary structures moved from point to point, but mail-order houses could be sent from Chicago to a new community for immediate erection. Here the foundation story seemed to come true. A French mining engineer in Cheyenne saw that "Houses arrive by the hundreds from Chicago, already made. . . . Do you want a palace, a cottage, a city or country home; do you want it in Doric, Tuscan, or Corinthian; of one or two stories, an attic, Mansard gables? Here you are! At your service!"[49] In North Platte and other Nebraska towns, "hotels for travelers would spring into existence in a day; a bank and an opera house would rise simultaneously side by side; stores and outfitting establishments of every variety would line the main streets with their quaint signs and emblems of trade. Mechanics and merchants would pour in from other parts of the road and with them would come the lawyer and the doctor. . . ."[50] The same process continued further west in Washington and Oregon, where new cities burst into life in grid patterns that took their orientation from the railroad line.[51]

Acceleration of social development was an essential ingredient of the railroad foundation story, which became shorthand for the effects of industrialization generally. The influential clergyman Josiah Strong, who surveyed a century of development from the vantage point of 1885, emphasized in his best-selling book *Our Country* that in the late eighteenth century people still lived in isolated communities. A person born in 1800, he reminded his readers, would not have known the steel plow, the telegraph, the railroad, or the steamship. As the culmination of his argument, Strong declared: "Ten years in the New West are, in their results, fully equal to half a century east of the Mississippi. There is there a tremendous rush of events which is startling, even in the nineteenth century. That

Figure 7.3
U.S. General Land Office, "Map of Franklin County, Arkansas" (1893). Courtesy Library of Congress.

western world in its progress is gathering momentum like a falling body. Vast regions have been settled before, but never before under the mighty whip and spur of electricity and steam."[52] The result was a process of settlement completely unknown to previous ages, even within the United States: "In the Middle States the farms were first taken, then the towns sprung up to supply their wants, and at length the railway connected it with the world; but in the West the order is reversed—first the railroad, then the town, then the farms. Settlement is, consequently, much more rapid, and the city stamps the country, instead of the country's stamping the city. It is the cities and towns which will frame state constitutions, make laws, create public opinion, establish social usages and fix standards of morals in the West."[53]

Strong was merely repeating what many others had said. "In the East," Albert Richardson had observed 20 years earlier, "railroads are built for the towns; on the border they build the towns."[54] From Strong's point of view, this urban dominance was worrisome, as was the imminent exhaustion of good farmland and the arrival of waves of new immigration. He sought to convince his readers that the character of the American West would be determined during the following 20 years, and that the West would eventually become the nation's dominant region.[55] Therefore, "the emergency created by the settlement of the West—a grand constellation of empires—is to be met by placing in the hand of every Christian agency there at work all the power that money can wield."[56] A railroad civilization was potentially sinful and in need of extensive missionary work.[57]

The very ease with which cities might rise at the side of the railroad suggested their fundamental instability and their weak social fabric. Such towns appeared or disappeared in a season. The station, as John Stilgoe has noted, was "a liminal zone" representing the portal through which everything passed to the larger world.[58] Like the river town that sprang to life whenever a steamboat arrived, the western community set its daily rhythm in accord with the railway timetable; it was most active when the steam engine raced into the depot. By implication, it was naturally slothful and indolent without the external stimulus. People with nothing else to do gravitated to the station, curious to know who and what might arrive there. If the train could supply new inhabitants, mail-order goods, and newspapers and magazines from distant cities, it could also suck the life out of a community by moving its repair shops elsewhere or by changing its timetable. Many early railroad towns grew rapidly for 10 years but then stagnated. The communities most likely to continue to grow had already been founded before the trains came.[59]

Cities west of the Alleghenies understood early that their success depended on expanding rail lines. Charles Cist wrote of the railways passing through Cincinnati: "On every side, toward every point, radiating lines from Cincinnati will penetrate the most fertile regions of America. They will connect the lakes with the rivers; they will bind ocean to ocean; they will bear the burdens of enormous harvests; develop the treasures of the disemboweled earth, and carry bread to laboring millions."[60] Cist went on to argue that Cincinnati was destined by its geography to become the West's dominant city. He was not the first to do so. A generation earlier, an author in *Western Monthly Review* had noted that three canals were nearing completion nearby and had declared: "Cincinnati will soon be the center of the 'celestial empire,' as the Chinese say, and instead of encountering the storms, the sea sickness, and the dangers of a passage from the gulf of Mexico to the Atlantic . . . the opulent Southern planters will take their families, their dogs and parrots, through a world of forests, from New Orleans to New York, giving us a call by the way. When they are more acquainted with us, their voyage will often terminate here."[61] Such rhetoric was common at celebrations held to celebrate the completion of new lines. Stripped to their essentials, such orations are foundation narratives.

In 1867, to mark the building of the transcontinental railroad and to profit from it, the American Publishing Company (also the publisher of Mark Twain's *The Innocents Abroad*) brought out a book by Albert Richardson titled *Beyond the Mississippi* for the subscription trade. It was subtitled "From the great river to the great ocean, life and adventure on the prairies, mountains, and pacific coast, with more than two hundred illustrations, from photographs and original sketches of the prairies, deserts, mountains, rivers, mines, cities, Indians, trappers, pioneers, and great natural curiosities of the new states and territories, 1857–1867." A popular verse from Whittier appeared just after the title page:

Behind the squaw's light birch canoe,
The steamer rocks and raves;
And city lots are staked for sale
Above old Indian graves.

I hear the tread of pioneers
Of nations yet to be—
The first low wash of waves where soon
Shall roll a human sea.

In the preface, Richardson prophesied: "Five years hence, the Orient will be our next-door neighbor. We shall hold the world's granary, the world's treasury, the world's highway."[62] He emphasized the West's "exhaustlessness and variety of resources" and its "grand curiosities and wonders," and concluded: "The Pacific Railroad hastens toward completion. We seem on the threshold of a destiny higher and better than any nation has yet fulfilled. And the great West is to rule us."[63] The first chapter begins with an 1857 railroad journey from St. Louis west, in which the line is treated metaphorically as "the entering wedge—the first deadly blow at her relics of barbarism."[64] The West is to rule the east, then, only after being tamed. Leaving the railroad at Jefferson, he boards a steamboat for Kansas City, whose many passengers include "the irrepressible agent of a new Kansas town, proving incontestably by statistics and diagrams that his will become the largest city west of New York" and "the eager-eyed speculator bound for the land sales, with wonderful stories of his uncle who became a millionaire from Chicago investments, or his wife's cousin, who made forty thousand dollars in six months upon Michigan pine-lands. . . ."[65]

Richardson saw everywhere a West in convulsive growth. He described the many new towns and the speculative mania of their founders and declared:

On paper, all these towns were magnificent. Their superbly lithographed maps adorned the walls of every place of resort. The stranger studying one of these, fancied the New Babylon surpassed only by its namesake of old. Its great parks, opera houses, churches, universities, railway depots and steamboat landings made New York and St. Louis insignificant in comparison. But if the newcomer had the unusual wisdom to visit the prophetic city before purchasing lots, he learned the difference between fact and fancy. The town might be composed of twenty buildings; or it might not contain a single human habitation. In most cases, however, he would find one or two rough cabins, with perhaps a tent and an Indian canoe on the river in front of the "levee." Any thing was marketable. Shares in interior towns of one or two shanties, sold readily for a hundred dollars. Wags proposed an act of Congress reserving some land for farming purposes before the whole Territory should be divided into city lots.[66]

Richardson concluded that "it was not a swindle, but a mania," and that the speculators—who continually reinvested their inflated profits only to lose them in the end—were "quite as insane as the rest."[67] The apparent truth of the technological foundation narrative was everywhere too strong for mere facts of geography. Indeed, Richardson himself fell into profitless speculation, and yet his book is suffused with opulent visions and gorgeous prophecies. The fruits and vegetables of his California are immense and require no irrigation. His Plains States have immense coal reserves. His Omaha, it seems inevitable, "will rise a great city, heart of a dense population, on the grand highway of travel and traffic for the whole globe."[68] The transcontinental railroad ensures the city's future, infusing it with "wonderful vigour; and I found the little capital of Nebraska the liveliest city in the United States."[69] Yet Richardson also knew that railroad agents "manifested an avarice for donations of lands and lots to themselves. . . . If the owners of any village refused to comply, they could run the cars by, establish the station on the bare prairies beyond, and kill the town by establishing a new one."[70] But the activities of swindlers and confidence men did not undermine his confidence or that of most Americans in the last half of the nineteenth century. Rapid growth, sudden fortunes, and overnight construction all seemed incontestable.

Dickens was quick to satirize this aspect of American development in *Martin Chuzzlewit*, whose hero invests in New Eden, an almost wholly imaginary town. A detailed map shows "banks, churches, cathedrals, market places, factories, hotels, stores, mansions, wharves, an exchange, a theatre, public buildings of all kinds"—even the office of the daily newspaper, the *Eden Stinger*.

Such satire had little effect on American behavior. The sense of inevitable growth stimulated by the railroads suffuses the opening pages of James Rusling's 1877 book *The Great West and Pacific Coast*. The whole area from New York to St. Louis is described as an already settled garden,[71] and the mental location of the West has shifted from Kansas and Missouri to the Rocky Mountains. An 1882 summary of the 1880 census commented on the centrality of railroads in promoting western settlement. In 1840, near the time of Dickens's visit, "except for some sparse settlements in Iowa Territory, all else to the north and west was essentially an unsettled blank."[72] Whereas canals had been important in the settlement of states as far west as Ohio and Indiana, the railroad's importance increased as the population moved west. The greater flatness of the land and the gradual increase in aridity made railroads the more adaptable, as well as the faster, mode of transport. "Many persons still live within a hundred miles of the greatest grain market of the world in ancient or in modern times, who recount to their children how they hauled wheat to Chicago with their teams, camped at night by the wayside, and, after a week or so spent in going and coming, reached home with a few yards of calico, or a few pounds of groceries, as the net proceeds of the journey. . . . The railroads have brought the grain fields and the pastures of the West to the doors of the factories of the East. They have stimulated the finer farming of the East, too." Now the most profitable agricultural land in the United States was in the East—for example, in Massachusetts, in Connecticut, and in New Jersey.[73] "The cities and the factory towns of Massachusetts buy eagerly every relishing fruit and every staple vegetable that can be coaxed out of the ground, even if the great supplies of meat and bread come from the fertile West."[74]

Because of railroads, many western cities saw themselves as the future centers of national commerce and culture. Omaha, for example, was laid out in 1854, incorporated in 1859, and had only 1,900 citizens in 1860. Many emigrants disembarked at this little town on the Missouri River to begin their trek westward. In the 1860s, Omaha became the railway center where the Union Pacific and Central Pacific connected. Its population reached 16,000 in 1870 and 30,500 in the following decade. By 1886 it had 70,000 people and expected to become the most important city on the Great Plains. Omaha's spectacular growth was almost entirely due to the construction of the transcontinental railroad. Its sheer existence seemed to verify the foundation story, assuring other cities that such growth was to be expected. The heavily illustrated 1889 book *Picturesque Sketches of American Progress* touted the spectacular growth of one city after another.[75] Minneapolis had grown from a population of 45 in 1845 to 129,000 in 1885, based on its enormous flour mills driven by the water power of the Mississippi and its extensive rail connections to eastern markets. Denver had fewer than 5,000 people in 1870 and more than 70,000 in 1885. Kansas City's population had tripled to more than 100,000 people in the same period. These instant cities, and many other towns that never grew, all expected to become the next Chicago.

A decade before Turner made his famous declaration about the end of
the frontier and the closing the West, the Census Bureau determined that
(not including swamp lands or Alaska) there were 700 million acres as yet
unsurveyed and unclaimed.[76] The railroad continued to expand rapidly
into these areas, growing by an astonishing 70,000 miles in the next 10
years to 163,597 miles of track. At the end of the nineteenth century, with
fewer than 50 motorcars on the nation's roads highways and before the
invention of the airplane, the monumental importance of the railroad to
the nation seemed obvious. No one could then imagine that the system
would reach its greatest extension—254,037 miles—in 1916 and would
then decline every year for the rest of the twentieth century.[77]

In 1900 the railroads were coextensive with America, holding it
together, fueling its cities, carrying its goods, and offering the fastest
passenger service in history. Between 1830 and 1900 the railroad had
grown from a few miles of track to become the central channel of American life and commerce. It seemed to exemplify every element of the technological creation story. The railroad company entered an undeveloped
region and used its powerful technology to transform it, creating new
towns and bringing new settlers. It made empty space into valuable land
where prosperous farms soon appeared. As Emerson Hough declared in
1905, "the railroad was not to depend upon the land, but the land upon
the railroad."[78] For the immigrant, Hough declared, "the railroad was
explorer, carrier, provider, thinker, heart, soul and intellect for [a] population that in another generation was to be American. No wonder these
folk stand and stare when the railway train goes by. It has been Providence
to them."[79]

The settlement process seemed almost endlessly repeatable, and the narrative was based on four interrelated premises, each of which seemed confirmed by experience. The first premise, that the land and resources were
abundant, seemed extravagantly fulfilled by the lands west of the Appalachians and by the provisions of the Homestead Act and its successors.
The second, that the market was free, was the political orthodoxy of
Republicans and Democrats alike. The third, that land could be divided on
geometrical principles and treated as an abstraction, had become the standard practice of government surveyors and had been accepted by buyers
and sellers as natural and normal. And the fourth premise, that of man's
mastery of force, was triumphantly exemplified by the railroad itself,

which conquered both space and time, demonstrating to passengers and to onlookers the power of human creation.

In 1836, Daniel Webster celebrated "the application of the power of steam to transportation and conveyance by sea and by land," asking "Who is so familiarized to the sight even now, as to look without wonder and amazement on the long train of cars, full of passengers and merchandise, drawn along our valleys, and the sides of our mountains themselves with a rapidity which holds competition with the winds?"[80] For Webster's contemporaries, such a train contrasted with a straggling line of wagons ponderously drawn by horses over poor roads. Whitman spoke for his generation when he proclaimed that the railroad engine was "type of the modern—emblem of motion and power—pulse of the continent."[81] The completion of the first rail line across the West to San Francisco in 1869 marked the apotheosis of a national vision. Whitman celebrated "the Pacific railroad surmounting every barrier":

I see continual trains of cars winding along the Platte carrying freight and passengers,
I hear locomotives rushing and roaring, and the shrill steam-whistle,
I hear the echoes reverberate through the grandest scenery in the world. . . .[82]

For Whitman, the train was a part of a larger "passage to India" that would link the Atlantic and the Pacific and place the United States at the center of world commerce. He saw this "marriage of continents, climates and oceans" as more than the completion of Columbus's voyages, for it intimated transcendent journeys to the stars. Whitman was hardly a real-estate speculator, yet he found the railroad an extravagant confirmation of their dreams of transformation. The force of the locomotive became symbolic of the power to remake the land and to found new communities. In the American imagination, the sod house on the prairie, far from being a marginal dwelling, stood at the future center of world civilization.

The railroad seemed the inevitable culmination of an evolutionary sense of progress. Frederick Jackson Turner argued that the United States had developed in an orderly sequence. He compared the continent to a vast page to be read from west to east, with the hunters and pioneers followed by settlers and then by internal improvements. For Turner and his readers, railroads seemed a part of an automatic unfolding of the national destiny: "The transformation by which the slender lines of the Indian trail became the trader's trace, and then a road, superseded by the turnpike and canal, and again replaced by the railroad is typical of the economic development of the United States."[83]

At century's end this nearly automatic, evolutionary optimism was linked to an industrial aesthetic. Americans were awed by the sheer scale of the railroads, with their bridges spanning even the largest rivers, their enormous freight yards and roundhouses, and their impressive stations. This system, together with the landscape of steel mills and factories that the railroad made possible, was understood in terms of the sublime. The new industrial aesthetic emphasized man's domination of nature, billowing clouds of smoke and steam, mountains of coal and slag, and the searing heat of furnaces—a landscape where everything was man-made.[84] This vision of a second creation that entirely replaced nature was promulgated in trade journals and in magazines such as *Century* and *World's Work*, whose authors proclaimed the sublimity of the industrial "wonderland," typically glimpsed through a train window at dusk or during the night. The aesthetic was also visualized in the work of Alfred Stieglitz and Alvin Langdon Coburn, who photographed railroad yards, steam engines, and factories.[85]

The industrial sublime effaced the details of individual lives, whether those of railroad magnates or those of navvies laying the rails, and represented the railway system as a powerful force transforming American life. It justified the foundation story of steam by presenting its landscape as an inevitability. The ever-acute Henry James realized this when he returned to the United States at the turn of the century. As he pointed out in *The American Scene*, there was no force visible in the American landscape capable of resisting the "bullying railway."[86] As Leo Marx concluded, in the America of 1903 there was "nothing in the visible landscape—no tradition, no standard, no institution—capable of standing up to the forces of which the railroad is a symbol."[87]

Yet there were a great many counter-narratives.

8

"The Route of Superior Desolation"

The good land you will certainly hear of from the magnificent circulars of railroad and emigration companies.
—J. H. Beadle, 1873[1]

Every train rolls on through dismal smoke and barbarous melancholy ruins, and the companies might well cry in their advertisements: "Come! Travel our way. Ours is the blackest. It is the only genuine Erebus route. The sky is black and the ground is black, and on either side there is a continuous border of black stumps and logs and blasted trees appealing to heaven for help as if still half alive, and their mute eloquence is most interestingly touching. The blackness is perfect."
—John Muir, 1897[2]

But steam has no soul.
—Emerson Hough, 1903[3]

Though Calhoun's conquest of space and Whitman's passage to India sounded grand in the abstract, the actual construction of canals and railroads and their use confronted Americans with new forms of political power, class conflict, accidents, land swindles, pollution, and unfamiliar environmental problems. Starting in the 1820s, the first counter-narratives to canals and railroads emphasized the political and economic dangers of internal improvements. Were the federal government to intervene in the states to fund roads, canals, and railroads, it would undermine state sovereignty. Jacksonians feared any concentrated economic power that received special privileges, such as the Bank of the United States. President Jackson eliminated the Bank and looked on any permanent institution fostered by the federal government with suspicion. After privately owned railroads were built, they were attacked as unjust monopolies that wielded too much economic power. Later critics emphasized the social costs of canals and railroads to labor and local communities. A third critique, increasingly important after the Civil War, focused on the railroad's destruction of the environment.

Ralph Waldo Emerson had celebrated the railroad, but he grew skeptical with age. As John Kasson notes, after the trip to Britain that led to the publication of *English Traits* in the 1850s, Emerson's "stance toward technological civilization grew more critical. Though in the flower of his transcendental revolt he had celebrated technology as a stimulus to creative vision, in his later career he emphasized more its tendency to debase the imagination."[4] In 1851, Emerson declared that he "could not accept the railroad and the telegraph in exchange for reason and charity."[5] This did not mean that he, or Thoreau for that matter, blindly rejected new technologies, but they did fear the subordination of imagination to materiality. In 1853, as I have already noted, Emerson wrote in his journal: "The age has an engine, but no engineer."[6]

In *The Machine in the Garden*, Leo Marx examines the literary resistance to industrialization and, more particularly, to the railroad as it invaded the rural landscape. His elegant and still influential argument is that all the writers of the American Renaissance were intensely aware of the railroad's intrusion into the rural world. Hawthorne meditated on the meaning of a locomotive's shrill whistle echoing through the forest and concluded that "it brings the noisy world into the midst of our slumbrous peace."[7] Emerson recorded the same experience: "I hear the whistle of the locomotive in the woods. . . . It is the voice of the civility of the Nineteenth Century saying, "Here I am." It is interrogative: it is prophetic: and this Cassandra is believed: Whew! Whew! Whew! How is real estate here in the swamp and wilderness? Ho for Boston! Whew! Whew! . . . I will plant a dozen houses on this pasture next moon, and a village anon. . . ."[8] Such passages recognize the foundation story—the railroad as the force behind settlement—but do not celebrate the railroad. Rather, they counterpose the machine to the natural world. "Their heightened sensitivity to the onset of the new industrial power can only be explained by the hold upon their minds of the pastoral ideal."[9] Thoreau provides an excellent example. He refers to the railroad often in *Walden*, even though the work is ostensibly about a retreat into nature. Marx notes that Thoreau "takes special pains to impress us with the 'cut' in the landscape made by the embankment" for the railroad, and that he treats this scar as "a wound inflicted upon the land by man's meddling, aggressive, rational intellect"—a wound that is healed, however, by the end of the book, through nature's urge to create form and design from the sandy sides of this "cut" as it thaws during the spring.[10] The heat of the sun causes the sand to assume various organic forms, and Thoreau feels that he is standing "in the laboratory of the Artist who made the world and me—had come to where he was still at work, sporting on this bank, and with excess of energy strewing his fresh designs about."[11] Yet the power of nature to heal industrial wounds seemed less certain with each passing year.

Attacks on canals and railroads increased as they became more familiar. The completion of a canal or a railroad was invariably greeted with civic ceremonies that included speeches, fireworks, and predictions of prosperity,[12] and the first journeys were thought to be things of wonder. But after a few years, boats and rail cars no longer amazed, and some began to find them squalid. As early as 1836, the magazine *Knickerbocker* carried two disparaging articles about travel on the Erie Canal. The correspondent found the food appalling, the crowding below decks insufferable, and the snoring and mosquitoes at night unbearable. "At last, morning dawns; you ascend into pure air, with hair unkempt, body and spirit unrefreshed, and show yourself to the people of some populous town into which you are entering, as you wash your face in canal water on deck. . . ."[13] The contrast between these cramped, slow boats and the "spacious cabins of the Hudson steamers" was palpable.[14] Likewise, railroad cars were smoky, poorly ventilated, and furnished with hard, uncomfortable seats. *The Atlantic* praised European trains and called the typical American railway car "a moving stable" that was overcrowded and poorly ventilated: "One's feet rest in an ice-bath of bitter air, and one's head reels in a burning, disoxygenized atmosphere."[15] Comfort gradually improved, but other problems persisted, including poor construction on the lines, abuse of land grants, exploitation of labor, accidents, and environmental degradation.

In theory railroads hastened the settlement of the West, but in practice they often distorted it. "Everywhere that a railroad was built or projected," Henry George argued in *Progress and Poverty,* "land was monopolized in anticipation, and the benefit of the improvement was discounted in increased land values. The speculative advance in rent thus outrunning the normal advance, production was checked, demand was decreased, and labor and capital were turned back from occupations more directly concerned with the land. . . ."[16] Furthermore, government policy prevented the sale of lands along a railroad right of way. The moment an official route map was registered with the Department of the Interior, it had "the policy of withdrawing from sale all the lands within the indemnity limits." Land on either side of the right of way for miles was kept off the market until the railroad had selected the sections it was entitled to. It often took years before the lands were selected, surveys were made, and all the paperwork was complete.[17] In a perfect world where information circulated instantaneously, keeping land off the market would have been inconvenient, because it would have retarded development. But because information circulated slowly, local land offices often sold real estate to settlers who much later found that the railroad had title to their claim. As a newspaper correspondent asked in 1874, what justice was there in a system that allowed "the local land offices to receive applications for homestead and preemption rights and to encourage settlements and improvements on the public domain" and then get "instructions from the General Land Office in Washington to drive the settlers away and turn their improved property over to some railroad corporation"?[18] In this narrative, before railroads built their tracks, they seized the right of way from homesteaders. The shifting land policies of different administrations and local variations in their implementation makes generalization difficult, but both law and practice favored the railroads more than individual settlers.[19] In a landmark case brought by a dispossessed homesteader, the Supreme Court ruled in favor of the Santa Fe Railroad.[20] The case amounted to official recognition that the railroad foundation narrative was falsified by the actual workings of the bureaucracy and the legal system.

During the boom in railway construction that occurred in the 1870s, "memorials and petitions poured into Congress from dispossessed settlers" of the West and from their legislatures.[21] For example, the Southern Pacific had invited settlers into California's San Joachin Valley in 1871, promising to sell them farmland for $2.50 an acre. Once the settlers had improved the land, however, the railroad evicted them. In the armed confrontation that followed, two deputy marshals and five ranchers died.

Seventeen settlers were indicted for "conspiracy." The case became widely known, and 47,000 people petitioned to President Rutherford B. Hayes to pardon the settlers.[22] Such petitions inverted the foundation narrative: The railroad drove pioneers away and seized their lands, rather than fostering settlement. The railroad did not open the West; it closed it to poor settlers who could not afford to pay the higher prices that property suddenly fetched along its routes. The railroad stimulated rampant land speculation and distorted normal development, driving prices up before the road was completed and, often as not, leading swiftly to depression thereafter. Henry George noted that "as the transcontinental railroad approached completion, instead of increased activity, symptoms of depression began to manifest themselves" and there "succeeded a period. . . during which wages and interest have steadily fallen."[23] Worst of all from the western point of view, railroads transformed settlement from an individualistic process to a centralized corporate practice. Each line advertised the lands along its routes, even translating booklets and sending sales agents to Europe. The railroad did not bring free-market development and prosperity; it brought immediate economic disaster for some, feverish land speculation, and long-term economic control for all.

In 1860, Captain R. B. Marcy, a western guide, published a handbook for emigrants, travelers, hunters, and soldiers. This was a practical manual explaining the different routes west, the best places to purchase supplies and provisions, what to take, how to pack, and how to ford a river, repair a wagon, treat saddle wounds, and cure rattlesnake bite. But Captain Marcy used his first page to issue a warning about the perfidy of those who lived along the major routes westward: "Information concerning these routes coming from strangers living or owning property near them, from agents of steam-boats or railways, or from other persons connected with the transportation companies, should be received with extreme caution, and never without corroborating evidence from disinterested sources."[24] Marcy did not elaborate further, so well known were the deceptions of the planners of cities and the railroad companies. Similarly, at the end of a chapter explaining in detail why most of the area west of the 100th meridian would always be desert, an 1873 account of the West concluded sarcastically: "The good land you will certainly hear of from the magnificent circulars of railroad and emigration companies."[25] In the

North American Review, General W. B. Hazen protested the way that railroads and land companies misrepresented their holdings to prospective settlers, who frequently bought worthless, arid land.[26]

Speculators promoted empty sites as though they were towns, and towns as though they were cities. A journalist traveling west on the newly completed Union Pacific Railroad stopped at the new community of Columbus, Nebraska. Its citizens were convinced that Columbus was "sure to be quite a metropolis, the great central city of this valley, and certainly the capital of the State and possibly of the Nation." They believed it stood at the geographical center of the United States. "Town lots were at handsome figures and advancing, and there was speculation in the eyes of real estate owners."[27] But the journalist was not impressed: "Beyond Columbus there were then no 'cities' for four hundred miles. Of all, which sprang up on the road, only two or three, survive in anything like their first greatness. A speculative and uncertain character attached to all of them. . . ." Indeed, "lots in the 'wickedest city,' Julesburg, which once sold readily for a thousand dollars, are now the habitations of the owls and prairie dogs. . . . The town lasted only five months, but was quite successful in establishing a graveyard."[28] During a land boom in Los Angeles County, lots sufficient for 2 million residents were laid out in more than 60 new towns; however, in 1889, after the speculative bubble burst, the actual number of new residents was only 2,351.[29]

In their 1873 novel *The Gilded Age,* Mark Twain and Charles Dudley Warner described how eastern investors connived with local speculators to plan a railroad, lobby Washington for land grants, and line their pockets from property sales.[30] An article in *The Nation* averred: "A doubt has risen in the public mind as to whether, after all, the policy of giving away the lands almost indiscriminately to great railroad corporations is wise. . . . In the West, especially, this doubt has almost developed into positive opposition. . . ."[31] Ending land grants became a prominent issue in local party platforms, and even President Ulysses S. Grant expressed some hostility to the system. An 1873 lithograph, "Anti-monopoly," depicts railroads as barriers to western settlement. Their tracks cut through the center of the image, and a smashed farmer's wagon underscores the difficulty of crossing them into the western lands beyond. The most prominent building is not a settler's cabin but a magnificent mansion whose fence blocks access

Figure 8.1
W. Bostwick, "Anti-Monopoly" (lithograph, 1873). Courtesy Library of Congress.

to the vast unfarmed areas behind it. The sign in the foreground, "Cheap Farms," can only be read as ironic commentary. In contrast to "Across the Continent" from just 5 years earlier, in "Anti-Monopoly" the railroad blocks the homesteader, prevents the settlement of the West, and fosters not democracy but oligarchy.

In Congress, which once had given away millions of acres to railroads, sarcastic speeches became common in 1871, including one address by James Proctor Knott that remains famous today. Knott ridiculed a bill that would cede land along the St. Croix River for a railroad:

Duluth is situated somewhere near the western end of Lake Superior; but as there is no dot or other mark indicating its exact location, I am unable to say whether it is actually confined to any particular spot or whether "it is just lying around there loose." I really cannot tell whether it is one of those ethereal creations of intellectual frostwork, more intangible than the rose-tinted clouds of a summer sunset; one of those airy exhalations of the speculator's brain, which I am told are ever flitting in the form of towns and cities along those lines of railroad, built with government subsidies, luring the unwary settler as the mirage of the desert lures the famishing traveler on, and ever on, until it fades away in the darkening horizon; or whether it is a real, bona fide, substantial city, all "staked off," with lots marked with their owners' names. . . .[32]

Applause and laughter frequently interrupted Knott's speech, and the railroad to Duluth, despite the city's brilliant future, its salubrious climate, and its central location, did not get its land grant.

Henry George was perhaps the most outspoken critic of railroad land grants. His 1871 book *Our Land and Land Policy* examined the published advertisement of the Northern Pacific Railroad. "The Directors show," George wrote, "that if they can get an average of but $2 per acre for their land, they can pay the whole cost of building and equipping the [rail]road and have a surplus of some $20,000,000 left. That is to say, the government might have built the tracks by merely raising the average price of their lands by $1 per acre" and still made a profit. Had the government done this, it would then have owned the railroad and been able to lease it to the corporation that charged "the lowest rates." "As it is," George continued, "the government has raised the price to settlers on one-half the land $1.25 an acre; the other half it has given to the Company to charge settlers just what it pleases; and then on this railroad which it has made the settlers pay for over and over again both government and settlers must pay for transportation just as though the road had been built by private means."[33]

It was not merely that the government gave away land worth more than what it received in return. George reasoned that without land grants railroads would be built in any case as soon as settlement reached a certain density. In terms of the greatest good to the greatest number, land grants to railroads hurt at least as many citizens as they helped. When "we get a railroad to precede settlement," George asserted, "if the first settlers gain at all, the others lose."[34] Because railroads charged more for land, often raising the price for undeveloped land as high as $13 an acre, it seemed that the "first advantage in having a railroad before its natural time, is offset by the subsequent retardation of settlement in their neighborhood which the land grant causes."[35] The idea that railroads would inevitably come when the rise of population justified such an investment underlay George's argument, which concluded: "We need not trouble ourselves about railroads; settlement will go on without them—as it went on in Ohio and Indiana, as it has gone on since our Aryan forefathers left the Asiatic cradle of the race on their long westward journey. Without giving away any of the land, railroads with every other appliance of civilization will come in their own good time."[36]

As a result of the outrage of Twain, George, and many others, including mainstream politicians, 1871 was the last year in which Congress enacted land grants. But in that single year 20 million acres were given away, much of it in the South. "There was not a party platform in 1872 which did not condemn the whole system and insist that the public domain be held for actual settlers."[37] The underlying belief in the historical inevitability of "Aryan" western expansion and settlement was not affected by reining in the railroads, however, and it remained an article of faith among politicians of almost every persuasion. Josiah Strong might have written the same as George, who declared: "We have subdued a Continent in a shorter time than many a palace and cathedral of the Old World was a building; in less than a century we have sprung to a first rank among the nations. . . . We are carrying westward the center of power and wealth, of luxury, learning and refinement, with more rapidity than it ever moved before."[38] From this point of view, railroads did not cause western development, which was inevitable. Rather, railroads were monopolies assisting that process. They required regulation, and their land grants had no place in a free market. By the 1870s, Americans realized that railroads might either accelerate or retard settlement, but not many questioned the automatic unfolding of technological creation.

The difficulties with railroads did not end once the practice of giving land grants had been halted. Both canals and railroads introduced an entirely new class of people into the regions they traversed. Most of these workmen were male, and they worried the middle class. Disconnected from any local community and often drunk and disorderly, boatmen and railroad workers quickly gained a reputation as rough, combative, and profane. By the 1840s, the Erie Canal employed 30,000 men, women, and boys. Most had unskilled jobs, notably the boys who drove the horses along the towpath, the women who did the domestic duties on board, and the day laborers who performed routine maintenance. The logic of capital investment rewarded continuous boat operation, even on Sundays, despite the protests of religious people. Passengers and shippers did not want to interrupt their journeys, and boat owners realized that faster trips increased their profits. Therefore a large class of laborers who often drank heavily worked on Sundays. In 1835, the Erie Canal averaged a tavern or grog shop every quarter-mile, all the way from Albany to Buffalo, providing 1,500 opportunities to purchase liquor. Estimates for other canals yielded similar results.[39] The sense that communities had been invaded by foreign elements further increased after Irishmen were recruited to do much of the canal digging. A largely Protestant population feared Catholics, and the poor sanitary conditions in Irish shantytowns seemed threatening to middle-class households.

Railroad construction workers moved more rapidly through the landscape than canal workers, and they often lived in specially constructed cars that rolled to a new location each night. This transient population was followed by saloonkeepers, prostitutes, and others who sought to make a profit from their daily wages. The historian of the Atchison, Topeka & Santa Fe noted that "drifting railroaders gained a reputation as a boastful, rough-speaking, hard-living, immoral lot."[40] The railroad towns seldom resembled the Currier and Ives image of the ideal American community. As John Reps summarized: "Like other end-of-track railroad towns, Cheyenne presented a rough face to the world with its saloons, gambling halls, and brothels. Tents and crude shacks mixed with more substantial structures, and its unpaved streets were either dusty or muddy depending on the state of the weather."[41] The predominant forms of recreation were drinking, gambling, whoring, and fighting.

As the canal and railroad workers gained unsavory reputations, the middle class confronted an entirely new narrative in which the appearance of new forms of transportation threatened the moral health of the community. Missionary societies were soon formed to minister to the canalers, who were denounced from middle-class pulpits as a degenerate class that threatened public order and undermined the Republican values essential to democracy.[42] The energies set loose by the technological conquest of space, paradoxically, turned out to be threatening. When several hundred businessmen signed petitions asking that the Erie Canal be closed on Sundays, they declared: "Free institutions will be sustained only by an intelligent and moral people. Make then this people ignorant and immoral, and you will destroy our prosperity, peace, and republic."[43] The canal, and later the railroad, released commercial energies in the pursuit of wealth, just as the dominant narrative said they would. But the conquest of space turned out to involve the construction of shabby, temporary communities, the palpable exploitation of men, women, and children, an increase in public drunkenness, the cultivation of immorality, and violation of the Christian Sabbath.

If these issues troubled the middle class, workers faced even more serious problems. Building canals and railroads was dangerous, wages could decline or disappear without warning, and a life of constant movement left men with no community to fall back on. These workers were arguably the first industrial proletariat in the United States, being cut off from the land and relying entirely on wages from large corporations for their subsistence. Once transportation services began, some of the men might find regular employment with the company, and in theory they might rise to higher positions. In practice, however, canals and railroads required a large and primarily unskilled work force for construction and a somewhat smaller and more highly skilled group for operation. Even those fortunate enough to make the transition from building the line to running it found railroad work dangerous and uncertain.

Nevertheless, because the railroads grew rapidly at the middle of the century, experienced "boomers" were in demand. The superintendent of the 1860 U.S. census, Joseph C. G. Kennedy, asked how the railway affected labor. "The first impression made on the popular mind by any great improvement in machinery or locomotion, after the admission of their beneficial effect, is that they will, in some way or other, diminish the demand for labor or for other machinery." But he concluded that experience of building railroads into the Midwest disproved such ideas: "It is now established, as a general principle, that machines facilitating labor increase the amount of labor required."[44] In the 1870s, however, the supply of labor consistently outpaced the demand, and railroads began to cut wages and reduce workers' autonomy.

Worsening labor relations culminated in the Great Strikes of 1877 and in "an extraordinary rise in strike activity under extremely adverse conditions" between 1881 and 1894.[45] During these strikes, the very towns that railroads had created were likely to side with the workers. In contrast, the wholesale merchants and politicians of a town that pre-existed the railroad, such as Burlington, Iowa, tended to lead public opinion to side with the railroad. The citizens of a town built by a railroad constantly felt its power and presence. Its merchants typically "rankled under the limits imposed by their dependence on a single railroad line. Their anti-monopoly sentiments provided the basis for an alliance with the community's railroad workers in their mutual difficulties with the outside railroad corporation."[46] If in the technological creation story the railroad was the community's creator and benefactor, those who lived in such towns often felt dominated and controlled. During the nationwide strikes of 1877, most railroad towns sided with the strikers, and angry crowds attacked and burned railroad property in major rail centers such as Baltimore and Pittsburgh.[47] Animosity toward the railroads was often greater among ordinary citizens in cities such as Albany, Syracuse, and Buffalo than among the workers themselves, and in some cases strikers actually protected railway property from an inflamed public.[48] Ordinary citizens were angry not only because railroad crossings impeded and endangered them in the increasingly congested central city. "A person who was not a railroad striker might have participated in crowd attacks against railroads for any number of mutually reinforcing reasons: anger over the railroad's killing or maiming of a friend, neighbor, or family member; smoldering opposition to the railroad's disruption of the social and economic fabric of streets and neighborhoods; support for a striking railroad worker; fear of the concentrated economic and political power of the railroads; animosity over a house set on fire from a passing locomotive."[49] Aldermen and city officials also were dissatisfied with railroads because they resisted city ordinances to reduce smoke pollution and requests to build tunnels and viaducts that separated pedestrians and wagons from the trains.

State and local authorities found it extremely difficult to exercise control over railroad corporations. In the 1850s, the businessmen of Pittsburgh complained of the exorbitant rates the Pennsylvania Railroad demanded, and they believed the railroad undermined the city's competitive position. As Michael Holt found, there was "a virulent anger against the railroads widespread among the local populace."[50] After the Civil War, public dissatisfaction increased. New York and Chicago, the nation's two most important railroad centers, voiced similar complaints of rate fixing among the apparently competing lines.[51] Horace Greeley gave voice to the public mood: "Very soon after the commencement of the railroad era the process of combination began, and has finally produced the present era of consolidation. . . . It is not too much to say, that in the railroad history of this country, the public has never, for a period of twelve consecutive months, been the gainer from railroad competition."[52] Railroads were private investments, run for private gain. Although some towns sprang to life along their routes, they did so only because the railroad's board of directors had decided to build a station there. A bypassed town fell into lassitude, and regions without service lost importance. The power of railroads began to resemble that of an absolute monarch. Greeley declared they were a more potent power "than royalty has ever seen." This dominance was strongest at the local and state levels. For example, Greeley noted that "the railway monopoly of New Jersey has for years held the political and financial control of that state, and levied an onerous tax upon all travel between the East and South."[53] Likewise, Greeley observed, railroads controlled the shipment and pricing of the most important ingredients of the industrial economy, coal, iron, and steel. Pennsylvania, Maryland, Massachusetts, "and various other states, east and west, find themselves . . . seemingly powerless in the grasp of railroad corporations their charters have called into being. With the growth of the principle of consolidation the evil is increasing."[54] Furthermore, railroads had vast assets. For example, they owned 16 percent of all the land in California as early as 1860.[55]

From the monopolistic realities of the 1870s, Greeley could look back critically at the earlier practice of giving land to the railroads as a spur to their development. These "land grants as first presented by the railroad had a certain air of being a legitimate business transaction. The land was wild, and the railroad proposed, by the building of the line, to bring it into the market for settlement. . . ."[56] But in practice the railways were awarded far more land than was necessary to finance construction, and by 1870 some railroads were being built primarily as a means to receive grants. Indeed, the land grant system encouraged promoters to build railways through places where no one lived, in order to obtain the land on either side of the tracks. When Greeley had traveled across Missouri in 1858, he had been surprised to find "infinitely less population and improvement" than he had expected. "Of course, this road was run so as to avoid the more settled districts, and thus to secure a larger allotment of the public lands. . . . I had not believed it possible to run a railroad through northern Missouri so as to strike so few settlements."[57] He found one section of level prairie 50 miles long virtually uninhabited and thought it "incredible that such land, in a state forty years old, could have remained unsettled till now."[58]

In many cases, railroads avoided towns that refused to grant them favors. In the 1870s the Southern Pacific "did not hesitate to change its route to avoid towns which refused it subsidies. Many of these deserted settlements were thus doomed to stagnation."[59] Even Los Angeles, in its early days, had to lobby in order to avoid being bypassed by the railroads. In Kansas immediately after the Civil War, anger at the railroads' successful land grabs led to the formation of a Neutral Land League and to attacks on railroad surveyors, destruction of railroad property, sacking of a pro-railway newspaper, burning of railway ties and equipment, and intimidation of construction parties.[60] By the 1870s the railroads were fervently hated throughout the plains states, providing a rallying point for discontented farmers who called for regulation and reform. The Grange, founded in 1867, became a political organization in the hard times of the 1870s and made railroad regulation a central issue. The major lines charged as much as double for an intrastate short haul, where they had no competition, as they did for long-haul service from Chicago to the East. People shipping identical goods over the same route to the same destination were charged different rates.[61] The railroads developed "preferential rates, rebates, drawbacks, under weighing, under classification, and numerous other devices" in dealing with the public.[62] Price differentials were often extreme. In Minnesota, for example, shippers from the town of Owatonna paid 2.6 cents per ton per mile, less than half the rate of 6 cents charged in nearby Rochester. The two towns were on the same line, but people in Owatonna had two railroads to choose from.[63] The Grange sponsored legislation to control freight rates, first in Illinois in 1871 and then in Minnesota, Iowa, Wisconsin, Nebraska, and Missouri.[64] Though they never entirely solved the problem, the older narrative of railroads as automatic stimulants to prosperity was being rewritten, first by popular opinion and later by legal institutions. In 1873 the Minnesota Supreme Court held that the legislature could regulate intrastate rates unless a railroad's charter specifically declared otherwise.[65] Nevertheless, Minnesotans vacillated between controlling railroads and stimulating railroad investment. Pursuing either policy eroded the sense that railroad development could be left to the invisible hand of the market. By the 1870s, state governments, like farmers and workers, were increasingly aware that railroads undermined the premises of the technological creation story. By 1890, Congress had been induced to rescind some of the land grants, typically in cases where lines had not been completed, and 25 million acres reverted to the public domain. Yet railroads held onto even more land, exploiting the timber, oil, and mineral resources, which by law they should have sold to settlers along their lines.[66] The market invaded by a railroad was not free. Instead of increasing access to what was perceived as the natural abundance in the geometrical space of the American West, the railroad controlled access to that abundance and exacted a heavy toll.

The railroad also was dangerous to its employees. The first American locomotive, the *Best Friend,* blew up in 1831 shortly after being put into use, and hundreds more explosions would follow during the rest of the century. Many train wrecks resulted from poor construction. Track was often flimsy or poorly laid, and bridges were built without strain calculations.[67] Cars were coupled and uncoupled manually, putting brakemen at risk of being crushed; their fatality rate was more than twice that of firemen or conductors.[68] Lack of common standards also caused many accidents. Signaling systems varied from one railroad to the next, not all freight cars had the same braking systems, and for decades each railroad ran according to its own clock. Railroads standardized time for themselves and for America as a whole in 1883,[69] but nevertheless during the 1880s fatalities on American lines were 50 percent higher than in Britain, and the accident rate twice as high.[70] Progressive reformers continually called for more safety devices and better pensions for those injured. "The death and disability roll of the employees of our American railroads," one reformer noted, "is a terrible indictment against the inhumanity of the service."[71] Safety had improved during the nineteenth century, but between June 1905 and June 1906 a total of 3,807 men were killed and 55,254 injured, not counting injuries to passengers. Critics noted that engines and rolling stock had increased in weight but still ran over the lighter rails that had been adequate in early days, and that workers often had such long hours that they could not attend properly to safety. The railroads successfully opposed a law designed to limit continuous work to 16 hours, and, compared with European practice, they employed too few men to inspect track.[72] For workers, the heroic narrative of railroad expansion driving national settlement became a story of peripatetic movement and danger.

Most worker fatalities were "small" accidents that attracted little attention from the press. In contrast, wrecks and explosions that killed passengers were headline news, and reporters gave the public graphic details of bodies disemboweled, roasted flesh, faces burned beyond recognition, legs crushed, and torsos cut in half. In 1838, a man wrote in his diary: "I never open a newspaper that does not contain some account of disasters and loss of life on railroads. They do a retail business in human slaughter. . . ."[73] This mayhem spurred mounting public outrage. Legislation and liability suits gradually forced the railroads to abandon some of their most dangerous practices, such as hiring inexperienced men, using a single track with occasional sidings to carry traffic in two directions, and running "by the book" (i.e., by the printed schedule) rather than using the telegraph to keep constant watch over the actual locations of trains. All these shoddy practices were common as late as 1871, when a rear-end collision in Revere, Massachusetts, killed 29 people and injured many more.[74] Charles Francis Adams editorialized against railroad accidents in *The Atlantic Monthly*, pointing out that Britain had a much more thorough and competent system for investigating such disasters.[75] Only in 1893 did the Railroad Safety Appliance Act require railroads to use the Westinghouse air brake, which soon reduced accidents by more than 60 percent. Yet as late as 1912 advertisements for artificial limbs were common in the *Railroad Trainman's Journal*.[76]

Accidents not only led to lawsuits from the bereaved; they also put the public as a whole in an adversarial position. Doctors discovered that accident victims suffered from nightmares and trauma. By the 1880s they had abandoned physiological explanations for these symptoms in favor of psychological explanations that focused on the passenger's sense of helplessness and the instantaneous terror of a crash.[77] Furthermore, by the middle of the nineteenth century medical investigators were convinced that the shaking and involuntary movements imposed on the body during ordinary railway journeys exhausted travelers, who in some cases developed "railway spine."[78] The passenger increasingly seemed the railway's passive victim. The terror and fascination of accidents found popular expression in countless ballads (including "Casey Jones") and in staged accidents at more than 150 fairs, where locomotives were smashed together at maximum speed.[79]

If railroads overcharged farmers, skimped on worker safety, and traumatized some passengers, they paid even less heed to the rights of Native Americans. Railroads devastated the environment that sustained the Sioux, for example. Not only did the engines spark frequent prairie fires, but the companies hired sharpshooters to slaughter the buffalo herds that might otherwise block the tracks. As most game disappeared and the buffalo came close to extinction, Native Americans lost their sources of food and clothing. But Native Americans were not simply victims of a powerful technology that came unbidden from outside. Some tribes sought to invest in and to control railroads, although to little avail. In the Indian Territory that later became Oklahoma, the Cherokee, Choctaw, and Chickasaw tribes were ready to embrace the railroad. The Cherokee sold some of their lands to invest in the Union Pacific's Southern Branch, and they secured two seats on the line's board of directors. In 1870, the Choctaw and the Chickasaw planned to build their own railroads, with outside technical assistance, but disagreement between the two tribes and resistance from Washington combined to kill this initiative.[80] Later, attempted cooperation between the tribes and railroads led to the sale of Indian lands, but the money was invested for the tribes in railway stock that was held in trust by the government. In such a scheme, railroads often supported the doctrine of tribal sovereignty, because assimilated Native Americans had a stronger legal position and could better look after their interests than "protected" Indians, whose rights were controlled by the Congress. Both the House and the Senate consistently sided with the railroads against the tribes. One Texas senator declared: "We will not be penned up; we will not be hindered." Another put it more grandly: "All this aesthetic talk and constitutional argument amounts to nothing in the face of the great fact that the people of the United States today are stopped in their imperial course toward the Southwest."[81] The needs of railroads were made identical with those of "the people," and decades of petitions and legal actions from the tribes of the Indian Territory failed to give Native Americans much voice in decision making.

Railways were also attacked because of their environmental effects. Residents along the first lines complained of the noise, smoke, and sparks from locomotives. During the initial excitement over railroads, such complaints had little effect on public policy. By the late nineteenth century, however, coal smoke from railroads was widely recognized as a problem. Although pollution came from many sources, locomotives obviously did not burn coal as efficiently as large stationary steam engines, and they did not have tall smokestacks that could pump their smoke into the upper atmosphere. By 1906 the American Medical Association recognized that exposure to coal smoke increased mortality rates for tuberculosis and pneumonia.[82] The nineteenth-century myth that smoke killed germs had also been laid to rest, and for more than a generation reformers worked to improve engine efficiency, to require the use of cleaner coal, or to switch to diesel or electric drive.[83]

Moreover, coal smoke was bad for the economy. A study sponsored by the city of Cleveland calculated that smoke cost every family an average of $44 a year for cleaning, painting, artificial lighting, and replacement of damaged goods. That was an unskilled laborer's wages for a month, and yet such estimates—there were many in other cities as well—did not include the damage to plants, animals, and human health. The Chicago department store owner Marshall Field estimated that the "soot tax" was larger than his real estate taxes. A Chicago study estimated that railroads contributed 22 percent of that city's smoke pollution, and its inspectors focused particularly on locomotives.[84] Many industries burned coal, but railroads drew more attention as they thundered into town and blackened adjacent regions. City ordinances attempted regulation and engineers developed more efficient furnaces, but rail traffic and its pollution continued to increase. Air quality did not improve in most cities until the middle of the twentieth century, and that improvement was due in good part to the substitution of diesel and electric locomotives. By that time, the areas along railway tracks had long since become gray wastelands where no one wished to live. In the master narrative, the railroad brought prosperity, but the right of way was a place of filthy air, dingy tenements, disease, and poverty.

Railways not only spewed out sulfurous smoke that endangered public health and devastated the urban areas near their tracks; they also endangered the natural environment because of their enormous demands for wood. In the single state of Ohio in 1870, for example, the 6,000 miles of track required 15 million ties. Ohio's railroads built 10,000 miles of wooden fencing, and they crossed 10 miles of wooden trestles and 16 miles of wooden bridges. All this wood rotted rapidly and had to be replaced every 5–7 years, long before the forests that had been cut down could grow back.[85] There was yearly demand in Ohio for enough wood to rebuild completely 1,000 miles of the system, with 2,640 cross-ties for every mile. Wherever a new rail line was built, it created "a mania for buying woodland, stripping it of its timber, and then selling it for agricultural purposes."[86]

Ohio was a thickly wooded state that could supply most of its own needs. Such was not the case farther to the west. When transcontinental railroads began to cross the treeless plains, they demanded millions of ties, miles of fence and telegraph poles, and lumber to build stations and other structures. They stimulated rapid expansion of the logging industry in Michigan, Wisconsin, and Minnesota. A federal report estimated that by 1883 about 3 million acres of land had been cleared to supply railroad ties alone; an additional 472,400 acres of timber were needed every year for routine track maintenance and replacement. These figures were based on the size of the existing network of 93,000 miles in 1880, but the network would more than double to 193,000 miles in 1900 and would continue expanding until 1916 (when the railroads reached their greatest extent: 254,000 miles of operating track). By this time, they were using between

20 and 25 percent of the nation's annual timber production,[87] and had become what one critic in 1897 called "the insatiable juggernaut of the vegetable world."[88] By the turn of the century, a widespread "depletion myth" expressed the fear that American forests would soon disappear. Theodore Roosevelt said: "If the present rate of forest destruction is allowed to continue, with nothing to offset it, a timber famine is inevitable."[89] The famine did not occur, in good part because railroads planted their own forests and cut their consumption by adopting wood treatments such as creosote so that ties lasted longer. Because the counter-narrative of depletion was so fervently believed during the last years of the nineteenth century, new practices prevented it from becoming an accurate prediction. Furthermore, railway trackage began to contract after 1917. Rather than a wood famine, much of the eastern United States was gradually reforested during the twentieth century.

Yet if the wood famine never materialized, the railroads in the West nevertheless destroyed much of the landscape that they claimed to make available to the tourist. As John Muir lamented, fires devastated more forest than wood-cutting crews. During construction, railroads typically cleared away trees and brush from along the right of way and left it beside the tracks, where it was eventually set ablaze by sparks from the engines. Muir witnessed many fires started in this way, "and nobody was in sight to prevent them from spreading . . . into the adjacent forests and burn the timber from hundreds of square miles."[90] He wrote a parody of the advertisements of the transcontinental lines whose "gorgeous many-colored folders" described "scenic routes":

"The route of superior desolation"—the smoke, dust, and ashes route—would be a more truthful description. Every train rolls on through dismal smoke and barbarous melancholy ruins, and the companies might well cry in their advertisements: "Come! Travel our way. Ours is the blackest. It is the only genuine Erebus route. The sky is black and the ground is black, and on either side there is a continuous border of black stumps and logs and blasted trees appealing to heaven for help as if still half alive, and their mute eloquence is most interestingly touching. The blackness is perfect. On account of the superior skill of our workman, advantages of climate, and the kind of trees, the charring is generally deeper along our line, and the ashes are deeper, and the confusion and desolation displayed can never be rivaled. No other route on this continent so fully illustrates the abomination of desolation." Such a claim would be reasonable, as each seems the worst, whatever route you chance to take.[91]

Shortly after Muir published his critique, Frank Norris's novel *The Octopus,* a full-scale indictment of the railroad's effects on American society, appeared. The central force in this novel is a railroad, a thinly disguised copy of Collis Huntington's Southern Pacific. Norris described how the railroad strangled wheat farmers, not only by manipulating rates but also through intimidation and violence. Driving the farmers from their homes and corrupting politicians to attain its ends, Norris's railroad is an amoral octopus seizing all within its grasp. "They swindle a nation of a hundred million and call it Financeering; they levy a blackmail and call it Commerce; they corrupt a legislature and call it Politics; they bribe a judge and call it Law; they hire blacklegs to carry out their plans and call it Organization; they prostitute the honor of a State and call it Competition. And this is America."[92] *The Atlantic* printed an equally caustic article that described how many a community "sees its shops closed, its mills silent, its streets growing up with weeds, its capital and its best talent insensibly drawn away to some neighboring city which enjoys better rates from the railroads, here is an evil as monstrous as it is insidious." In 1894, it seemed obvious that the country was "covered with dying villages and towns whose expanding life has been stifled by railway discriminations."[93]

Norris, Muir, and others of the Progressive Era who attacked the railroads for their selfish pursuit of wealth and power were writing in the tradition of what Sacvan Bercovitch has called the "American jeremiad."[94] The railroad's original promise that it would foster development and democracy was followed by its "declension" or spectacular failure to fulfill its social mission. The railroad had stimulated wasteful land speculation, created ghost towns, undermined and destroyed existing communities by denying them service, driven some pioneers off their claims, slowed settlement of some areas by charging high prices for both freight and land, and caused forest fires and environmental devastation. And yet the railroad might still serve its original purpose if it were better regulated (or owned) by the state. Thus the critique of the railroad often ended in a hopeful prophecy of what it could be in the future. The public admired the idea of railroads, but it no longer loved the corporations that ran them. The admiration runs through Emerson Hough's popular book *The Way to the West,* yet even Hough was compelled to admit in 1905 that "once we depended upon it; now it rules us almost without argument."[95] He complained that the railroads drove up food and coal prices for ordinary Americans because they were "non-competitive transportation." "The iron trails," he continued, "are built over the hearthfires of America."[96] "Steam will establish our doctrines and our tariffs," he complained. "But steam has no soul."[97]

"When the term 'soulless corporation' was first coined," Theodore Dreiser wrote in *Harper's New Monthly Magazine*, "it was used to describe the nature of those largest of the then existing commercial organizations, the railroads." As an experienced journalist, Dreiser knew the popular attitude. The railroads "were declared to be all that the dictionary of iniquity involves—dark, sinister, dishonest associations which robbed the people 'right and left.' . . . They were, as the public press continually averred, bribe-givers, land-grabbers, political corruptionists, hard-fisted extortionists, thieves. . . ." The railroads were universally suspected, and "by the masses of the people they are still viewed with suspicion, and everything which they undertake to do is thought to be the evidence of a scheme whereby the people are to be worsted, and the railroad strengthened in its opulent despotism."[98] Dreiser went on to argue that some of the western railroads were adopting a more enlightened approach to the public, based on the realization by sponsoring educational lectures on farming and using their telegraphs to help with marketing they could increase the prosperity of their customers and thereby generate more freight traffic.

Both readers who accepted Dreiser's argument and those who remained deeply distrustful of the railroad knew that it was a colossus. It was the largest business in the United States, the most powerful technical system, the fastest and most convenient form of transportation, and a state and national political force. Despite all the counter-narratives, and despite the protests from Native Americans, farmers, small town merchants, and railway workers, the foundation story, with its evolutionary sense of progress, remained dominant. Most jeremiads were directed far less against the railroads than against the corporate monopolies that controlled them.

Henry James, however, assailed the railroad itself in the final pages of *The American Scene*.[99] Having ridden trains across the nation for the better part of 2 years, James abhorred "the general pretension to charm, the general conquest of nature and space, affirmed, immediately round about

you, by the general pretension of the Pullman, the great monotonous rum-
ble of which seems forever to say to you: 'See what I'm making of all this—
see what I'm making, what I'm making!'" He could only respond: "'I see
what you are *not* making, of what you are so vividly not.'" James imagines that if he "were one of the painted savages you have dispossessed, or
even some tough reactionary trying to emulate him" he would find beauty
"in the solitude you have ravaged, and I should owe you my grudge for
every disfigurement and every violence, for every wound with which you
have caused the face of the land to bleed."

But James realizes that, as a comfortable passenger on a train, he cannot pretend to that Native American critique. So he accepts "your ravage"
and instead complains to the railroad: "You touch the great lonely land—
as one feels it still to be—only to plant upon it some ugliness about which,
never dreaming of the grace of apology or contrition, you then proceed
to brag. . . ."[100] In the master narrative, the railroad creates new communities all along the line. However, James finds these towns endless rep-
etitions, and he asks the railroad why it must "multiply the perpetrations
you call 'places'" and then give them "some name as senseless, mostly,
as themselves."[101] Along the railroad's right of way, James wonders, "Is
the germ of anything finely human, of anything agreeably or successfully
social, supposedly planted in conditions of such endless stretching and
such boundless spreading as shall appear finally to minister but to the triumph of the superficial and the apotheosis of the raw?" As James sits
"by the great square of plate-glass through which the missionary Pullman
appeared to invite me to admire the achievements it proclaimed," he realizes that "it was in this respect the great symbolic agent; it seemed to stand
for the irresponsibility behind it. . . ."[102]

When business historians came to write about railroads, they did not, like
James, evoke the possibility that Native Americans or early settlers had
grounds for complaint, nor did they see railroads as the "apotheosis of
the raw." Instead, most of the early writers accepted the foundation narrative. Joseph Schumpeter argued in *Business Cycles* that "the western and
middle western parts [of the United States] were, economically speaking,
created by the railroad."[103] During the 1940s and the 1950s, most historians agreed with this assessment, regardless of their political perspective.

In the 1960s, this orthodoxy came under attack when Albert Fishlow examined "this sequence of zero population, railroads, and then economic development," which by then had "become an implicit ideal type of construction ahead of demand."[104] Echoing the criticisms made in the 1870s, Fishlow argued against the historical reality of this sequence. He found that "few areas of the country penetrated by railroad in the 1850s were frontier even in 1840, let alone without population." Instead, Fishlow found, "railroad promotion in already settled areas sparked anticipatory population movement to less settled areas. As a consequence, demand was already there when railroads were ultimately built farther west."[105] Fishlow argued that the lines in the region from Michigan and Ohio to Missouri and north to Minnesota were built piecemeal, starting in the more settled areas, and that they stimulated migration ahead of construction. They therefore represented a "rational exploitation of opportunity,"[106] and were not built in advance of or to stimulate demand. Fishlow bolstered his argument that the railroads had customers from the beginning by comparing the new lines with those in the East and found that those to the West were equally profitable virtually from the start.[107] If Schumpeter had dressed up the foundation narrative in economic theory, Fishlow rewrote it to make consumer demand the stimulus to capital investment and railway construction.

Yet no sooner had Fishlow suggested a new orthodoxy than Robert Fogel attacked the notion that national development required railroads. In 1960, Fogel had concluded that the Union Pacific Railroad had an average rate of return on the costs of its construction of 11.6 percent, and that it "was a highly profitable venture that should have been taken up unaided by private enterprise."[108] In other words, the huge land grants had not been necessary. Four years later, in *Railroads and American Economic Growth*, Fogel declared that canals alone would have been adequate to underpin economic growth. Examining 37 proposed canal lines in the trans-Mississippi West, he argued that they could have brought "almost all of the agricultural land in the Midwest within 40 straight-line miles of a navigable waterway."[109] The topography of the generally flat region was not a major problem, and no proposed canal had as much rise and fall as four working New York canals.[110] He concluded: "Cheap transportation rather than railroads was the necessary condition for the emergence of the North

Central states as the granary of the nation. . . . The combination of wagon and water transportation could have provided a relatively good substitute for the fabled iron horse."[111] Fogel completely reversed the railroad foundation story. What in 1840 had seemed the stimulus to agriculture, the founder of cities, and the veritable engine of progress had become an unnecessary investment.

This argument was "subsequently refuted on almost every conceivable point."[112] Fogel aspired to scientific exactitude where the available data were incomplete and where the innumerable forces and interests that might benefit from a railroad made precision far more difficult than he suggested. He treated the railroads primarily as carriers of freight, and non-perishable freight at that, largely ignoring the economic advantages of celerity and year-round operation that canals could not offer. Historians in the 1960s were much taken with new quantitative methods, which they thought could reveal patterns in the past that had remained hidden. Indeed, Fogel later "proved" that slavery had been much more profitable than hitherto suspected—another argument that was later refuted. One might perform a similar analysis to "prove" that the Internet is of marginal utility or even unnecessary because the postal service is adequate to carry the mail. Fogel's theories were widely appreciated, however, because he used then new quantitative methods to overturn what had become the standard wisdom of the railroad foundation narrative. In his hands the engine of progress became wasteful and unnecessary, while the canal was reestablished as the mother of cities. In effect, Fogel resurrected John C. Calhoun's 1817 program of internal improvements, which had called for roads and canals but had not foreseen the railroad. He claimed that American western settlement and industrialization could have been based on the mill and the canal. In rejecting the inevitability of the railroad or even the need for it, he denied one version of the technological creation story in favor of earlier second-creation narratives.

In the middle 1960s, when Fogel's book appeared, the railroad had lost its centrality in American economic life. As John Stilgoe has argued, between c. 1880 and 1930 the "metropolitan corridor objectified the ordered life, the life of the engineered future. Industrial zones, small-town depots, railway gardens, suburbs, even backyard vegetable gardens and lawns all drew characteristics from the railroad. . . . For one half-century moment, the nation created a new sort of environment characterized by technically controlled order."[113] But as trains lost passengers to automobiles and freight to trucks, the metropolitan corridor was broken, and settlement and new roads sprawled in new directions. As weeds grew along the metropolitan corridor, a nostalgic narrative emerged. For some people the steam train became virtually a cult—societies of enthusiasts lovingly restored famous engines and patronized special sections in bookstores that contained glossy volumes of photographs, technical details, maps, and old timetables. For others, constructing meticulous model-train layouts offered a way to preserve the landscape of railroads.

In 1923, one of the earliest canal and railway companies, the Delaware and Hudson, marked its centenary. Appropriately, the event was held on successive days in Scranton (the center of the anthracite fields) and in New York (the chief market). One of the speakers remarked that Julius Caesar, had he accompanied George Washington to New York for his inauguration as president in 1789, would have found most of the objects around him easily comprehensible:

Some changes there were; the outer garments were different. . . . The carriages were different, but their mechanism and the principles involved were familiar to him in the Roman chariot. The city and the occasion might well have reminded him rather of a celebration in some outlying province than one in Rome itself. How great a contrast would today confront these two historic men. The buildings outstripping anything then or previously known; the elevators to be seen through the windows rapidly moving, a mode of transportation quite unique; at the cross streets glimpses to be had of the surface cars on Broadway, and trains moving on elevated roads, the streets thronged with automobiles moved by a power not indicated, through a mechanism not understood, faintly indicate the long list of changes that makes the material progress of this time more significant than all that which has gone before.[114]

Despite this stunning technical progress, however, the president of the Delaware and Hudson was not sanguine about the future, for railroads in the 1920s were suffering economically. He located the source of the problem in excessive government regulation and frequent denial of requests for rate increases, and he concluded dourly: "Railroading is no longer a business, it has become a calamity."[115] In the following decades railroads would fight back against the automobile with faster rains and improved Pullman service, but they would nevertheless lose money on most routes. Eventually they would sell profitless passenger services to the federal government, which would unify them into Amtrak.

The decline of the railroads also was evident in world's fair exhibits. Displays of steam engines and railroads had been vast and exhaustive at Philadelphia (1876), Chicago (1894), and St. Louis (1904), and remained quite important at San Francisco (1915). During this 40-year period, railroads also carried visitors to the fairs and around their perimeters. The history of the railroad then seemed to be nothing less than the history of the expansion and settlement of the United States. During the 1930s, however, nostalgia began to creep into the exhibits, not least because the railroad corporations began celebrating their centenaries. At the Century of Progress International Exposition (Chicago, 1933), the transportation exhibits included three locomotives from before 1836, the Tom Thumb (Baltimore and Ohio), the DeWitt Clinton (Mohawk & Hudson, later New York Central), and the Thomas Jefferson (Winchester & Potomac). Also on display were "the old 'Pioneer,' first locomotive ever to run out of Chicago" and the first sleeping car ever built, "a little wooden car with open platforms and crude berths, that looks a bit humble as it stands between two great modern Pullmans, all of aluminum, and stream-lined, which are the last word in sleeping car construction for 1933."[116] Such contrasts throughout the exhibit were intended to show how far railroads had advanced.

Nevertheless, the railroads became more marginal to each successive fair, both as a way to get to the site and as an exhibit. At New York's World of Tomorrow (1939) the massive railroad exhibits covered 17 acres. One of them visualized the foundation story, showing how the railroad moved into and transformed empty space. Located in an eight-story building, it featured a mountainous landscape where was enacted "the progress of railway building—forest clearing for a right of way; ore brought down the mountains to smelters, factories, and fabricating plants; logs cut and floated to saw mills; raw materials converted and assembled into the finished product." A nearby exhibit, Railroads at Work, featured a 160-foot-wide diorama in which 500 engines and cars traversed 3,800 feet of miniature track (and 70,000 miniature ties), winding their way through "cities, towns, villages, farms and factories."[117] However, neither this

Figure 8.2
Edward Hopper, "Hotel by a Railroad" (1952). Courtesy Hirshhorn Museum and Sculpture Garden, Smithsonian Institution. Gift of Joseph H. Hirshhorn foundation.

enormous model train layout nor Railroads on Parade (with music by Kurt Weill) proved as popular as General Motors' Futurama, an enormous miniaturized landscape depicting a high-tech world of 1960 where trains had disappeared and radio-controlled cars whisked passengers along limited-access highways.[118] As the public preference for Futurama suggested, the future belonged to the automobile and the airplane, while the railroad increasingly belonged to the past. With each passing exposition, the railroad was more likely to be imagined as an earlier stage in transportation that began with the chariot and ended with rockets to the stars. These representations of transportation history erased the enormous social tensions and environmental problems the railway system had created. Reduced to a miniaturized system, the train seemed a charming remnant of history.

As railroads ceased to be the central mode of passenger transportation, the landscapes they left behind seemed to have little redeeming social value. John Updike's sharp eye noted one characteristic scene:

. . . the laborers of old hand-dug a great trench to bring the railroad tracks into the city, tracks disused now, and the cut, walled in limestone, a pit for tossing beer cans and soda bottles down into, whole garbage bags even, mattresses; Brewer was always a tough town, a railroad town, these blocks along the tracks full of tough men, bleary hoboes who'd offer to blow you for a quarter, sooty hotels where card games went on for days, bars whose front windows were cracked from the vibrations of the trains going past, the mile-long trains of coal cars pulling right across Weiser, stopping all traffic. . . .[119]

This damaged and sooty landscape was the subject of a number of Edward Hopper's paintings of the 1920s and the 1930s. In some of these paintings, tracks are in the foreground; house and neighborhoods in the background lack such conventions of landscape as lawns and shrubbery. In others, people on a train or a subway car stare past one another without acknowledgment.[120] In "Hotel by a Railroad" (1952), a seated woman reads her book, oblivious to a man who stares expressionless out the window while smoking.[121] Outside are the railroad tracks, a curtained window, and a blank wall. The view is cut off, empty of nature or signs of humanity. The mirror between the man and the woman likewise reflects nothing. It is an emotional and physical cul-de-sac. Nothing could be further from "westward the course of empire takes its way."

However, railroads remain a vital part of the American transportation network. In 1990, freight trains still carried more than trucks, even though 90,000 miles of track had disappeared since 1920.[122] Yet because American railways carry so few passengers, they exist at the margins of public consciousness. Some of the great passenger stations, like those in Omaha and Kansas City, long stood unused and empty. Others have been torn down. In Washington and Indianapolis the central stations have been converted to shopping malls. Gone are both the dynamism of the technological creation story and most of the angry social criticism of railroads. As happened with the canal and the mill, the nostalgic rewriting of this foundation narrative has transformed the train into an emblem of the past.

9

"Conquered Rivers Are Better Servants than Wild Clouds"

Conquered rivers are better servants than wild clouds.
—John Wesley Powell, 1890[1]

It is the fortune of Arid America to be so palpably crude material that it can not be used at all, save upon the divine terms.
—William Smythe, 1905[2]

When white settlers and railroads moved westward and entered drier regions, they modified the technological foundation narrative accordingly. As before, surveyors carved the public lands into a vast checkerboard for sale to speculators and farmers. As before, railroads were regarded as the economic lifeline that opened a region to development. As before, farmers were expected to improve the land. The arid plains, once rejected as forever unsuited to agriculture, were to be transformed. Earlier settlers in the East had drained swamps and logged entire counties, drying out the land so that they could farm it. The West at first seemed useless to farmers conditioned by such previous experiences, and the area west of the 97th meridian was viewed with skepticism until after the Civil War. Attempts to settle the high plains met with only limited success, and many gave up on the attempt, concluding that the area was too dry.[3]

Gradually, a new idea emerged. On the plains, the theory went, settlers would plant trees rather than cut them down. Their plowing would activate the earth and stimulate the hydraulic cycle. The idea that "rain followed the plow" was justified by scientists, promoted by land speculators, embraced by settlers, explained in editorials, and spread far and wide by the railroads.

Further west, homesteading the desert through irrigation was the logical extension of Manifest Destiny. Like the clearing of eastern forests or the damming of streams to build mills, by the 1880s irrigation was depicted as an inevitable development that opened the land to full use and promoted prosperity. In the most extreme form of the story, the aridity of the West was considered an act of providence. Just as Zachariah Allen saw New England's lack of coal as a sign of God's intention that water power (rather than steam) was to be the basis for a factory system, western irrigation enthusiasts such as William Smythe wrote of "man's partnership with God,"[4] suggesting that the arid West was an unfinished creation waiting for Americans to complete the work of the Almighty. And just as early canal and railroad towns promoted themselves as inevitable centers of national commerce, the irrigated West proclaimed itself the future center of the United States. "All through that region," Secretary of the Interior John W. Noble declared in 1893, "much of which is now arid and not populated, will be a population as dense as the Aztecs ever had in their palmiest days in Mexico and Central America. Irrigation is the magic wand which is to bring about these great changes."[5]

The narrative of irrigation, like earlier technological foundation stories, was justified by future pastures, contrasted with present wastelands. In 1894 a correspondent in *McClure's* declared: "Millions of acres of land are lying idle in western Kansas and Nebraska, in Colorado, Wyoming, Utah, Nevada, Idaho, Montana, Arizona, New Mexico, and California, wanting only the magic touch of water to make them bloom into a flower-garden, and yet producing nothing but lean coyotes, sun-dogs, and scenery."[6] By the 1890s the desirability of irrigation seemed obvious, though one had to distinguish between two narratives, one individualistic and the other statist. The first focused on private initiatives, and it recalled the self-reliance of the pioneer with the axe. This story was most common east of the Rocky Mountains. The settler had only to divert a stream or install a windmill-driven pump, and water would fructify his land. In contrast, the second form of the story acknowledged a need for collective action to build dams and canals, and it eventually became the dominant narrative in the Colorado River Basin and California. This variant began with the investment of state or federal money and engineering expertise to prepare the way for irrigation. Yet the water was still to be consumed by individual farmers. The second irrigation story resembled that of the railroad, for each called for large investments before public lands could become valuable.

In either form of the narrative, mastery of irrigation would make barren land bloom, raise property values, and become the basis for new communities. The underlying structure of both variants of the irrigation story recapitulated the fundamental technological creation story. The land was regarded as empty space waiting for the white man to develop its latent potential. Settlers arrived and transformed the land using a new technology that made possible the foundation of new communities. The individualistic variant was essentially a rewriting of the story of the axe. The pioneer did not have to clear the treeless plains, but instead had to fructify the land by drawing water down from the skies or up from the earth. It was still the story of free land that anyone could farm with small-scale technologies. The second variant (treated in the next chapter) told of desert land that could be farmed only after large-scale state investment. In either case, a Lockean individual farmer was expected to earn the right to own the land by mixing his labor with it. Together, the two irrigation narratives recapitulated the development of the technological creation story from the axe to the railroad, from individual to corporate effort, and from muscle power to mechanization.

Before the individualistic variant of the irrigation story could emerge, Americans had to redefine the Great Plains and the Mountain West. When Europeans arrived in the New World, they assumed that the vast, treeless prairie was incapable of supporting agriculture. A pantheon of influential people declared it a wasteland, including Washington Irving, Zebulon Pike, James Fenimore Cooper, and the scientist Joseph Henry. In 1856, Henry concluded that the plains were "a barren waste . . . of little value to the agriculturalist."[7] Yet even as he wrote these words, the first waves of settlers were pushing beyond the Mississippi. These settlers had adopted a far more hopeful vision of the region, and they reimagined it as an untouched space that could be transformed into a vast garden.[8] Men seemed particularly prone to seeing the land as virginal and waiting to be taken.[9] Bitter experience would soon teach them that beyond the 98th meridian was a land of little rain.

The unusually wet years between 1878 and 1887 encouraged homesteaders to believe that rain and hence civilization followed the plow.
Railroad promoters trumpeted this message throughout the nation. The
Rock Island Railroad insisted that Kansas was "the garden spot of the
world. . . . Because it will grow anything that any other country will grow
and with less work. Because it rains here more than in any other place, and
just at the right time."[10] Near the end of this rainy period, the Department
of Agriculture commissioned Richard Hinton to prepare a report on irrigation. He sent letters to "irrigators, arboriculturists, engineers, landowners, colonists, and all other persons known to be actively interested."[11]
Most of those who responded wrote enthusiastically about the possibilities for extensive irrigation. Hinton's report came close to embracing the
notion that "rain follows the plow," albeit in a variant form. He admitted
that precipitation had not been increasing in many areas, but nevertheless
he concluded that "where settlement and cultivation have progressed to
any marked degree, and especially where the latter has been aided by irrigation, there has been a decided increase of terrene humidity. Springs have
increased in volume. The running waters are more regular in their flow and
quantity. The increase in some places is a very noticeable phenomenon, as
that of Salt Lake, for instance . . . in California, Utah, and Colorado . . .
wherever irrigation has been longest applied the necessity for the use of
water by its means has diminished, owing to the seepage from the ditches,
and that capillary attraction. . . ."[12] This improvement was partly the
result of deforestation, but Hinton argued that the process of humidifica-
tion was enhanced by the substitution of crops for native grasses. Furthermore, the plains area experienced "the movement westward, with the
movement of population, of an increased rainfall. This precipitation is
likened by the State Engineer of Colorado to a wall pressed westward."[13]
The report cited studies by Professor F. H. Snow of Kansas State University which "proved" that rainfall had increased 25 percent in the dry parts
of his state as a result of settlement. After adding to this the possibilities of
irrigation from the Arkansas River and other sources, the development of
artesian wells, and the systematic programs that rewarded homesteaders
for planting trees, Hinton concluded: "Western Nebraska and Eastern
Wyoming will show a steady climatic change."[14] Professors at local state
universities declared that the appearance of desert was a temporary effect

of not cultivating the soil. The accuracy of this new version of the second-creation narrative seemed to have been demonstrated by experience, for in the early 1880s white Americans were apparently transforming the entire West into a vast garden, in a process that promised to be self-reinforcing.

Yet even as the Hinton Report appeared, a long drought began that showed it to be erroneous. With heartbreaking faith, "tens of thousands expended all their money and the most precious years of their lives in discovering what could not be done in the semi-arid region."[15] Yet if the rain was unreliable, experiments with irrigation had already begun. In 1871 Horace Greeley was a powerful advocate. Speaking to Texas farmers, he declaimed: "In the great Future which Science and Human Energy are preparing, artesian wells, bored to depths of a thousand to fifteen hundred feet, will be sunk on every arid plain, and near the head of every capacious valley wherein water is deficient, to enable the strong currents that flow from subjacent mountain or elevated plateau [to] rise by gravitation to the surface and fruitfully over spread hundreds of acres, instead of uselessly coursing in darkness beneath." He predicted that "the wide, misnamed 'Desert' at either foot of the Rocky Mountains, will yet be transformed into the verdurous, plenteous feeding ground of innumerable cattle and sheep by irrigation." The new farming would require little muscle power. Greeley recalled the "rude pioneer, wrestling stubbornly with the giant forest" whose "fields are subdued and tilled, his crops produced and secured, almost wholly by dint of the strength in his good right arm."[16] But that foundation story belonged to the past. The pioneers' "enlightened descendant and successor" now commanded greater forces: "Water, wind, steam, supply the needed power; his task is to mould and guide that power to benevolent ends."[17]

The audience knew that this was not merely rhetoric, for Greeley himself had encouraged the creation of a new irrigation-based community: Greeley, Colorado. Success in that state suggested how much could be done. "Prior to 1860," Hinton observed, "the practice of irrigation in Colorado was confined to a few scattered Mexican settlements in the southern part of the Territory, with an imitation, but little improved, by the few American settlers in other parts of the Territory on the bottom lands lying immediately alongside the streams. The ditches were small and short [and the irrigation was] confined to small patches."[18] Greeley visited the region in 1859 and became convinced that the higher tablelands could also be productive. He used his New York newspaper to promote the idea, and after the Civil War he located a site on the Cache la Poudra River north of Denver. In what was conceived as a cooperative venture, 687 people subscribed for membership, and hundreds of settlers poured onto the 12,000 acres that had been purchased. They erected a town, built extensive irrigation works at a cost of more than $400,000, and discovered through trial and error what trees and crops would grow there.[19] Their example inspired others, and by the middle 1880s Colorado had canals of "between 900 and 1,000 miles, and the land susceptible of being irrigated 1,700,000 acres."[20]

Irrigation was fast becoming the foundation narrative of the arid West, and in 1873 a convention was held in Denver to consider the prospects. As one correspondent summarized the situation for the *New York Times,* the 99th meridian of longitude defined the point where the prairie ended and the Great Plains began: "West of this line lies one-half of the area of the United States, all of which, excepting a small strip on the shores of the Pacific, is without sufficient rain-fall for the cultivation of the soil. . . . Here are one million square miles of barren country, and the question is, What shall we do with it?"[21] The journalist went on to praise the systematic irrigation practiced in Utah and to compare it with the uncoordinated efforts in Colorado, where little had yet been done: "According to careful estimates, Colorado has a water supply sufficient to irrigate 6,000,000 acres, an arable area which, in Egypt, in the times of the Ptolemies, supplied food for 8,000,000 people. . . ." Embracing the grandiose rhetoric of the convention, the *Times* corespondent concluded that "the mountain streams, if turned into proper channels, [would] irrigate the greater part of the Plains, both east and west of the mountains."[22]

The reporter had swallowed the most optimistic creation story, which would persist for a generation. Even the aged Walt Whitman took up the new gospel of irrigation. In *Specimen Days* he declared that the western plains had a "sure future destiny" as "the inexhaustible lands of wheat, maize, wool, flax, coal, iron, beef and pork, butter and cheese, apples and grapes—land of ten thousand virgin farms—to the eye at present wild and unproductive—yet experts say that upon it when irrigated may easily be grown enough wheat to feed the world."[23] In this new avatar of the foundation narrative, the individual needed more tools that an axe and a plow, and more knowledge than the farmer in the well-watered East. But the system, once installed, seemed perpetual, so long as the windmills turned or the winter snows melted each spring and filled the reservoirs. Irrigation seemed to exemplify the idea of a man-made second nature.

Most visitors to the West were impressed by the success of the Mormons, who developed irrigation on a large scale in Utah, one of the driest areas in North America. A typical account dating from 1866 condemned their religion but praised the people as "industrious, frugal, and thrifty," noting that "by their wonderful system of irrigation, they have converted the desert there into a garden, and literally made the wilderness, 'bloom and blossom as the rose.'" By that date the Mormons had already constructed more than 1,000 miles of canals and ditches, watering 150,000 acres. Because of their careful methods and the double cropping possible in the mild climate, "one man cannot well manage over ten or twelve acres per year; nor is more necessary for an ordinary family."[24] This lesson was lost on most early irrigators, however, and even Congress long missed the point, giving inappropriately large land grants. Mormons also realized the need for central planning and systematic control in order to make the most of their water. Dams and ditches were supervised and regulated by the Church, and an outside observer concluded that "the whole institution of Mormonism—polygamy and all—apart from its theological aspects, impresses you rather as a gigantic organization for collecting and consolidating a population, and thus settling up a Territory rapidly."[25]

As white settlement pressed into drier regions, Congress passed the Desert Land Act of 1877, which quadrupled the provisions of the Homestead Act and gave title to 640 acres of arid land to any settler who claimed and irrigated at least 80 acres within 3 years and paid $1.25 per acre.[26] This act did not include the western portions of states such as Kansas and Nebraska or even Colorado (until amended in 1891), but it did apply to the Dakota Territory, Wyoming, New Mexico, and the rest of the West. Though well intentioned, the Desert Land Act did not encourage settlement as much as was hoped. Experience would show that 80 irrigated acres was more land than almost any family could manage to develop in 3 years, and that the $800 it cost to claim 640 acres was beyond the means of most.

In contrast, locations that permitted direct irrigation from rivers and streams were quickly exploited. In Nebraska, a large water wheel driven by the current on a branch of Hat Creek lifted water 30 feet to serve nearby farms.[27] In western Kansas, starting in 1879, some farmers diverted water from the Arkansas River into their fields in response to a 14-month drought.[28] Writing for *Harper's Weekly,* Hinton dramatized the transformation of Garden City between 1880 (when it was merely "a score of log cabins, or prairie dugouts, with a frame store building") and 1888 (when its population had swelled to 8,000 and it was a center of regional development).[29] It seemed a paradigmatic example of the irrigation creation story, transforming a region with only 18 inches of rainfall a year into a rich farming area. Some years were much wetter than others, however. When rain was sufficient, farmers refused to contract for water, leaving the local irrigation companies teetering on the brink of bankruptcy and unable to keep their ditches in repair. When dry years came and farmers again wanted water, it was not always deliverable.[30] In 1889 one of the ditch systems near Garden City gave way and flooded the railway tracks and other lands. Irregular availability of water, poor engineering, and rampant land speculation brought irrigation into disrepute, and "by the 1890s irrigation was at a low ebb in the Garden City area."[31]

Elsewhere on the high plains there were few easy opportunities to divert streams into fields. However, some farmers discovered that the water they needed was often right beneath their feet, if they could only drill down and pump it up. Few early irrigators had steam-driven pumps, for they were expensive and required expert attention. Gasoline engines and electric pumps were still displayed as novelties at world's fairs in the 1890s, and they did not become a practical option until the early twentieth century. In the nineteenth century, windmills offered the most common way to raise water, and by 1880 there were 69 factories producing them as ready-to-assemble kits. The windmill offered a new version of the technological creation story. Again, white Americans entered a "virgin" region, and again they were to develop it using machines that drew upon local conditions—in this case, the underground aquifer and the persistent winds. Garden City not only drew water from the Arkansas River; the local farmers also built private reservoirs fed by 150 windmills.[32] The early irrigators used gravity flow to move the water from reservoirs into fields. This new second-creation narrative was spectacularly confirmed in western Kansas, where settlers discovered that water lay as little as eight feel below the surface. Smythe summarized: "It was found in the Arkansas Valley of western Kansas that water could be obtained by shallow wells. . . . This is raised by hundreds of windmills into hundreds of small reservoirs constructed at the highest point of each farm. The uniform eastward slope of the plains is seven feet to the mile. The indefatigable Kansas wind keeps the mills in active operation, and the reservoirs are always full of water, which is drawn off, as it is required. . . ."[33] Smythe, an indefatigable organizer of irrigation congresses and associations, estimated that the cost of such a farm, including reservoirs and ditches but exclusive of the farmer's labor, was $20 per acre, a modest sum in exchange for "perpetual guaranty of sufficient 'rain.'"[34]

An irrigated farm often consisted of no more than 20 intensely cultivated acres. Cottonwood trees planted around fields protected them from the drying wind and the scorching sun. The reservoirs could also be used as sources of fish in summer and ice in winter. The appeal of such small farms was due in part to their independence. "These small individual pumping-plants have certain advantages over the canal systems which prevail elsewhere. The irrigator has no entangling alliances with companies or cooperative associations, and is able to manage the water supply without deferring to the convenience of others, or yielding obedience to rules and regulations. . . ."[35] Thus Smythe told the individualistic version of the technological creation story, which he had first heard as a journalist in Nebraska.

Farmers invented new kinds of windmills that they built themselves from scraps of wood and inexpensive odds and ends, such as burlap bags for sails. One windmill type looked like a small Ferris wheel lying on the ground and rotating horizontally. Another resembled an overshot water wheel. In 1900 *Scientific American* reported that these ingenious devices could be built for $1.50 to $150. They not only pumped water but also saved the labor of "running a grindstone, churning, working a feed grinder, corn-sheller, the wood-saw, and other farm machinery."[36] The largest of these homemade mills reportedly could irrigate 10 acres. Thus individual technological ingenuity seemed to provide the basis for developing the land. Usually, however, a windmill could irrigate 8 acres at best, and less if it served the household, the livestock, or a separate vegetable garden. Though one farmer managed to irrigate 50 acres using 10 windmills, most results were less impressive. Another farmer found he needed 24 windmills to water 40 acres.[37] No one seems to have come very close to using the wind to irrigate the 160 acres obtainable under the Homestead Act. In Garden City, for example, the windmills of all the farms together did not pump water onto more than 1,000 acres as late as 1905, while 40 times as much land was watered by older methods. Artesian wells, which were deeper, required powerful pumps that cost several thousand dollars; such expenditures were out of the question for most farmers until after World War I.

Early success convinced William Smythe and other proponents of irrigation that "the very lands which refused to yield a return for the industry of the first settlers will sustain the densest population in the future and give the most absolute assurance of permanent prosperity."[38] In this new Jeffersonian vision, the land would be irrigated and farmed intensively in small parcels. Such farming appeared possible not only in western Kansas and Nebraska, where it started, but also in west Texas, eastern Montana, and the Dakotas, where artesian wells could be used to draw on enormous underground reserves. The conclusion seemed clear: "From Canada to Mexico the revolution in the Great Plains is now in full tide. It is the most dramatic page in the history of American irrigation. It has saved an enormous district from lapsing into a condition of semi-barbarism. It has not only made human life secure, but revolutionized the industrial and social economy of the locality."[39]

By the early 1890s, settlers in all parts of the Great Plains were experimenting with irrigation. They held excited meetings, gave speeches about artesian wells, read government reports on techniques for building earthen dams, and invested in canal systems and windmill-powered pumping stations. Individual farmers dug a tracery of ditches that used the apparently providential fact of the persistent gentle slope of the whole continent toward the east to provide a gravity-based feeder system that would make the dry land blossom. In 1894, at virtually the same time that Frederick Jackson Turner was declaring the frontier closed, Thomas B. Reed declared in a speech: "The same power which wastes millions on the Mississippi [River] can be utilized to make the desert blossom with the homes of men, for whom and for all of us the now blighted soil will bring forth the fruits of the Garden of Eden."[40]

The contrast between barren land and irrigated luxuriance repeatedly called forth Edenic metaphors, although strictly speaking the biblical garden had required no labor and by definition existed before any technology. In 1894, a western journalist presented the case for irrigation this way: "An irrigated farm is the only 'sure thing' farm on the face of the Earth. Here a man may, for the small sum of one dollar an acre, make it rain or shine, on any or all of his acres, when he wills. An irrigated farm never wears out."[41] A clergyman from Brooklyn, New York, after seeing the irrigated sections of Colorado, returned to Brooklyn and preached a sermon full of Edenic imagery that decried the fact that "we have allowed to lie waste, given up to the rattlesnake, prairie dog, and bat, land enough to support whole nations of industrious population."[42]

The same conversion to the religion of irrigation took place in California and Arizona. Early Anglo-American settlers had assumed, as did the Census Bureau, that irrigation was a marginal possibility. In 1860, California's aridity seemed unalterable. Former governor John Downey noted that the dry land supported millions of cattle and horses, "inferior stock which have become a nuisance."[43] Owing to the lack of a railroad connection to the East, there was no market for such livestock. Little of the land, he argued, could be cultivated: "This is not owing to any want of fertility, but to the absence of rains in the summer, and the scarcity of water for irrigation on a large scale." "We need never hope for a dense population such as will swarm the great Northwest," Downey erroneously predicted.[44] Raising sheep appeared to be the best way to make use of California's land. Downey's inaccurate assumptions and conclusions were widely shared.

When John Hittel devoted 85 pages to a survey of agriculture in California in 1863, he concluded that 30 million acres of arable land would forever remain fallow because of "the impossibility of irrigation."[45] Only 1 million acres were then in production, and Hittel considered the soil inferior to that of all the eastern and midwestern states. California then had only one-tenth as much land under cultivation as Ohio.[46] Despite the mild and sunny climate, Hittel thought crop rotation was "impossible on the greater part of the land"[47] because it was arid. "Most of the land in Los Angeles, San Bernardino, and San Diego counties," he wrote, "is sandy and dry, and very little of it is cultivated."[48] In eastern California, he recognized only a remote possibility "that a considerable tract of land will be rendered fit for tillage by turning the Colorado into the low part of the desert." Yet such a project required such immense resources that it did not appear feasible. Only in a few valleys, Hittel emphasized, could farming flourish.[49] The Census Bureau too expected irrigation to develop only in a few specific river valleys. Smaller streams seemed more promising than large rivers. For example, it was estimated that 3.5 million acres of Arizona could be irrigated,[50] and, in fact, 5 years later Arizona had "about 700 miles of irrigation works."[51] The Colorado River, in contrast, could "not be used to serve any large quantity of land, owing to its bluff, high banks, and canyon walls, and its slight fall where it flows through open country."[52] Although many realized that California was producing some high-quality produce,[53] most observers expected mineral resources to be the chief source of western development.[54]

A few, however, were becoming excited about Californian irrigation. In 1851 the Mormons briefly established a colony in San Bernardino, laying out an irrigation system and a town with a population of 2,000 before selling out and retreating to Utah.[55] Successful irrigation also had begun in Anaheim, where in 1857 a group of 50 German settlers began a colony of 20-acre farms. This colony set the pattern for "the mutual water companies that became dominant in southern California."[56] In 1870, by which time 60,000 acres were being irrigated, a judge from Tennessee, John Wesley North, formed an association that purchased 4,000 acres of dry land, which it planned to irrigate by means of a $50,000 canal that would deliver water from the Santa Ana River. North's town, San Bernardino, proved a success. The following year, other investors spent $1.3 million on a 70-mile canal to carry water from the San Joaquin River.[57] Another colony of settlers came from Indiana and established a town that became Pasadena. Both communities grew oranges,[58] and in 1877 the first railway car of oranges was shipped east. An 1883 guidebook depicted the San Bernardino area as an earthly paradise, in "a beautiful valley, with picturesque mountains on three sides. . . . It is supplied with water by artesian wells, and all parts of the town are embowered in fruit and ornamental tress. Fruits of all kinds are grown here, and oranges and lemons are produced in great abundance."[59] With such reports widely circulating, the next census summary was more alert to the possibilities of irrigation, noting that, when water is dependably available, "the cultivator is entirely independent of the weather; droughts have no terrors for him." In addition, "the waters of irrigation come from the mountains charged with fertilizing material, which by evaporation they leave in the fields. Irrigated land thus needs no manure."[60]

Early enthusiasts for irrigation expected not government-sponsored projects but profit-making private initiatives, like those in Greeley and Pasadena. Albert Richardson cited these successes to project the irrigation foundation story into the future: "In time, simple and cheap machinery for irrigation from wells will doubtless be introduced. Then the great American Desert will become a thing of the past; and the thousands upon thousands of miles of sage-brush and grease-wood dwarf-cedar and cactus, sand and alkali . . . will yield barley oats and fruit as profusely as the Mississippi valley produces corn and hay."[61] Note how in this passage the invention and the widespread use of "simple and cheap machinery" for irrigation are taken for granted. Those who believe in a technological creation story assume that the needed tools and natural resources will be available, so that the narrative can automatically unfold.

There were many problems with this narrative as a description of any possible reality, and none realized its inadequacies better than the director of the United States Geological Survey, John Wesley Powell. The most obvious difficulty was not the lack of cheap and efficient pumping machinery but the lack of enough water to pump. Powell estimated in 1874 that no more than 3 percent of the West could be irrigated.[62] He had traveled the region extensively and studied the amount of water available. Powell never believed that rainfall would increase with settlement, and in 1902 his hydraulic engineer for the Geological Survey stated categorically: "It has not been possible to detect any progressive increase or diminution in the amount of precipitation when records extending over thirty or forty years are had."[63] For a quarter of a century, Powell argued for an alternative vision of what western agricultural would look like. His sober narrative helped educate the public about what was possible, undercutting the boosterism and bogus science that envisioned the entire West as farmland.

Powell's views first appeared in detail in his 1878 *Report on the Lands of the Arid Region of the United States*. His charts and statistics showed how little water was available in the West. Powell also wrote against the still widely held belief that the climate of the West was changing in response to settlement and irrigation. As evidence, many pointed out that the Great Salt Lake had risen between 1850 and 1870 despite the fact that the Mormons withdrew irrigation water from its tributaries.[64] Powell was "frequently told . . . that wherever and whenever a settlement was established, there followed in a few years an increase of the water supply."[65] Likewise, an entomologist who had observed conditions in Colorado from its settlement on found "there has been a gradual increase of moisture . . . of a permanent nature, and not periodical. . . . It is in some way connected with the settlement of the country."[66]

In reply, Powell agreed that the Great Salt Lake had grown larger, but attributed this to deforestation and grazing, each of which reduced the soil's retention of water. Powell accepted the fact that the water level had risen, but attributed it to temporary causes: "The destruction of forests, which has been immense in this country for the past fifteen years; the cropping of the grasses, and the treading of the soil by cattle; the destruction of the beaver dams, causing a drainage of the ponds; the clearing of drift-wood from stream channels; the drainage of upland meadows, and many other slight modifications, all conspire to increase the accumulation of water in the streams, and all this is added to the supply of water to be used in irrigation."[67] Such gains were illusory, as they came at the expense of drying out the watershed.

The arable lands so greatly exceeded the available water, wrote Powell, that "monopoly of land need not be feared." Indeed, "the question for legislators to solve is to devise some practical means by which water rights may be distributed among individual farmers and water monopolies prevented."[68] Powell saw that the common law, which had developed in well-watered England and the eastern United States, was not an adequate framework for development. Not only was land valueless without water rights, but "the water has no value in its natural channel. In general the water cannot be used for irrigation on the lands immediately contiguous to the streams—i.e. the flood plains or bottom valleys."[69] The water was needed higher up, on the "bench lands." And, Powell anticipated, "all the waters of all the arid lands will eventually be taken from their natural channels, and they can be utilized only to the extent to which they are thus removed."[70] The useful lands and the water were seldom contiguous; dams and diversion canals often had to begin miles from the fields to be irrigated.

Because of these circumstances, Powell argued, the common laws of water use, "if applied to the Arid Region, practically prohibit the growth of its most important industries,"[71] farming and mining. The new conditions had already resulted in the separation of water rights and land rights. Concentration of water rights in a few hands was possible, in which case "the whole agriculture of the country will be tributary thereto." The central question became whether water rights should be owned by farmers or by the water companies controlling the canals. The problem could be solved by forming cooperative water companies that would include all the local farmers. It also could be solved by government regulation of water monopolies, or even by outright government ownership. Powell did not propose any of these solutions, however, instead arguing that water rights ought to be legally attached "to the land where used, not to the individual or company constructing the canals."[72] The farmer was to retain this right by using it. Powell's solution required first that a detailed survey and classification of the western lands be conducted, then that settlement be limited to land with water rights. Furthermore, the shape of the land would have to be different: "The lands along the streams are not valuable for agricultural purposes in continuous bodies or squares, but only in irrigible tracts governed by the levels of the meandering canals which carry the water for irrigation. . . ."[73] Powell thus attacked both the common law governing the use of streams and the geometrical division of space that had become an unquestioned practice in American expansion.

If Powell rejected the naive assumptions that the needed water and the irrigation technologies were easily available, he did champion irrigation. In 1890 he inspired readers of *Century* with a vision of western transformation: "The people are constructing cities and towns, erecting factories, and constructing railroads, and great industries of many kinds are already developed." The growing population had to be fed, "and the men of the West are too enterprising and too industrious to beg bread from the farms of the East."[74] Indeed, "already they have redeemed more than six million acres of this land; already they are engaged in warfare with the rivers, and have won the first battles." The solitary pioneer with his axe had been replaced by "an army of men" on a "campaign—not for blood, but for bounty; not for plunder, but for prosperity."[75] Powell embraced Henry Carey's attacks on Malthus and Ricardo, and like Emerson he believed that the best agricultural lands were the hardest to put into production. He declared: "The arid lands of the West, last to be redeemed by the methods first discovered in [ancient] civilization, are the best agricultural lands of the continent."[76]

Just as mills had made logging profitable in the East, dams would make agriculture even more profitable in the arid states than it was in the Midwest. Powell emphasized that, despite the limited potential farmland in the West, "the limit in quantity has compensation in quality"[77]: "An acre of western land, practically worthless without irrigation, when the works are constructed to supply it with water at once acquires a value marvelous to the men of the East. In new California, settled but yesterday, cultivated lands command better prices than in Massachusetts or Maryland."[78] Like most other writers on irrigation, Powell conceived of this project as "the redemption of the arid region."

Critical though he was of western boosterism, Powell's narrative remained that of the foundation story. In each case, Americans entered a new region, deployed the appropriate technology, reinvented the landscape, and created a prosperous new community. Irrigation was a new tool to be used in rewriting the familiar story to fit western conditions. Powell may have disproved meteorological claims, he may have argued for the need to rethink the common law, and he may have pointed out the absurdity of land division based on geometry rather than water levels, but he did not write a full counter-narrative, nor did he reject the technological creation story that had developed over generations. Ultimately, he was only recognizing environmental limits within which an irrigation narrative might still describe reality.

Like Powell, many others argued for state involvement in systematic, large-scale irrigation. The California legislature began to move beyond piecemeal local efforts by appointing William Hall as state engineer in 1878. For the next 11 years, with wavering support from the legislature, Hall conducted extensive surveys and produced reports and histories that private entrepreneurs found useful for decades after. Hall argued for the formation of irrigation districts that owned and operated waterworks, and something like this did pass into California law.[79] But Hall was unable to sell to the individualists of the 1880s a new technological narrative in which the state was to take a major role.

Yet even within limits of private capital and partial control of watershed areas, irrigation communities could be developed as a coordinated whole. Two Canadians in southern California, the Chaffey brothers, showed how. In 1881 they purchased 1,000 acres of land and water rights, built a concrete pipe system, and organized a mutual water company. One share in this company was permanently attached to every acre, and the land was divided into small plots that a single family could manage. Each property was further supplied with electric light and the telephone, making the new community of Etiwanda one of the most modern in the United States. The project's immediate financial and agricultural success encouraged the Chaffeys to build the larger community of Ontario.[80] They understood that, unlike homesteading in the Midwest, irrigation required an extensive infrastructure and an overall design. Their towns immediately served as models for how water could transform the region as a whole. Indeed, two decades later, at the St. Louis World's Fair, a model of Ontario was presented as a paradigm of western development.

There were other success stories. *Scientific American* reported on the economics of irrigation in 1891, translating the foundation story into facts and figures. It found that each acre to be irrigated, after being purchased from the government for $1.25, had to be grubbed, leveled, and plowed, at an average cost of $14.85. After that, "the average first cost of bringing the water to land in Utah" was $10.55 per acre. These expenses of $26.65 compared quite favorably with the $19 per acre average annual value of crops. Farms with such yields were worth $84.25 per acre. If a farmer sold immediately after improving his land, he could realize a 200 percent profit.[81] The essential elements of the foundation story were thus authenticated. No matter who recounted or acted out the irrigation narrative, it was more overtly technological than the narratives based on the pioneer with his axe. Yet in exchange for this greater investment, the farmer would become independent of the weather. Bumper crops were guaranteed. Little wonder that irrigation was regarded as a farming bonanza, comparable to boomtown real estate. Even with little federal assistance, by the end of the century 100,000 farmers had used streams, small rivers, and wells to irrigate 7.26 million acres. In 1889 their crops were worth $84.4 million[82]—$20 million more than their irrigation systems had cost. Colorado was the most extensively irrigated state, with 1.6 million acres, followed by California with 1.45 million. Yet California farmers were producing more than twice as much value per acre, because many of them grew expensive fruits and vegetables for nearby urban markets.

These small irrigation projects whetted the appetite for larger projects, some of which could be realized by joint stock companies.[83] The Census Bureau found that the "first great 'boom' in irrigation construction occurred in the late eighties and early nineties, when many large enterprises were undertaken by promoters who hoped to profit by the increase in land values created by irrigation."[84] Their platting of new towns and new farms was modeled on railroad practices, and the mania for land speculation reappeared. Unlike railroads, irrigation projects demanded longer-term investments, and many feasible schemes went bankrupt because they failed to produce a rapid enough return. It often took two decades before an irrigation district was fully working. By 1894 the private boom was largely over, even though Congress passed the Carey Act (which gave each western state 1 million acres in public lands, to be used to stimulate irrigation). Little acreage was irrigated as a result of the Carey Act until after 1900. Nevertheless, between 1889 and 1899 individuals added a yearly average of 389,000 new acres to the irrigation system.[85]

The press hailed these developments as portents of an enormous change imminent in the West. Leading the chorus of boosters was William Smythe, whose many articles in national magazines during the 1890s were expanded into a book titled *The Conquest of Arid America* (1905). Smythe saw God's intentions in the landscape, and declared: "He depends on man, working in partnership with Him and in harmony with the laws of the universe, to bring the world to completion."[86] This was particularly the case in the desert lands of the West. "There are conditions in Arid America which make men peculiarly conscious of their partnership with God or Universal Purpose. They seem, indeed, to begin where God left off and to go forward with the actual material creation of the world. Here, the English-speaking race entered into a new environment. Nature had done what it would, then withdrawn and left its unfinished task to the ingenuity of man." The apparent wasteland of the West was not a land that God had forgotten. Rather, "above all other sections Arid America is the God-remembered land" and the "place where man should become supremely alive to his divinity."[87] In the desert, men would be "driven by the club of necessity into a brotherhood of labor" and forced to develop "the conscious partnership of the universe." Rather than despair at the sight of western aridity, those imbued with the idea of second creation believed that if they followed the "torrential stream to its mountain sources" they would "discover the reservoir sites which nature provided at the right elevation to command the valley and to furnish power with which to bring the hidden water from the bowels of the earth."[88]

Smythe saw technology as a means to spiritual growth and economic prosperity: "The man who works intelligently in creating his irrigated farm with the raw materials of land and water, knows that in this smaller sphere he is engaged in finishing the world. He feels himself to be an instrument in the process of evolution."[89] Just as Zachariah Allen had once proclaimed New England's lack of coal a blessing, Smythe found the arid West an ideal landscape for the realization of the second creation: "It is the fortune of Arid America to be so palpably crude material that it can not be used at all, save upon the divine terms."[90] If Smythe was the most strident of the irrigation visionaries, the editor of the movement's first journal, and the convener of its first national conferences, he was by no means alone. What Kevin Starr observed of California was true of the West as a whole: "Prophesying the bringing of water to the desert and to the cities on the plain, they saw themselves embarked upon a work of social redemption biblical in metaphor and suggestion."[91] Indeed, the Rio Grande Western Railroad even distributed maps showing "the striking similarity between Palestine and Salt Lake City, Utah."[92] It seemed that irrigation had built a new Jerusalem.

Three novels imbued with the rhetoric of second creation, retold the history of irrigation in California. The first, Frank Lewis Nason's *The Vision of Elijah Berl* (1905) was loosely modeled on the success of Riverside, California, whose irrigation system had opened in 1871. Nason's story was a classic tale of technological origins. As in the narratives of the axe, the mill, and the railroad, human innovation transformed a landscape, making it productive and enriching the founders. Mary Austin's *The Ford* was more nuanced; it looked affectionately at the ranches and the pastoral life

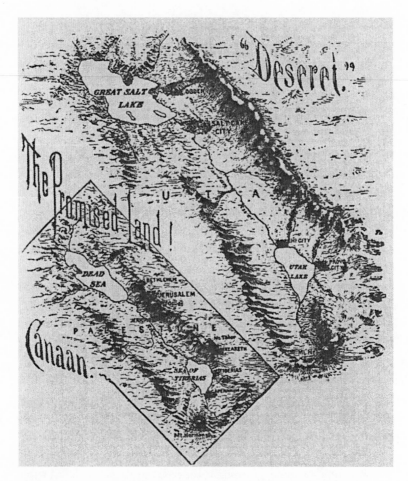

Figure 9.1
Map showing similarity between Promised Land and Salt Lake City area. Distributed by Rio Grande Western Railroad; reprinted in Smyth, *The Conquest of Arid America.*

which irrigation replaced. Harold Bell Wright's *The Winning of Barbara Worth* (1911) examined the shift from such small-scale and local developments to the larger investments and organizational requirements needed to divert a part of the Colorado River into the California desert.

Nason's book opens with a rancher offering a young engineer partnership in a project he has studied for years. The young man muses: "The building of a great storage dam in the mountains, the layout of canals that should lead the stored waters to the sun-parched deserts; this was an engineer's work, and he was an engineer." He imagines "the bare brown hillsides clothed in verdure and teeming with prosperity."[93] The irrigation scheme is presented as a sober investment, in contrast to land speculations in a nearby town. The engineer makes careful surveys and cost estimates; his partner, Elijah, proves that oranges will grow in the local climate, and lines up investors. He even develops "a variety of seedless orange which had been hitherto unknown." (Indeed, California fruit growers had done this.[94]) Elijah begins to purchase lands from poor Mexican farmers that will increase in value from $1 to $1,000 per acre if the irrigation work succeeds. The focus of the story is not the building of the dam, canals, and irrigation system but Elijah's corruption by land speculation and financial intrigue. In contrast, the engineer's work is "inspiring, exhilarating." He conceives it less as "the conquest of Nature" than as "higher Nature asserting itself, selecting and assimilating that which had hitherto been uncalled into active existence."[95] The canyon in the San Bernardino Mountains is presented as a "natural reservoir" awaiting human use. The engineer tells a friend: "We are really only doing here what nature herself did and then undid. You can see that this valley was once a great natural lake."[96] In the novel, as in Smythe's rhetoric, man and nature are in partnership, and the engineering work completes the design latent in the landscape.

"Man" was understood to be white. To Smythe the promoter and Nason the novelist, Mexican-Americans seemed shiftless, lazy, and unimaginative. They had failed to realize the land's potential, and even when the irrigation works are underway right in front of them they do not grasp the change. Like the Native Americans in the narratives of the axe, mill, canal, and the railroad, the dispossessed are shadowy figures whose lives are unexplored and whose voices remain mute. Neither Smythe nor Nason expresses sympathy for their plight. As the latter puts it, before irrigation California was "a country which had been neither developed nor appreciated by its natives and early settlers."[97]

In *The Ford*, Mary Austin focuses primarily on these precursors. In her pastoral vision, whites and Mexican-Americans live harmoniously in Arcadian simplicity. Austin's novel depicts two forms of energy exploitation: drilling for oil (which leads farmers into land speculation and get-rich-quick schemes) and irrigation. Oil wells offer spectacular rewards to a few people in the valley. In contrast, the sheep and cattle ranchers in the mountains only gradually awaken to the possibilities of irrigating their land. Austin describes "a little river, but swift and full, beginning with the best intentions of turning mills or whirring dynamos, with the happiest possibilities of watering fields and nursing orchards, but, discouraged at last by the long neglect of man, becoming like all wasted things, a mere pest of mud and malaria."[98] As in Smythe, the river is given intentions. The landscape wants to be a partner with man. The river is not merely available to run mills and dynamos; it wants to run them, just as it wants to irrigate farmland. The stream has feelings. It becomes "discouraged" and wastes away into disease and mud. "Not but what it did its best with such opportunities as were offered it. The Caliente Ditch, taken out above the branding-pens, watered a great green oasis of alfalfa, and the 'Town Ditch' turned the original purlieus of a Spanish roadhouse into a green, murmurous hive. . . ."[99]

In Austin's novel, as in Nason's, natural resources are waiting to be used to their full potential. The little river is presented as young and unfulfilled, and it remains a potential force more than an actual one until the last pages of the novel. In a characteristic moment, two characters "[mark] the shining waste of the river, and [hear] as always the clear call of the empty land to be put to human use."[100] They "hear" a gendered call for help from a feminized landscape. In another scene, a father looks "at that almost untouched valley as a man might at his young wife, seeing her in his mind's eye in full matronly perfection with all her children about her."[101] The same image of a man marrying the land and fructifying her appears elsewhere in the novel. During a drought, a developer named Elwood begins to buy options on land, whose "voice . . . had spoken to him in terms of canals, highways, towns, so that what to the Pierra Longans [the local residents] had been in the nature of an enslavement, had become to Elwood the clear call to realization. It was as if they had cherished all these years, in the hope of what the valley might become, a very noble and lovely lady, too exalted for any of them to mate with, but who yet might be persuaded to look favorably on this more accomplished suitor."[102] Austin sees irrigation as the consummation of humanity's marriage with the earth. She reimagines the early phase of irrigation, telling a story in which enterprising individuals confront a problem presented by nature and resolve it in partnership with the land.

In *The Land of Little Rain,* Austin wrote: "It is the proper destiny of every considerable stream in the West to become an irrigating ditch. It would seem the streams are willing. They go as far as they can, or dare, toward the tillable lands in their own boulder fenced gullies—but how much farther in the man-made waterways."[103] By giving the river agency, Austin made it a historical actor. Man and nature work in partnership, man helping the river to complete what it would like to do. This partnership is local and is based on an intimate long-term relationship between humans and the landscape. *The Land of Little Rain* idealizes not individualistic homesteaders arriving in the desert from the East, but a Mexican-American agricultural community, El Pueblo de Las Uvas. Its agriculture is based on irrigation, but not aggressively so; some of the water seeps away into *tulares* (bulrushes) and much of it is taken by cottonwood trees. Every house has a garden plot, and there grow "in the damp borders of the irrigating ditches clumps of *yerba santa,* horehound, catnip, and skienard, wholesome herbs and curative."[104] The town is based on subsistence, not growth. Nominally a part of the United States, it celebrates the Mexican rather than the American Independence Day, and in no way does it resemble an Anglo town: "At Las Uvas every house is a piece of earth—thick walled, whitewashed adobe that keeps the even temperature of a cave. . . ."[105] Ending *The Land of Little Rain* with a portrait of this town, Austin presents it as the ideal that results from cooperation between man and nature.

Austin's second creation is quietly subversive of the Anglo-American technological creation story, for it is small-scale, communal, and largely self-sufficient. In contrast to new Anglo communities carved out of "virgin land," Las Uvas originated with a silver mine, which drew people into the region but then closed. An accidental by-product of this failure, the town has been built not by driving Yankee enterprise but by "those too idle, too poor to move, or too easily content with El Pueblo de Las Uvas."[106] Austin evoked an Arcadian existence that was being swept aside by irrigation. If she still saw water's "destiny" in an irrigation ditch, she wanted it to be a ditch owned by the local community, not by a private corporation or the federal government.

Both Austin and Nason saw irrigation as a "natural" extension of creation, and both celebrated projects initiated by local residents. In contrast, Harold Bell Wright's 1911 novel *The Winning of Barbara Worth* describes the transitional moment when eastern corporate capital enters the scene and competes with local entrepreneurs. (In 1926 the book, a best seller, was made into a feature-length Hollywood film starring Vilma Banky, Ronald Colman, and the then-unknown Gary Cooper.) Still in print today, *The Winning of Barbara Worth* is based on the Imperial Valley project, which diverted water from the Colorado River into the desert of eastern California. Though Wright's heroic engineer does struggle against the rampaging Colorado, the major conflict in the novel is not between man and nature but between two forms of capitalism.

Wright's novel dramatized the enthusiasm for controlling the Colorado River, which had begun as early as 1853, when a railway survey party determined that it was possible to irrigate the Colorado Desert of eastern California.[107] In 1875, Eric Bergland, a West Point engineer, studied the desert area of southeastern California on the western side of the river, near the Mexican border.[108] Bergland had spotted the most lucrative opportunity for private enterprise, and in the 1890s an underfinanced and quarreling team of investors in the California Development Company discovered a way to take Colorado River water through an old channel into the million acres of its former delta.[109] Under the direction of the same George Chaffey who had created Etiwanda and Ontario, a gravity-feed canal that carried water 60 miles into the desert was built at a cost of only $100,000. It began to silt up almost before settlers had their first crops, but it opened the sluice gates to development and, soon after, to disaster.

The investors bought little desert land, for the value inhered in the water. Homesteaders who flocked into the area could obtain water only if they were ready to sign contracts that demanded regular payments to the Irrigation Company. This scheme demonstrated the accuracy of Powell's early realization that water rights alone created value in land and that ownership of these rights created political and economic power. Although it seemed that the California Development Company had made possible a new form of homesteading based on irrigation, in practice ownership of land rapidly concentrated in a few hands. Often homesteaders overextended themselves, for the leveling of desert in a fiercely hot desert demanded much of any farmer and his horses. Many acquired more land (and more debt) than they could manage. A single year of poor agricultural prices, higher interest rates, or unexpected personal expenses could force

Figure 9.2
"Subduing the Desert, Imperial Valley, California," c. 1913. Courtesy Library of Congress.

a family into bankruptcy. Sixty years after the project began, 90 percent of the more than 1 million irrigated acres in Imperial Valley would belong to only 800 individuals and corporations.[110]

This long-term result contradicted the Jeffersonian rhetoric about family farms; however, such a perspective was impossible in Wright's novel, which celebrated the recent completion of the irrigation project. Wright's fictional Irrigation Company was going "back into that land of death to save that land from itself."[111] There seemed to be no contradiction between redeeming the landscape and being "the advance force of a mighty army ordered ahead by Good Business—the master passion of the race."[112] The surveying party feels that soon the "main army would move along the way they had marked to meet the strength of the barren waste with the strength of the great river and take for the race the wealth of the land."[113] In this effort, the central figure is Jefferson Worth, the local banker, of whom Wright admiringly declares: "He was Capital—Money—Business incarnate."[114] With a name that evokes both Thomas Jefferson and wealth, Jefferson Worth is the "good capitalist" who invests money not merely to make a profit but also to improve society.

Worth seeks to transform "hundreds of thousands of acres equal in richness of the soil to the famous delta lands of the Nile."[115] The sequence he imagines is precisely that of the foundation story. First come the surveyors, then irrigation canals, then farmers who level the ground. "The fierce desert life would give way to the herds and flocks and the home life of the farmer." To help this land to grow further, "the railroad would stretch its steel strength into this new world, towns and cities would come to be where now was only solitude and desolation, and out from this world-old treasure house vast wealth would pour to enrich the peoples of the earth. The wealth of an empire lay under the banker's eye, and Capital held the key."[116] When news comes that the railroad will be built, "from every side the swelling flood of life poured in. Every section of the new lands felt the influence of the rush . . . every tract was seized by incoming settlers. Townsite companies quickly laid out new towns, while in the towns already established new business blocks and dwellings sprang up as if some Aladdin had rubbed his lamp."[117] Thus the irrigation narrative incorporated the railroad foundation story, just as the railroad narrative had absorbed the stories of the axe and the mill.

The Winning of Barbara Worth marginalizes the engineering problems involved, focusing on the struggle between capitalists. Yet selling water was not the same thing as controlling the river. In 1905 the Colorado changed course and discharged most of its flow—90,000 cubic feet of water per second—into the Imperial Valley, much of which lay below the level of the river. It began to refill an ancient lake, the Salton Sea. The California Development Company failed four times to shift the Colorado back into its channel to the Pacific. Verging on bankruptcy, it was forced to sell out to the Southern Pacific Railroad, which in 1907 spent an additional $3 million to contain the rampaging river. Two years of uncontrolled flooding demonstrated conclusively that private enterprise lacked the expertise and the resources required to construct and maintain large water projects. Wright's novel devotes comparatively few pages to this crisis, however, and does not draw this conclusion. Instead, it suggests that failure to control the Colorado revealed the inability of eastern capital to understand western conditions.

Future dams and canals would not be private matters. Even before the spectacular mistakes of the California Development Company, the federal government had entered the irrigation business. In the twentieth century, the dominant role would be played not by local capitalists, such as Jefferson Worth, but by the Bureau of Reclamation.

10

Water Monopoly: Federal Irrigation and Factories in the Field

Tomorrow the Colorado River will be utilized to the very last drop. Its water will convert thousands of additional acres of sagebrush desert to flourishing farms and beautiful homes for servicemen, industrial workers, and native farmers who seek to build permanently in the West.
—Department of the Interior, 1946[1]

The land fell into fewer hands, the number of dispossessed increased, and every effort of the great owners was directed at repression.
—John Steinbeck, 1939[2]

As these contradictory statements suggest, both the narrative of state-sponsored irrigation and its counter-narratives belong largely to the twentieth century. As with the canals and railroads of earlier generations, the enormous expense was matched by the gigantic scale of ambition: to subdue, coordinate, and develop entire river systems. Many factors recalled railway development, notably intensive political lobbying, land speculation, lack of respect for Native Americans' rights, and concentration of economic power in a few hands. Even more than a railroad, an irrigation company made land valuable. Indeed, railroads themselves owned extensive land in dry areas that could be irrigated, and the same capitalists often invested in both. Yet even railroads balked at the high initial costs of building large dams. Furthermore, the larger the project, the more state and local authorities unavoidably became involved. Both economics and politics underlay the second form of the irrigation narrative, in which the state (rather than individuals) initiated the process of transforming the land.

As had been the case with the axe, the mill, the canal, and the railroad, counter-narratives immediately challenged state-funded projects. Three different forms emerged. The first looked at the human costs of irrigation, demonstrating the gap between the rhetoric of family homesteading and the reality of agri-business based on huge land-holdings. The second emphasized the displacement of white ranchers, Native Americans, and Hispanic Americans. The third focused primarily on environmental problems.

In the 1890s, irrigation remained a mirage throughout much of the West, an unrealized possibility that shimmered in the desert sun. The success in Mormon Utah and in Southern California's fruit-growing communities suggested that Americans could conquer much larger areas if they dammed the major rivers that flowed "unused" to the sea, notably the Colorado, the Rio Grande, the Sacramento, the Columbia, and their many tributaries. As early as 1891, the United States Geological Survey had examined potential reservoir sites and concluded that substantial areas could not be tapped through local or individual efforts. "Simple farmers' ditches," *Scientific American* observed, "are totally inadequate, but competent engineering skill must be called upon to collect and distribute a material proportion of the immense supplies of hitherto unused water often coursing in destructive floods from our great Western mountain system."[3]

Once investors and promoters realized that they did not command the necessary resources, they held a series of National Irrigation Congresses and lobbied the federal government for help. The congress of 1893 proposed various schemes to transfer federal land and resources into state or private hands. After the 1894 congress (held in Denver), a local newspaper concluded that investors wanted to "gobble the arid lands and make of them vast landed estates with the ill concealed object of turning the Western farmer into a tenant or a peon." It appeared that the National Irrigation League "was organized for the sole purpose of manufacturing public opinion as a lever to be used on Congress to force the cession of the arid lands."[4] Such populist suspicions, widespread at the time, were not groundless. The *Chicago Times* also feared private monopolies; it called for the federal and state governments to build dams, canals, and irrigation systems and then rent them to farmers.[5]

Yet the National Irrigation League was a more complex organization than its opponents asserted. Its promoters and speculators also included idealists, such as William Smythe, who had founded it. Smythe advocated government control of water rights and small family farms of 40 acres for people of modest means. Several cooperative communities were modeled on Smythe's ideas, including the town of San Ysidro on the Mexican border near San Diego.[6] Some members of the National Irrigation League were speculators bent on achieving the monopolies anathema to populists, but others sincerely believed in irrigated homesteading. A centerpiece of California's exhibit at the Chicago World's Fair of 1893 was "a model of the great irrigation system that has reclaimed the desert" in the Central Valley.[7] Westerners wanted to reinvent their landscapes, and they demanded first state and then federal assistance.

Before government stepped in, however, a landmark Supreme Court Case was decided in 1896. The legal issue was the constitutionality of California's Wright Act, unanimously passed by the state legislature in 1887.[8] Supported by the small farmers who expected to benefit from it, the Wright Act provided that local governments could create irrigation districts "upon application of a majority of the owners of lands susceptible of a uniform mode of irrigation from a common source." In practice, this meant that many small landholders outvoted the large owners, which looked like democracy to the former but seemed "communistic" to some of the latter. When an irrigation district was created, government could exercise the right of eminent domain to acquire the water rights and land needed to build dams and canals, often taking some property from large landowners who had voted against the idea. To create a district required an election in which two-thirds of all property owners, including non-farmers, had to vote in favor. The expectation was that pooling individual water rights would rationalize and extend the existing piecemeal irrigation systems.

The Supreme Court of California found the Wright Act constitutional. The Ninth Circuit Court reversed this decision, however, because it permitted the taking of property to benefit some landowners but not all. Every citizen had to pay taxes to support the water districts, but only irrigators would be assured water. The Supreme Court reversed the Circuit Court's decision, however, and held that the existence of millions of acres of arid land in California made their exploitation a public use, even though this would not be true in a state with better rainfall. In effect, the Supreme Court found that the law varied depending on the weather. Confiscation of property to benefit private irrigation became legal. The *San Francisco Call* was delighted: "Irrigation is now declared to be constitutional wherever it is necessary for the cultivation of the soil, and the legislature and the courts of each State are the judges of that necessity."[9] The *Chicago Times Herald* accurately predicted: "This decision will stimulate similar legislation in all the arid and semi-arid region."[10] In the following two decades most western states passed laws that resembled the Wright Act, despite the fact that only twelve of the first fifty water districts formed in California remained solvent by 1910.[11]

In accepting this law, the Supreme Court admitted the inaccuracy of the individualistic irrigation story. It declared that farmers were unable to develop irrigation on their own, and needed state assistance or cooperative organization. The *St. Paul Globe* concluded: "This is one of the cases where the State has to step in self-defense. The work is too great for the individual. No man but a millionaire could afford to construct the great ditches and canals necessary to bring water from some distant point to his particular holding." The danger lay not in the state exercising its powers of eminent domain to acquire water rights. Rather, it lay in the grasping monopolies which attempted "to gobble every water-course and water-right within its reach and hold them at extortionate prices. . . . Inasmuch as the supply [of water] is absolutely limited, and as it would take millions of dollars to contest supremacy, this amounts to putting the whole future of the arid belt into the hands of greedy corporations."[12] Irrigation districts were widely approved as a progressive reform that would encourage family farms, protect the farmer's access to water, and curb monopoly. In practice, however, many water districts were poorly run. They had difficulty selling their bonds, which they often traded for water rights, leaving them underfinanced. Administrators frequently lacked engineering expertise or experience with large projects, and construction typically cost more than estimated. When litigation and legal fees were added to their expenses, little wonder that in 1910 almost none of $8 million in debt had been paid off.[13] The Wright Act had been declared constitutional, but that did not make it practical.

Meanwhile, the scale of private irrigation projects continued to grow. Speculators and large landowners enlisted new allies—notably James Hill, whose railroad passed through the arid and largely undeveloped northern plains, and the Union Pacific, Southern Pacific, and Santa Fe railroads, which served the even drier Southwest. Together they quietly funded a major lobbying effort during the 1890s, working together with western senators and congressmen to gain passage of the Reclamation Act of 1902.[14] After the public backlash of the 1870s against land giveaways, the railroads had learned not to be too vocal in supporting schemes that benefited them. Instead, like proponents of the Wright Act, their public relations emphasized the need for family farms. They called for expansion of the United States within its own borders, rather than for the acquisition of imperial possessions abroad. By 1902 the National Irrigation League had learned to sell its message by evoking the "narrow limitations of city life" that sapped the American character: "The man who rears his sons and daughters in the rural life of our irrigation empire will give them a better chance to become useful men and women than boys and girls will have when raised in the city."[15] The foundation narrative of irrigation was presented as homesteading with guaranteed water plus comforts unknown to earlier pioneers, such as the telephone and electricity.

Theodore Roosevelt's speeches articulated the irrigation creation story. Roosevelt, who had traveled widely in the West and had lived for a time in the Dakotas, was acutely aware of the problems of drought. In December 1901, in his first message to Congress as president, he declared:

In the arid region it is water, not land, which measures production. The western half of the United States would sustain a population greater than that of our whole country today, if the waters that now run waste were saved and used for irrigation. The forest and water problems are perhaps the vital internal questions of the United States. . . .

Great storage works are necessary to equalize the flow of streams and to save the floodwaters. Their construction has been conclusively shown to be an undertaking too vast for private effort. Nor can it be best accomplished by the individual states alone. Far-reaching interstate problems are involved; and the resources of single states would often be inadequate. It is properly a national function, at least in some of its features. It is as right for the national government to make the streams and rivers of the arid region useful by engineering works for water storage as to make useful the rivers and harbors of the humid region by engineering works of another kind.[16]

Roosevelt wanted to put water within the reach of homesteaders, and he worked with a coalition of western congressmen from both parties to create the Reclamation Service. Its very name suggested the idea of a second creation. "Lost" land was to be reclaimed for human use. The 1902 law established the Reclamation Fund, which could be "used in the examination and survey for and the construction and maintenance of irrigation works for the storage, diversion, and development of waters for the reclamation of arid and semiarid lands"[17] under the direction of the Secretary of the Interior. The scope was broad, including dam construction, canals, irrigation ditches, and artesian wells.

Though enthusiasts saw federal dam building as a way to project Jefferson's nation of small farmers into a new geographic context, western water development in practice seldom sustained yeoman farmers. The start-up cost of irrigation farming was high, the land accessible to water was limited, and political influence and economic advantage were closely linked. Irrigation favored those with capital and good political connections, and it typically encouraged large landholders or cooperative groups. These problems had been evident before, but the Reclamation Bureau was expected to solve them and open a new era for the West. In 1905, *National Geographic* enthusiastically reported:

There is now lying in the Treasury vaults the sum of nearly $30,000,000, which is reserved exclusively for the government irrigation projects of the West. This immense sum has been realized during the past three and one-half years from the sale of public lands, and the amount is increasing daily at a very rapid rate. Work has already been begun on eight great projects which will make gardens of nearly one million acres, an area equal to the State of Rhode Island and probably capable of generously supporting a population of several million people.[18]

The illustrated article described the "giant dam 240 feet high and costing $3,000,000" planned across the Salt River to irrigate 200,000 acres near Phoenix. Photographs showed other proposed dam sites in Colorado, Idaho, and North Dakota. It was emphasized that the costs would be repaid in only 10 years and that "government irrigation works cost the people considerably less than works built by private corporations."[19] The anonymous author concluded enthusiastically: "The reclamation law is working admirably. It is elastic and equally fair to all sections."[20] (Just how it was fair to farmers in Ohio or Virginia was not specified.)

The Salt River Valley in Arizona exemplifies both the problems of scale that private irrigation had encountered and their solution through federal intervention. The Salt River was well suited to irrigation. It fell rapidly, and water could be diverted at almost any point. White settlers had built twelve diversion canals between 1868 and 1889, but their capacity was far greater than the river could supply during the hot months when water was most needed. The potential solution to periodic shortages lay upstream, where the river ran through a deep gorge that could be dammed, but the high cost and the engineering problems were too great for local investors.

Immediately after its formation, the Reclamation Service [later the Bureau of Reclamation] planned a large dam on the Salt River, to be named after Roosevelt. When a flood damaged the existing canal system in 1905, the Service bought out the private interests and developed a more comprehensive system. Roosevelt Dam impounded a lake that covered 16,832 acres and irrigated a larger area than previously served, as well as supplying farmers with electricity.[21] The Salt River Project demonstrated both the advantages of large-scale planning and the expertise of the Bureau of Reclamation. Federal planning and engineering had replaced uncoordinated and underfunded private efforts.[22]

In this new version of the foundation story, government intervened only to provide an essential technical infrastructure; it then stepped back to allow market forces a free hand in calling the second creation into being. The structure of the story was otherwise the same as it had been when a solitary man entered the woods with his axe. In each case the farmer was expected to improve the land and become prosperous and independent. Yet the earlier pioneer in Ohio, Kentucky, Indiana, or Missouri had not paid for every drop of the water he needed, but could reasonably assume enough rain would fall. Homesteading by irrigation created debt and dependency. It required a large number of benevolent government agencies, which would build dams, regulate water supplies, protect forests, and otherwise supervise the environment. The work of these agencies was interdependent. For example, civil engineers knew that forests held water and helped regulate stream flow. Conservation of wooded areas not only became government policy in the new Bureau of Forestry, but it was good policy for water conservation as well. Though at their inception these agencies had a utilitarian vision of their role, their very presence curbed individual initiatives. The irrigator, unlike the homesteader in Illinois or Indiana, was largely dependent on federal agencies, even if the rhetoric of settlement remained that of independent homesteading.

Figure 10.1
Roosevelt Dam and Lake, c. 1915. Courtesy Library of Congress.

Irrigation on a regional scale also required rethinking the ways in which homesteading could be carried out. Cutting up the land into perfect squares made little sense in terms of watershed. Echoing John Wesley Powell a generation before, Frederick H. Newell commented: "The rectangular system of division of the public lands, while one of the most beneficial measures leading to the settlement of the Ohio and Mississippi Valleys, has been found to be detrimental to the best growth of the western two-fifths of the United States. . . . Attention has been concentrated upon land titles, and great care has been exercised in the survey and marking of boundaries . . . while the water, which alone gives value, has hardly been considered, and the rights to its use have often been left to be adjusted largely by local or temporary expedients."[23] The Bureau of Reclamation tried to remedy this, but existing land divisions and water laws interfered with its task. The national survey had chopped up the West into units too small for grazing but too large for irrigation, and the inherited common law had been devised to ensure equity in the well-watered world of rural England.

A single family could prosper on far less than 160 irrigated acres. In the 1890s, Powell had advocated a maximum of 80 acres and had calculated that a million acres of irrigated land would be sufficient for 12,500 families.[24] Something resembling this vision was realized in a few places, for example the Huntley Project in Montana and the Snake River Project in Idaho. At these sites the Bureau managed to limit landownership to smaller plots. After the Bureau completed the Snake River Project, in 1920, its 630-mile canal system served 121,000 acres that produced alfalfa, potatoes, sugar beets, and garden crops, and made it possible to build up a dairy industry. The dam also supplied the farmers with hydroelectric light and power, at a time when 90 percent of American farmers still did not have electricity.[25] But such tightly knit communities of small farmers seldom resulted from the Bureau's activities.

The first lines in Newell's work on irrigation read: "Home-making is the aim of this book; the reclamation of places now waste and desolate and the creation there of fruitful farms, each tilled by its owner, is its object."[26] Newell was certain that the arid portion of the United States "under good management is capable of sustaining a large population." However, "as the control of the vacant public lands is now tending, these areas are not being made available for the creation of the largest number of homes."[27] Putting it another way, he declared: "It is unquestionably a duty of the highest citizenship to enable a hundred homes of independent farmers to exist, rather than one or two great stock ranches, controlled by non-residents, furnishing employment only to nomadic herders."[28] The question was rhetorical, for most Americans disliked absentee landlords and had no use for nomads. Given "underdeveloped" resources, government should intervene in the interest of "good management" to create a large, permanent class of farmers. The technological creation story had always been justified by future farmers, who would automatically appear, but in its earlier versions they were more self-sufficient and they owned their technologies.

An irrigator could seldom own the technology, however, and government-sponsored irrigation took on more expensive projects. Between 1899 and 1919, the extent of irrigated lands increased 250 percent, from

7.74 million acres to 19.2 million,[29] and the number of irrigated farms doubled to 231,541.[30] However, the cost of federal irrigation water was "about two and one-half times the general average."[31] Between 1900 and 1910 the total cost of irrigation shot up from $66 million to $307 million, an increase of almost 500 percent.[32] The Census Bureau dourly reported: "An effect of the increased cost of a water supply is to limit the class of persons to be drawn upon as settlers for new lands. . . . Persons with little or no capital, who might have settled on irrigated lands under the old conditions, can not do so. . . ." Developers before 1900 had been quite rational in their choice of sites, and they had been able to bring water to the land at a cost between $8 and $9 per acre.[33] The Bureau of Reclamation, in contrast, did not develop gravity feed canals at the bottoms of small valleys; it built "expensive storage works, high diversion dams, difficult tunnels, or long, expensive canal work upon side hills, where large investment was necessary before any water was brought to the land."[34] According to its own estimate, these delivered water at a cost of $55 per acre, 700 percent more than the small-scale, private projects before 1900.[35] An irrigation program intended to promote family farming was "practically limited to the well-to-do."[36] In California, an acre of unimproved land quadrupled in price between 1900 and 1920, to $116, and land ready to irrigate cost much more. A government sponsored project near Chico spent $80 per acre to level the land and prepare it for irrigation, and only accepted settlers who had a minimum of $1,500 in capital.[37] By 1925, experts concluded that farmers needed at least $5,000 to purchase and prepare 40-acre farms.[38] Congress had established a maximum of 160 acres that any individual could irrigate from publicly supplied water, but the Bureau's interpretations of this limit were so relaxed that much larger farmers acquired water at subsidized prices.[39] While few small farmers could afford to be irrigators, large farmers, even those with wells and pumping equipment,[40] often received water from the Reclamation Service. The law intended to promote homesteading instead helped concentrate wealth in a few hands.

Large landholdings preceded irrigation, and many areas, particularly in California, had never been open to small farmers. As early as 1871, just 516 people owned 8.6 million acres of the best California real estate, and much of the rest had fallen under the control of railroad companies.[41]

Lands were monopolized before settlement and developed as enormous ranches, not as the homesteads rhetorically evoked when lobbying Congress. Henry George complained in 1871: "The water of California . . . is threatened with monopoly . . . and already we are told that all the water of a large section of the State is the property of a corporation of San Francisco capitalists."[42] California was "not a country of farms, but a country of plantations and estates. Agriculture is a speculation. The farm houses, as a class, are unpainted frame shanties, without garden or flower or tree."[43] A large class of itinerant *vaqueros* stood in striking contrast to the (often absentee) landowners.

Furthermore, the federal government often did a poor job of selling the lands it had opened for homesteading. By 1917 the Bureau of Reclamation was proud of building 21 projects spread through all parts of the arid West, but some sites went unsold and others never were fully developed. Many would-be irrigators bought too much land. As Arthur Powell Davis admitted, 160 acres was too many for one family to cultivate, and even allotments half that large were often ruinous in practice. Davis estimated that a typical farmer managed to cultivate only 25 acres, but had to pay water charges for 65. Because he had too much land, "he is almost sure to cultivate more than he can properly level and provide with irrigating ditches, so that he obtains inferior results even from the acreage which he does cultivate."[44] Families who settled in anticipation of irrigation commonly spent most of their money before the water flowed into their ditches. Attempting to hold onto large tracts often ended in financial ruin. Overextended families then had to sell out, and water and wealth were soon concentrated in a few hands.

As water costs rose, the competitiveness of western agriculture fell, and settlers went bankrupt. In 1915 an irrigation economist at the Department of Agriculture admitted: "Irrigated lands cannot be expected to repay directly the cost of irrigation works, with interest, as is ordinarily expected of investments generally. Past experience demonstrates, however, that if the loss to original investors is overlooked, irrigation has been a decided success."[45] Irrigation, the argument ran, lost money in the short run but yielded benefits "indefinitely" and not for a mere 20 years.[46] "It amounts to the state making an investment on which it will get returns for all time."[47] By the 1920s, payments for irrigation water were millions of dollars in arrears, and Commissioner Elwood Mead focused on completing existing projects, decentralizing their administration, and forcing farmers to pay operating costs.[48]

Although reclamation did not pay for itself, it intensified in the 1930s. The Army Corps of Engineers and the Bureau of Reclamation had developed parallel programs that often competed for the same projects. Gradually, they adopted different versions of the idea of "multiple uses." This doctrine, which did not take much time to become popular in California, justified dams not only for irrigation but also to control floods, to generate electricity, to create recreation areas, to improve water transportation, and to stimulate regional economies. If the Bureau of Reclamation initially focused on irrigation and flood control, it soon saw the advantages of hydroelectric dams for the operation of pumping systems and began to sell surplus power to farmers. "By the 1920s, power revenues increasingly constituted a significant factor in the economics of reclamation,"[49] and generating electricity became an integral part of watershed development plans. The Tennessee Valley Authority also embraced the doctrine of "multiple use" during the 1930s. In practice, "multiple use" made it possible for government accountants to lower the amount farmers had to pay for irrigation water, since a dam also improved navigation, protected communities from flooding, and provided facilities for swimming, boating, and fishing. "Multiple use" meant something different to the Army Corps of Engineers, which gave a higher priority to river transportation than did the Bureau of Reclamation, whose dams ended any possibility of sailing up and down the Colorado. In either case, "multiple use" was a narrative that justified the plans of Washington bureaus, even if they made no economic sense from a strictly agricultural standpoint. "Multiple use" shifted the focus away from the farmer's creation story. Instead, the government projects reengineered entire regions.

The most famous of these irrigation projects began with Hoover Dam, planned in the 1920s and built in the 1930s. At completion, it was the largest man-made structure on earth, containing 6.9 million tons of concrete. Its intake towers are as tall as skyscrapers. In building it, the government took the first step toward controlling the entire Colorado River. The dam impounded the extremely irregular flow of the river in a vast lake. It regulated discharges and made it safe to construct a series of smaller downstream dams to siphon off water for irrigation in Arizona and California. (A good deal of the water eventually would be pumped over the mountains to reach Los Angeles and sold there for less than it cost to deliver.)

President Franklin D. Roosevelt came to Nevada to dedicate Hoover Dam. On the last day of September 1935, he celebrated "the completion of the greatest dam in the world" and "the creation of the largest artificial lake in the world" as achievements of the New Deal. The president's speech was an exemplary statement of the technological foundation story. He began by emphasizing that the site had been barren and uninhabited:

Ten years ago the place where we are gathered was an unpeopled, forbidding desert. In the bottom of a gloomy canyon, whose precipitous walls rose to a height of more than a thousand feet, flowed a turbulent, dangerous river. The mountains on either side of the canyon were difficult of access with neither road nor trail, and their rocks were protected by neither trees nor grass from the blazing sun. The site of Boulder City was a cactus-covered waste. The transformation wrought here in these years is a twentieth-century marvel.[50]

As with the private construction of canals or railroads, intervention in the natural order was expected to open up new land to agriculture, stimulate new towns, and provide surplus electrical power, for the "millions of others who will come to dwell here in future generations."[51] Engineering had remedied nature's inadequacies, and the dam stood as "another great achievement of American resourcefulness, American skill and determination."[52] Before government had intervened, "the mighty waters of the Colorado were running unused to the sea. Today we translate them into a great national possession."[53]

After World War II the Reclamation Service set out to complete this series of dams and canals. It emphatically endorsed the improvement of nature in a "Comprehensive Report on the Development of the Water Resources of the Colorado River Basin for Irrigation, Power Production, and Other Beneficial Uses in Arizona, California, Colorado, Nevada, New Mexico, Utah, and Wyoming." The report was subtitled "A Natural Menace Becomes a National Resource." The foreword began: "Yesterday the Colorado River was a natural menace. Unharnessed, it tore through deserts, flooded fields, and ravaged villages. It drained water from the mountains and plains, rushed it through sun-baked thirsty lands, and dumped it into the Pacific Ocean—a treasure lost forever. Man was on the defensive. He sat helplessly by to watch the Colorado River waste itself, or attempted in vain to halt its destruction."[54] The Colorado was socially maladjusted. It was a thief on the rampage, misusing water in floods, denying it to "thirsty lands," then throwing its treasure away. In this drama, the river robs the land, and man is helpless. Presenting the Colorado as a misguided being suggested the possibility of reform. The Report continues: "Today this mighty river is recognized as a national resource. It is a life giver, a powerful producer, a great constructive force. Although only partly harnessed by Boulder Dam and other ingenious structures, the Colorado River is doing a gigantic job." Its energies have been rechanneled: "Its water is providing opportunities for many new homes and for the growing of crops that help to feed this nation and the world. Its power is lighting homes and cities and turning the wheels of industry. Its destructive floods are being reduced. Its muddy waters are being cleared for irrigation and other uses."[55] Rationality had controlled and clarified the stream and made it work for America.

Government engineering projects, like the foundation stories of the axe, the mill, the canal, and the railroad, projected automatic economic development into the future. The aforementioned report also celebrated the final result as man's second creation:

Tomorrow the Colorado River will be utilized to the very last drop. Its water will convert thousands of additional acres of sagebrush desert to flourishing farms and beautiful homes for servicemen, industrial workers, and native farmers who seek to build permanently in the West. Its terrifying energy will be harnessed completely, to do an even bigger job in building the bulwarks for peace. Here is a job so great in its possibilities that only a nation of free people have the vision to know that it can be done and that it must be done. The Colorado River is their heritage.[56]

Transforming the river into "heritage" meant more than conceiving of it in utilitarian terms. Building America and its "bulwarks for peace" through water control meant more than creating beautiful farms and homes for servicemen. Only a free people, it seemed, could sustain such a technological vision and see it as a historical necessity. Harnessing its "terrifying energy" was a part of the American tradition of conceiving natural objects as resources available for a reason. Americans tended to assume that natural facts were providential, including the absence of coal but abundance of water power in New England, the opening through the Appalachians for the Erie Canal, the gentle west-to-east gradient on the plains (so convenient for irrigation), and the presence of the Colorado River in the southwestern desert.

The Colorado was more than wild nature to be subdued or a wasted energy source to be rationally developed. Its complete transformation was a part of America's historical fate, an actualization of its Manifest Destiny. Only a free people could see that they must create a second nature. To realize their freedom, they had to recognize, paradoxically, that they had no choice. The river was "theirs" not only in the sense of its location (primarily) within the United States, but also in a teleological sense. Its domination was a part of the larger technological creation story of western expansion. Though the rest of the official report does not maintain a lofty rhetorical tone, all of its statistics, maps, and projections were predicated on white migration into this region as a foreordained necessity. Similar plans had been expressed in one way or another by every Secretary of the Interior since the 1890s. Second creation seemed to be the logical and inescapable destiny of a free people.

Yet, as many critics soon realized, the result contradicted the technological foundation story. Expensive dams intended to serve homesteaders primarily supplied inexpensive water to large landowners and electricity for distant cities. Federal irrigation was not individualistic but collective, not small-scale but massive. It was controlled not at the private, the local, or state level, but in Washington. It was not guided by the invisible hand of the market; it required planning and central control. Instead of fostering farming communities, it often assisted monopolies and large estates. Instead of many small farms of roughly equal size, irrigation reinforced the concentration of land in a few hands. Thus, the experience of state irrigation could not easily be assimilated to the earlier foundation narratives, which emphasized individual action. A federal bureaucracy replaced the solitary pioneer with an axe or the individual miller. The government, not an individual or a private group, entered an undeveloped region and deployed new technology to transform the land for multiple uses.

The underlying presuppositions of the irrigation story contradicted earlier technological creation stories. The irrigation narrative began with water scarcity, not with the natural abundance assumed in the narratives of the axe, mill, and railroad. Because water was scarce, the geometrical division of land into uniform parcels no longer made economic sense, because western spaces were far less uniform and interchangeable than those in Illinois or Iowa. Nor was irrigation a part of a free market in the sense that Americans had understood it in 1830 or 1860. A tiny minority controlled most of the good land, the federal government and water companies controlled the water, and the railroads controlled shipping. Farmers who managed to survive had few illusions of sturdy independence in a free market. The populist revolt expressed widespread frustration that land, water, and transportation monopolies thwarted individual effort. Thus the irrigation story violated three of the unspoken assumptions that underlay the earlier foundation narratives. All that remained was the conception of increasing human control over natural forces in a Newtonian universe of cause and effect. The irrigation story relied on mastery and rechanneling of enormous forces as the way to overcome water scarcity. It redefined the free market to include the Bureau of Reclamation and the Army Corps of Engineers, and it rewrote the geometry of the national survey to accommodate dams, canals, ditches, and farms much larger than 160 acres. If economic growth occurred once whole regions were irrigated, the profits did not trickle down to many individual farmers.

The class of landless farm workers grew, but its composition changed. At first, owners recruited white men from the East, but they abandoned the fields once they realized they had little chance of becoming independent farmers. In the late nineteenth century, Chinese and later Japanese workers were recruited from abroad, but the white majority so resented and feared them that restrictive immigration laws were passed to cut off this supply. Other Asian workers came instead, but gradually growers turned primarily to Mexican labor. Racial inequality reinforced economic inequality. California in particular had a great mass of migrant workers.[57] By the end of the 1930s, popular counter-narratives exposed the failure of federal programs to act out the foundation story. Carey McWilliams wrote one of the most persuasive attacks, *Factories in the Field: The Story of Migratory Farm Labor in California*. It detailed the monopoly of land and water; the importation of foreign workers from Japan, China, the Philippines and Mexico; and then the gradual replacement of that labor during the 1930s by poor whites. So long as the migrants had been people of other races, their poor housing, low wages, and unstable lives had been scandalous but soon forgotten. But when old-stock white Americans were reduced to homeless poverty, the nation took notice.

In *The Grapes of Wrath,* John Steinbeck portrayed the plight of independent farmers who went bankrupt in the Southwest, migrated to California, and found that the only work available was day labor on huge irrigated farms. The Okies were victims of the irrigation foundation story twice over. First, they had homesteaded in the belief that rain would follow the plow west of the 99th meridian. Settling on the western plains after the drought of the 1890s, they had enjoyed a generation of modest prosperity before the blistering heat returned in the 1930s. At the same time, the shift from horses to tractors changed the scale of farming, making many tenant farmers superfluous. Bankruptcy and dust storms forced them off their land, disproving the earlier form of the technological creation story in which rain and prosperity followed the plow. Then they had believed it possible to acquire irrigated farms in the West. Piling their possessions into antiquated cars, almost 700,000 trekked to California in the 1930s.[58] Under the auspices of the Farm Security Administration, photographers captured the plight of these people in such images as Dorothea Lange's "Migrant Mother." Lange documented the harsh conditions in the Imperial Valley, where many migrant laborers lived in shacks and tents, often drinking water from irrigation ditches. That hard-working white men and women could lose their homes and become landless wanderers was further proof that the irrigation narrative had failed in the arid West. When Steinbeck's migrants reached California, they were astonished to find that there was no land to rent or homestead, that one man might own a million acres, and that, owing to overproduction, good land might lie fallow for years:

> . . . a homeless hungry man, driving the roads with his wife beside him and his thin children in the back seat, could look at the fallow fields which might produce food but not profit, and that men could know how a fallow field is a sin and the unused land a crime against the thin children. And such a man drove along the roads and knew temptation at every field, and knew the lust to take those fields and make them grow enough for this children. . . . The fields goaded him, and the company ditches with good water flowing were a goad to him.[59]

At the end of the 1930s, poverty in the face of plenty seemed to create a proto-revolutionary situation. If such landless farmers ever joined together, Steinbeck wrote, "the land will be theirs, and all the gas, all the rifles in the world won't stop them. And the great owners, who had become through their holdings both more and less than men, ran to their destruction, and used every means that in the long run would destroy them. Every little means, every violence, every raid on a Hooverville, every deputy swaggering through a ragged camp put off the day a little and cemented the inevitability of the day."[60] The foundation story had assumed that individual farmers controlled new and powerful technologies and that they had access to the land. But the counter-narratives of the Great Depression showed that these individuals had lost access to the soil. Farm Security

Figure 10.2
Dorothea Lange, "Migratory Labor Housing during Carrot Harvest, near Holtville, Imperial Valley, California." Courtesy Library of Congress.

Administration photographs, government reports and hearings, popular novels, and the songs of Woody Guthrie depicted the farmers as having succumbed to the natural forces of drought and dust storms. The technologies they owned were puny before the powerful new technologies of tractors and vast irrigation projects. The counter-narrative declared that their strength lay in their numbers. Steinbeck put it bluntly: "When property accumulates in too few hands it is taken away. . . . When a majority of the people are hungry and cold they will take by force what they need. . . . Repression works only to strengthen and knit the repressed."[61] Like the counter-narrative to the mill narrative, the new story also seemed inevitable. The oppressed, whether workers or farmers, would unite and take control of the means of production.

Native Americans and resident Hispanics also had a counter-narrative to tell about irrigation. Frank Waters's 1941 novel *People of the Valley* inverts all the elements of the foundation story. His Anglo-Americans are outsiders who enter not an uninhabited desert but a traditional Hispanic farming community. They gradually acquire land, at first quietly, then by legal trickery, and finally by force. These newcomers possess powerful new technologies, which they use to build a dam. The locals call these technologies collectively "the Máquina." Waters recounts the arrival of the construction crew and their equipment from the viewpoint of the local residents:

These mountains would never echo the sound of a train, but the Máquina assaulted them with more than a far-off, thin and muted shriek. It clanged and jangled, bumped, rattled and banged. Past El Alto de Talco and El Alto de los Herreras, through Santa Gertrudes, Cañoncito, over La Corriera and Trameras. To the tall jutting cliffs it rumbled ponderously, stopped and spread from wall to mountain wall.

Such a Máquina! It was iron and wood, steel and stone and concrete, bolts and long cables. It whistled, shrieked, snorted and clamored with blow and shout; it smelled of gasoline and crude oil, whisky and good coffee. It was terrible with power. It was wonderful with the new and strange. It was unreal.[62]

"The machina" transforms the landscape, ripping apart the old boundaries and inscribing the valley within the geometrical order of the grid. This can be seen in the precision of the new road:

Unlike the old one, which crawled along the side, following the contour of the mountain wall, the new one cut straight up the valley. Nothing stood before it, neither fences, adobes, log corrals, nor fresh plowed fields, a hogback of slate, a deep arroyo. It was a furrow that split men's families, fields and hearts. Its geometrical precision taught the unteacheable, that the shortest distance between two points is a straight line; this struck terror into their souls. And in its wake lumbered huge trucks and scrapers, winch-driven buckets and concrete mixers.[63]

The Hispanic community resists. People refuse to leave their property after the courts have appropriated it for construction; they move only on the last day, when threatened by a sheriff with a shotgun. Even then they are willing to go only because their leader, Maria del Monte, advises them not to resist. Maria—an old woman, descended from the first settlers, who spent her youth herding goats high above the valley—has won respect through a lifetime of giving good advice and providing traditional herbal medicine. Maria resists the dam as long as possible, but then resettles the evicted farmers in a remote, untouched valley that she recalls from her youth. Marginalized by the forces of progress, the Hispanic community departs, evicted by a massive dam that will enrich the land's new white owners.

Waters's novel does not emphasize the dam and irrigation; it focuses almost entirely on the world that existed before construction began. Like Louise Erdrich's *Tracks,* it concerns a landscape and a way of life that pre-exist a new technology. Both novels evoke a land ethic. Maria fights the dam "not opposing it with any reason, but calling forth the feeling for the land it would supplant."[64] She stirs up people with pointed remarks about their own property. She tells a farmer that he has a beautiful piece of land. But, she says, ". . . measure it and find comfort in it while you may. Soon it will be nothing but a lake. You will have to make your living fishing, though in winter you will probably starve."[65] By focusing on the eviction of the Hispanic farmers from the valley, Waters reverses the dam's meaning. It leads not to development but to dispossession. It does not create community; it destroys it. As in *Tracks,* the underlying sense is not of progress but of loss.

The ending to Waters's book was not a possible scenario for most Native Americans or Hispanics. When white Americans encroached on their lands, they seldom had a well-watered hidden valley to retreat to. In some cases no escape was possible. For example, whites slaughtered the Paiute tribe in Owens Valley and seized the land they had successfully irrigated.[66] Native Americans who survived in the rest of California retained little of their land. Across the West, the counter-narrative could be documented in tribe after tribe.

Nevertheless, tribes on reservations still retained federal rights to water, and they often had legal priority over other users. In the 1908 case *Winters v. the United States,* the Supreme Court established that tribes had the rights to water on their reservations, including rivers and streams that pass through or lie on their boundaries.[67] Because the water and land of the reservations had been deeded to the tribes with the expectation that they would change their way of life and become agricultural peoples—in effect, because they were expected to accept the Jeffersonian vision of family farming—these rights persisted even if they were not exercised. Despite this legal safeguard, however, the Bureau of the Interior often failed to respect the Supreme Court's ruling. In 1973 the National Water Commission's own publication admitted that government policy "was pursued with little or no regard for Indian water rights and the *Winters* doctrine" and that "with the encouragement, or at least the cooperation, of the Secretary of the Interior—the very office entrusted with protection of all Indian rights—many large irrigation projects were constructed on streams that flowed through or bordered on Indian reservations, some-times above and more often below the Reservations."[68]

Many federal construction projects took place on Native American lands. The Army Corps of Engineers forced the Mandan, the Hidstsa, and the Arikara off their reservations along the Missouri River, gave them scat-tered lands of inferior quality, and denied them even the right to log the lands to be flooded or to fish in the resulting reservoir.[69] They had lived on excellent bottomland for more than a century, but the Corps of Engineers chose to flood them out rather than disturb newer white communities. For most Native Americans, the irrigation narrative was untrue from beginning to end. The land was not originally empty, the decision to become farmers was not voluntary, and the government often did not pro-tect those who successfully converted to agriculture. Irrigation completed the destruction of their community.

When Native Americans who had lost land and water rights through state and federal encroachments sought legal redress, they usually found themselves ensnared in a tangle of jurisdictional disputes. By the 1970s, when the Navajo, the Hopi, the Apache, and other tribes were positioned to win back some of the water needed to irrigate their lands, Congress passed the McCarren Amendment.[70] The ostensible purpose of this law was to consolidate legal disputes over water rights so that they could be decided in a comprehensive way in one forum. This was to be done at the state, not the federal, level. Although the new law made no mention of Native American water rights, state governments immediately interpreted it to mean that the tribes were prohibited from going directly to federal court. A divided United States Supreme Court ultimately sustained this interpretation after almost a decade of litigation. Justices Stevens and Blackmun dissented, citing a Senate Report which had concluded: "There is great hesitancy on the part of tribes to use State courts. This reluctance is founded partially on the traditional fear that tribes have had of the states in which their reservations are situated. Additionally, the federal courts have more expertise in deciding questions involving treaties with the federal government, as well as interpreting the relevant body of federal law that has developed over the years."[71] The majority decision forced the tribes to undergo a long and costly process at the state and then the federal level, delaying a final allocation of their water rights for as much as a generation. The dispute over jurisdiction and the proper implementation of the McCarren Amendment distracted attention from the flawed character of the legislation. Not only did it mock Native Americans' water rights; it also made untenable assumptions about the relationship between land and water. It ignored the movement of water between states and assumed that each state had a neatly delimited supply to apportion among claimants. It gave jurisdiction to many separate state courts, which could not see the problem as a whole. (In the arid West, each state is a large geometrically defined space created without reference to watersheds. The ideology of the grid had become so fundamental to the perception of the western landscape that neither Congress nor the Supreme Court could imagine an alternative.)

The most defenseless victims of the irrigation narrative, however, were not migrant workers, dispossessed landholders, or even Native Americans trapped on waterless reservations, but fish, birds, and other wildlife. Agricultural reclamation destroyed the wetlands that many species require to survive. The delta of the Colorado River has largely dried up. In Utah, Sevier Lake has disappeared as more than a million acre-feet have annually been siphoned off for irrigation. In California there were more than 4 million acres of wetlands in 1850; only 10 percent remained in 1977.[72] But might irrigated fields be substitutes for wetlands? Unfortunately not. Any living thing on irrigated land is at risk from pollution in much of the West. In the heady days when irrigation was new and men such as Chaffey, Hall, Powell, and Smythe vigorously promoted it, they could not foresee the long-term environmental consequences. To them, irrigation seemed to be nothing more than rain that could be turned on and off. It seemed to have no adverse effects. No one seems to have realized that impurities in water could cause problems. The chief danger seemed rather to be one of too much water. By 1917, over-irrigation appeared to be a temporary problem that could be cured by drainage works.[73]

It took nearly a century for scientists to discover the dangers posed when evaporating irrigation water left behind poisonous minerals such as selenium, boron, arsenic, and mercury. Even before it is used for irrigation, the water behind many major dams is already polluted. The 200-mile-long Lake Powell behind Glen Canyon Dam is severely polluted. Its fish contain four times the level of mercury considered safe for human consumption, and selenium "has accumulated in some fish in amounts to cause their reproductive failure.[74] By 1981 many parts of the San Joaquin Valley of California had concentrations of selenium that far exceeded the survival limit for aquatic life.[75] Polluted water also drained into wildlife reserves and marshlands, poisoning fish and wildlife. By 1980, 40 percent of all bird nests contained eggs with dead embryos, and 20 percent of the birds that hatched were deformed.[76] At the Kesterson Reserve the damage was considered irreversible.[77] And this was not an isolated problem, for selenium is common in much of the arid West, from the Dakotas and Montana to the Imperial Valley. Drinking water that contains more than 10 parts per billion (ppb) is deemed unsafe, but irrigation runoff commonly contains 1,500 ppb and sometimes contains 6,800 ppb.[78] Farmers add pesticides and fertilizers to the water, further increasing its toxicity.

Such pollution is concentrated in some wildlife refuges. At the Bureau of Reclamation's Newlands Project, by 1987 the ecological system had collapsed. A reporter found "dead fish by the uncountable millions are washing up along the gooey shoreline, bobbing across the surface or decaying on the bottom, where bloating gases soon will pop great fetid masses more of them to the surface. Duck carcasses dot the shore. . . . Knight-crested herons, egrets, grebes, geese, cormorants—all are represented among the carcasses."[79] The main causes of the devastation were high concentrations of boron, selenium, and mercury in the water. Members of the Paiute tribe were also victims of the water pollution, and Pyramid Lake, on which they had relied for fish and game, was effectively destroyed. Its once-abundant large Lahonton cutthroat trout became extinct, and other local species were endangered. The tribe sued the federal government and local irrigators, some of whom, they contended, had no legal rights to the waters they were taking.[80] Likewise, some irrigated farms in California's Central Valley became deadly wastelands; others must lie fallow for years before they can be cultivated again.

Pollution in the few wetlands remaining in the arid West harms migratory birds from Hudson's Bay to Mexico. No less than "20 percent of all North American waterfowl use California wetlands as a stopover or winter residence,"[81] and millions of other birds migrate through Colorado, Arizona, Utah, and New Mexico. Thousands of snow geese from Alaska stop at Tule Lake and similar wildlife refuges. If they are poisoned, the loss affects Eskimo hunters and the ecological system they inhabit 2,000 miles away. Thus, Tule Lake cannot be treated as a local problem; it is a part of a continental support system for migratory birds.[82] The competing and cooperating interests at work have not acted out a technological creation story. Rather, from the viewpoint of the wildlife, it has been a narrative of destruction.

As late as 1980, historians writing about irrigation in the arid West still worked with a modified version of William Smythe's triumphant and idealistic vision of the desert as an unfinished creation waiting for human intervention. Arthur Maas and Raymond Andersen toned down the rhetoric in their book *. . . and the Desert Shall Rejoice,* but the message remained much the same. In the 1980s, Donald Worster attacked such complacent and progressive views in *Rivers of Empire: Water, Aridity, and the Growth of the American West,* a work inspired by the work of Karl Wittfogel and the Frankfurt School. Wittfogel had argued that irrigation was by its nature a centralizing, bureaucratic, and inherently undemocratic structure of power. In *Oriental Despotism: A Comparative Study of*

Total Power he focused on previous irrigation societies and, using a comparative method, argued that elites inevitably dominated societies based on irrigation, as control of water was translated into economic and political power. Worster applied this thesis to the American West, contrasting the rhetoric of individualism and the family farm with the reality that land and water were often concentrated in the hands of a few whose interests increasingly intertwined with those of politicians, or, in the case of the Mormons, with the interests of both church and state. He argued that federal water monopolies had created an authoritarian water regime that served special interests. His account echoed the stories of dispossessed settlers, disappointed homesteaders, and migrant workers, for whom the rosy promises of irrigation had never been fulfilled. Worster also emphasized the possible alternative land ethics of Native Americans, Hispanic Americans, and early environmental writers. In contrast with Smythe's incomplete, barren landscape waiting for humans to transform it into a garden, Mary Austin, Joseph Wood Krutch, and Willa Cather had seen the desert as a complete and spiritually satisfying world. Worster synthesized the counter-narratives to western irrigation and reintroduced these forgotten voices into the debate.

Marc Reisner's *Cadillac Desert* was just as critical of federal irrigation, but it focused primarily on the economic injustice and ecological absurdity of some of the dam building that took place after World War II. In the upper Colorado basin, systems were built that cost $2,000 or more per irrigated acre, six times the value of the best Illinois farmland, and yet all these irrigated acres could grow was cattle feed. Putting such land under water raised its market value to only $150 per acre, and it produced crops already in surplus nation-wide.[83] Reisner detailed how the Corps of Army Engineers and the Bureau of Reclamation invented new accounting methods that justified almost any dam project. By the 1960s, the Bureau of Reclamation alone had "nineteen thousand-odd employees [who] merely planned projects, supervised projects, and looked for new projects to build" while contracting out all the physical work to private firms.[84] The Corps of Engineers sometimes "stole" their projects, in addition to launching many more of their own. And when they could not outmaneuver one another, the Corps and the Bureau cooperated, perhaps most notably in a grandiose project to build a series of dams on the Missouri

River and at the same time spend half a billion dollars on its river chan-nel.[85] Reisner worked in the best tradition of muckraking journalism to expose the enormous federal subsidies to rich farmers and the ecological destruction of many areas. "Glen Canyon is gone," he wrote. "The Col-orado Delta is dead. The Missouri bottomlands have disappeared. Nine out of ten acres of wetlands in California have vanished, and with them millions of migratory birds. The great salmon runs on the Columbia, the Sacramento, the San Joaquin, and dozens of tributaries are diminished or extinct."[86] The indictment was savage: "illegal subsidies enrich big farm-ers, whose excess production depresses crop prices nationwide and whose waste of cheap water creates an environmental calamity."[87]

In the 1990s, others modified the counter-narratives of Worster and Reisner. Donald Pisani, Richard White, and William Cronon saw not only capitalist elites and exploitation of nature but also competing special interests and attempts to conserve resources.[88] Their narratives of western water included not only instrumental technologies but also unanticipated developments. Their irrigation narratives took account of a constantly interacting and adjusting system that included nature, politics, economics, and social values. If Worster used Wittfogel's theory to frame a century's counter-narratives, they synthesized the perspectives of government irri-gation engineers, agribusiness, smaller farmers, Native Americans, fisher-men, tourists, and wildlife managers.

In 1987, the Bureau of Reclamation claimed it had undergone a change in philosophy. It vowed to shift its major mission from development to con-servation. The agency that had long championed federal irrigation pub-licly recognized that there were limits to growth. Superficially, the elements of the technological foundation narrative were still in place. New technol-ogies were used to transform a region, land values increased, and a new landscape, or a second creation, emerged. But the government, not private enterprise, had paid for and controlled the technology. In practice the costly irrigation projects were seldom self-financing, and the "invisible hand" of a free market would never have built some of them. The govern-ment expected that construction and maintenance costs would be recov-ered through water charges, but western farmers seldom repaid the full costs. The new lands opened up were not a safety valve for labor, because they seldom were subdivided into family farms.[89] Instead, state-sponsored irrigation reinforced pre-existing land monopolies. Hydroelectric dams and irrigation systems increased the mastery of force, but they did not increase prosperity for all.

If one considered only the agricultural production, the promise of the irrigation narrative had been extravagantly fulfilled in the millions of acres farmed in arid Arizona, Utah, Colorado, and California, as well as in many areas of the Great Plains. But the story of reclamation as technological creation hid a great deal. If one analyzed government irrigation as an economic question a century after the founding of the Bureau of Reclamation, then some of it had been foolish, spending far more to deliver water than the farmers could ever repay. Some of the irrigation had also been illegal, seizing lands and waters that belonged to Native Americans. A great deal of it had not benefited the small farmers who initially supported it but instead had made a few large farmers quite wealthy. The environmental havoc for some species of fish and migratory birds had been catastrophic. The Bureau of Reclamation had shown just how costly second creation could be.

11

Progress, or Entropy?

In 1900, second-creation stories pervaded the American sense of the past. The economic and social history of the United States seemed to be summarized by the successive states of civilization suggested by the axe, the water-driven factory, the canal, the railroad, and irrigation. White Americans had used these technologies to move from the forests of the East to the deserts of the West in little more than a century. Frederick Jackson Turner and the many historians he inspired saw the moving line of the frontier as the first in successive waves of development that surged across the continent.[1] The popular writer Emerson Hough began *The Way to the West* with a chapter on the axe and ended with one on the railroad. Many Americans saw no reason to halt the project of national expansion. John Fiske spoke for many when he declared in *Harper's Magazine* in 1885 that Anglo-Saxons and their institutions would expand until "every land on the earth's surface that is not already the seat of an old civilization shall become English in its language, in its religion, in its political habits and traditions, and to a predominant extent in the blood of its people."[2] Josiah Strong had expressed similar views in his best-seller *Our Country*, which contained an entire chapter on "The Anglo-Saxon and the World's Future." To such writers, the foundation story applied beyond the borders of the United States, and neither "Manifest Destiny" nor second creation would be halted.

Henry Adams (1839–1918) knew the foundation stories, but from the same facts he drew different conclusions. In his childhood, the axe and the log cabin had been potent political symbols that Jacksonians and then Whigs had used to undermine any patrician's claim to national office. Adams had been born into the world of New England mills, and he knew that much of the wealth of his native state depended on them. His uncle Edward Everett, his grandfather John Adams, and his father all had championed internal improvements, embracing the canal and the railroad. Later, Henry and his brother Charles had written muckraking articles against the railroads' misuse of their power. Adams had traveled in the arid West and knew of its irrigation projects. He had seen the prodigious displays of steam engines and dynamos at more than one international exposition. Given his birth, education, and experiences, Henry Adams might have been expected to write the history of the United States as the triumph of these new technologies. But he did not, for he recognized the ideological limits of the technological foundation story. As Macherey noted, Adams was able to question that story "as to what it does not and cannot say." He was able to explore "those silences for which it has been made." He saw that "the order which it professes is merely an imagined order, projected on to disorder."[3] For Adams, second creation no longer made sense. He believed in entropy, not natural abundance, in inexorable decline, not progress.

The historian most frequently referred to in *The Education of Henry Adams* is Edward Gibbon, whom Adams periodically mimicked during visits to Rome by sitting in the same spot where Gibbon first meditated on the empire's decline and fall. The vast ruins of Rome offered a powerful contrast to each of the central elements of the American technological foundation narratives. In Rome, space was not abstractly geometrical but densely historical. In Rome, the certainty of progress was mocked by every vista. In Rome, one could not posit natural abundance as the normal historical condition. In Rome, the market was clearly a social construction that expanded and contracted with the borders of empire. By aligning himself with Gibbon, Adams emphasized that he looked at the United States and its narratives of technological progress from a comparative, European perspective. He would start not with the North American colonies of the seventeenth century, but with the rise and fall of ancient civilizations.[4] He would trace a much longer story, the United States appearing only in the last chapters.

This longer story was organized according to scientific ideas. Adams was far more at home with scientific theory than most historians. His personal library and marginalia reveal that he read about science not only in English but also in French and German. He was familiar with Haeckel's *The Riddle of the Universe,* Hertz's *Electric Waves,* Poincaré's *La Physique Moderne, son Évolution,* and Findlay's *The Phase Rule and Its Applications.* After reading widely in science, he concluded that the apparent progress of the nineteenth century was illusory. He then created a large temporal framework that began before mankind existed and intersected with his own experience.

During his final years, Adams became convinced that a fundamental break in history had occurred during his own lifetime. He tried to characterize the rupture in a letter to a friend: "A world so different from that of my childhood or middle-life can't belong to the same scheme. It shifts from one motive to another, without sequence. Any mathematician will say that the chances against such a rupture of continuity were a million to one— that it has been impossible. Out of a medieval, primitive, crawling infant of 1838, to find oneself a howling, steaming, exploding, Marconiing, radiumating, automobiling maniac of 1904 exceeds belief."[5] He sensed the contrast particularly when, in France, he purchased an automobile, which he used to visit medieval churches. Yet the sense of incongruity went far deeper. Even as a Harvard undergraduate, "the only teaching that appealed to his imagination was a course of lectures by Louis Agassiz on the Glacial Period and Palaeontology, which had more influence on his curiosity than the rest of college instruction."[6] Geology taught that the world had not begun a mere 6,000 years before in Eden, but millions of years earlier. It was not a happy tale of progress, but a series of catastrophes and extinctions, including the inexorable movements of vast sheets of ice across North America. In the 1860s, while in London assisting his father during his service as the American ambassador, Adams read Darwin and became personally acquainted with Sir Charles Lyell, whose work he discussed in the *North American Review.* Adams later would count among his closest friends the geologist Clarence King and the physicist Samuel Langley.[7]

Henry Adams's theory of history hinged on reconciling the incompatibility between evolutionary theory and the laws of thermodynamics. The concept of force had long been a puzzle to philosophers and physicists, but had seemed indispensable. During Adams's lifetime, however, the concept of force had been discarded.[8] Experimental investigations of the relationships among magnetism, electricity, gravitation, and other phenomena had shown that the concept of force was riddled with inconsistencies. It was inseparable from two other ideas rejected by late-nineteenth century science: the mechanistic view of the universe and the belief in the conservation of force.[9] In the place of force, thermodynamics substituted the quite different concept of energy and yoked it to the disturbing new idea of entropy.

Before the abandonment of the concept of force, many had tried to unify the work on heat, gravity, electricity, and mechanics into a general theory of the conservation of force. By the 1840s, at least 14 scientists claimed to have achieved a proof. They fervently wished to extend Newton's third law, which stated that for every action there is always opposed an equal reaction, to demonstrate the complete and absolute conservation of force as it passed from one form to another (for example, from heat to steam to rotary motion).[10] To achieve a theoretical unification, however, it proved necessary to jettison the concept of force itself. The "first law of thermodynamics" articulated by Lord Kelvin in 1861 was that the sum total of kinetic and potential *energy* in the universe always remains constant. But the second law recognized that the universe was nevertheless running down, as some of this energy was effectively lost. Adams realized that because thermodynamics posited entropy as an unavoidable by-product of all transformations of energy, it undercut the belief in progress. Over the long term, humanity's access to energy could only decrease.

However, during the nineteenth century, most Americans still thought in terms of force, not energy, and force underlay both technological foundation stories and popular evolutionary theories. When the widely admired Herbert Spencer proclaimed that force was "the ultimate of ultimates" he was not working with a clearly defined term, but few of his American readers (or listeners) were students of thermodynamics. Spencer argued in his *First Principles*: "We come down then finally to Force, as the ultimate of ultimates. Though Space, Time, Matter, and Motion, are apparently all necessary data of intelligence, yet a psychological analysis shows us that these are either built up of, or abstracted from, experiences of Force. Matter and Motion, as we know them, are differently conditioned manifestations of Force."[11] Though a scientist might have objected to this formulation, Spencer proved extremely popular. He spoke in terms that most people could understand, and he presented human history as a tale of evolutionary progress.[12]

Adams and Spencer were not the only ones who attempted to sum up nineteenth-century experience in terms of energy or force. Others attempting something similar included prominent inventors such as Nikola Tesla and Thomas Edison, and scientists such as Lord Kelvin. By situating Adams in relation to these figures and in contrast to his brother Brooks, one can see *The Education of Henry Adams* (particularly the chapter "The Dynamo and the Virgin") as a part of a comprehensive meta-narrative in which Adams destroyed the fundamental assumptions of technological foundation stories. Adams overturned the whole system of stories based on force.[13] Instead, he adopted the newer concept of energy and sought to integrate thermodynamics into history.

Adams's narrative was partly shaped by his politics. He "had, in a half-hearted way, struggled all his life against State Street, bankers, capitalism."[14] He had done so not in solidarity with the working class, nor as an agrarian, nor as any form of socialist, but as an anachronistic eighteenth-century gentleman, with a respect for Republican civic virtue, who distrusted the popular politicians who had emerged after the 1820s. Adams had known scores of such politicians personally, and he could appreciate their "efficiency," but he could not imitate or admire them. To him they ultimately represented the failure of social evolution (a failure epitomized by the election of Ulysses S. Grant to the presidency) and the impossibility of making sense of history in terms of individuals. Paradoxically, Adams, who had spent most of his life in close observation of diplomats and politicians, and who might have been expected to see history as the story of personal relationships, factions, parties, and serendipity, instead found that close scrutiny of the political world revealed inconsistency and chaos. Order only seemed to exist on an abstract, impersonal level that had nothing to do with private circumstances and particular events. That viewpoint partially explains why he tells the story of his education as a series of failures to make sense of the world, whether as diplomat, amateur scientist, historian, novelist, editor, professor, or journalist. In ordinary terms, Adams had been a brilliant success in each of these roles.[15] But from the viewpoint of his 1905 persona in the autobiography, each attempt at education had been doomed to failure by a lack of sufficient information and by the limitations of his personal perspective. Hence the need for third person impersonality in *The Education*: he wanted to remain in a more abstract realm where his life would be only a fact of history and not an end in itself. Adams, the author, needed to minimize his successes in order to construct a narrative of wandering self-education that could lead to a new theory of history.[16] Thus, he deprecated his early life and formal education, his European travel and informal education, the diplomatic work he did for his father, his early publications, and his teaching at Harvard.[17] For Adams, entropy explained both private and national experience.

From the time of John Quincy Adams and Edward Everett to the time of Josiah Strong and Teddy Roosevelt, most Americans viewed the western expansion of the United States and its simultaneous industrialization as evidence of progress and social evolution. These were narratives of *force,* about building a home in the wilderness, establishing a town around a mill, opening up the backwoods by steamboat, canal, and railroad, or making the desert bloom. They were about the physical conquest of space as directed by men. In Henry Adams's words: "Since the year 1830, when the great development of physical energies began, all school-teaching has learned to take for granted that man's progress in mental energy is measured by his capture of physical forces, amounting to some fifty million steam horse-power from coal, and at least as much more from chemical and elementary sources."[18] To most Americans, the control of physical power seemed to measure society's advance. This partly explains the popularity of Spencer, who was widely read in the United States.[19] A great deal of power was required to settle the nation, Adams emphasized: "On the new scale of power, merely to make the continent habitable for civilized people would require an immediate outlay that would have bankrupted the world."[20] With this hyperbole Adams restated what Strong and many others also had realized in the 1880s. However, there was a crucial difference: Adams did not assume that man was in control of the process.

The earlier narratives of force were about founding a place, about establishing a rooted relation to the world. Or at least so it first appeared. Yet the clearing in the woods proved to be temporary, for the frontiersman expected soon to move on, and the log cabin was an instant house, more generic than personal. Likewise, the mill site was developed for a succession of different purposes. After the trees were removed, the sawmill was taken to a new location and the water power devoted to other purposes. The river towns and canal towns and railway towns were thrown up in a few years, but if they did not attract many settlers their inhabitants restlessly moved elsewhere. By the end of the nineteenth century, the stories of technological foundation, if they were ostensibly about establishing a rooted relationship to the world, could be seen as descriptions not of America's settlement but of its unsettlement. Adams was acutely aware that he also had become unsettled, no longer at home in Quincy, Boston, New York, or London, but only in that most transient of communities, Washington, where the residents changed after every election. Adams shuttled between the Capitol and the expatriate community in Paris, with frequent visits elsewhere. Nor should this unsettlement be surprising, for his personal story, like the technological foundation narratives, was shaped in relation to the emergence of an industrial economy. However much narratives of force appeared to be establishing the foundations for society as a whole, they in fact reexpressed the right of each restless individual continually to (re)establish a (temporary) home.

In the larger trajectory of these stories, starting with the clearing made by an axe and ending with the irrigated fields in the arid West, the centrality of the individual was eroding as the power and complexity of technological systems increased. This loss of autonomy is so familiar a historical theme as to require little emphasis itself. Yet, in curious defiance of logic, the American sense of individualism seemed to expand as its possibility diminished. The deployment of mechanical systems had become so complex and costly that the ordinary individual could not avoid depending on many others. Yet the individualistic creed declared just the opposite. So the meta-history of energy narratives is about the erosion of community, about the diminution of the individual, about the increasing complexity of the machine, and about the unintended loss of the very autonomy celebrated in the narrative of the axe and clearing. Yet it is also about the simultaneous development of individualistic ideology and what can only be regarded as an illusion of self-reliance.

In explaining the persistence of this illusion, one must not overlook an obvious point: the power available to each person had increased enormously as Americans shifted from muscle power to water, then to steam, and, at century's end, to oil and electricity. It was this large fact that pre-occupied Adams, who was unquestionably the most prescient thinker on the subject in the United States in 1900. Changes in quantity, when large, have qualitative effects, and Adams understood that the concepts of force and energy implied quite different kinds of narratives. Nineteenth-century narratives of force were straightforward tales of cause and effect. Born into a world of water wheels beginning to give way to steam, Adams looked back toward muscle power, which had built the medieval cathedrals, and traced the increasing control of force from that point forward. Adams could personally recollect a time when horses were the primary mode of transportation and communication. He could vividly recall when the water-driven factories of Lowell were still innovations, and for him the steam engine was the central driving force of a new civilization. He had grown up in a world where the metaphors and mental habits of force were the norm and watched as steam power and then electrification trans-formed the appearance of cities. This early familiarity with all the major forms of power might have suggested a neat evolutionary order. Indeed, in *The Education* the idea of a "sequence of force" serves as an underlying structure in history, instead of the more traditional approaches that depended on wars, politics, and the achievements of individuals.[21] Yet *The Education* is not about orderly cause and effect. Rather, it is a twentieth-century energy narrative that operates in several temporal dimensions, with shifts in emphasis from the conquest of space to the acceleration of movement and the conquest of time. The sciences Adams preferred to read implied not a mere historical sequence of forces (that was the narrative of progress), but ruptures in continuity and unexpected swerves in develop-ment. One could observe the discontinuities, Adams believed, by starting a millennium before his own day.

As Cecelia Tichi has trenchantly shown, Adams's interest in the Middle Ages was not a retreat from his interest in the twentieth century or its ener-gies but an extension of it.[22] Adams saw monasteries and cathedrals as organic architectural forms, as successful solutions to design problems, and as elements of an architectural tradition that linked these buildings to the great railroad stations, world's fairs, and skyscrapers of his own time. He not only knew these modern buildings; he knew many of their archi-tects too. Far from being an anti-modernist escaping from the Protestant industrial United States into the distant Catholic past during the twilight of his career, as Jackson Lears would have it,[23] during the 1870s Adams had been a professor of medieval history at Harvard. In the last two decades of his life he learned to see that period in a new way. Tichi therefore rejects the idea that there was an irreconcilable opposition between the dynamo and the Virgin in the most famous chapter of Adams's *Education*. As analyzed by Leo Marx, these two symbols were a dichotomy: "On the one side lines up heaven, beauty, religion, and reproduction; on the other hell, utility, science, and production."[24] For Tichi, in contrast, the two symbols are points on a historical line.

Yet both Tichi and Marx are correct. Although Tichi shows that Adams saw a clear historical relationship between the thirteenth century and the twentieth, that relation was ultimately less one of continuity than one of rupture. The narrative was built on discontinuity. Adams sought to connect his own age to the medieval period in many ways, all without success. One of the most famous passages in his autobiography reads as follows: "Satisfied that the sequence of men led to nothing and that the sequence of their society could lead no further, while the mere sequence of time was artificial, and the sequence of thought was chaos, he turned at last to the sequence of force; and thus it happened that, after ten years' pursuit, he found himself lying in the Gallery of Machines at the Great Exposition of 1900, his historical neck broken by the sudden irruption of forces totally new."[25] These new forces included the automobile ("a nightmare at one hundred kilometers an hour, almost as destructive as the electric tram which was only ten years older"[26]) and radium, but were most centrally electrical, and "to Adams the dynamo became a symbol of infinity."[27] There was a fundamental rupture between steam power and electrical current: "The break of continuity amounted to abysmal fracture."[28]

For Adams, human history did not consist of continuities, and it did not record evolution. Adams felt that an abyss had opened between the dynamo and the virgin, but other ruptures were noted in his speculations about energy and history, particularly "Letter to American Teachers of History" and "The Rule of Phase Applied to History." In these late works Adams wrote about energy and history from three related perspectives: that of physics, focusing on the gradual decay of energy according to the second law of thermodynamics, that of geology, focusing on Darwinian theory as it pertained to human evolution, and that of chemistry, focusing on the slightly less than 1,000 years from the building of the gothic cathedrals until his own day. The possible conclusions changed depending on whether he focused on the long-term tendency of the universe to lose energy in an unavoidable entropic decline, the middle range, consisting of the fragmentary geologic record of life's emergence, or the short-term history of human mastery of energy since the Middle Ages. Before examining in detail the three temporalities in Adams's thought, it is instructive to look at other narratives of technology and human society that were articulated between 1895 and 1914 by Robert Thurston, Thomas Edison, Nikola Tesla, and Brooks Adams.

The most optimistic of these were Thurston and Edison, both of whom speculated on the synergies between evolution and new inventions. In interviews Edison had a flair for dramatic statements, which reporters often embellished, about how humans were entering a phase of rapid intellectual evolution.[29] Edison apparently believed that his lighting system stimulated mental growth, because people were active and awake for more hours of the day.[30] Women in particular, he once suggested, would become more intelligent as a direct result of a widening sphere of experience made possible by the electric light, photograph, telephone, film, and other technologies. "The exercise of women's brains will build for them new fibers, new involutions, and new folds. . . . More and more she will be pushed, and more and more she will advance herself. . . . This development of woman through the evolution of mechanics will . . . be the quickest which the world has ever seen."[31] Edison also speculated that, as humans evolved, entirely new senses as useful as sight and hearing might develop, expanding their sensory horizon.[32] Such narratives argued that industrialization directly stimulated evolutionary processes, speeding up their tempo in a self-reinforcing process. Like Etzler's utopian projections of the 1830s or the early pronouncements about railroads or irrigation, these were stories about how new technologies would accelerate progress.[33]

Many shared this belief in rapid progress, including Robert Thurston, a professor at Cornell University, a former president of the American Society of Mechanical Engineers, and one of the best-known engineers of his day. Adams is likely to have read Thurston's 1895 article in the *North American Review*, since he had once edited the journal and had been associated with it for decades. The general public encountered the condensed version of Thurston's ideas in the *Literary Digest*.[34] Thurston set out literally to chart American progress through diagrams that illustrated the rate of material advance. His attempt to translate historical trends into mathematical formulas is highly suggestive of Adams's later project. Nor was Thurston the first to try to reduce progress to mathematical certainty, even in the pages of the *North American Review.*[35] Thurston concluded that American wealth grew more rapidly in some areas than others. The total value of cattle grew slowly, for example, and that of factory buildings only a little faster; the value of land and houses increased more rapidly. In any of these areas, growth could be expressed as a straight line. But the value of "sundries" rose far more rapidly, as a curved line. By sundries Thurston meant what are now called consumer goods, such as clothing, appliances, and home furnishings. Intimately familiar with the workings of textile mills and other factories, in just 20 years Thurston had seen production per mill hand rise almost 30 percent while working hours dropped and real wages rose by 20 percent.[36] Similar gains throughout the nineteenth century had radically improved people's lives, as they had more time and more money to spend. Thurston asserted that the curve for sundries "best shows the trend of our modern progress in all material civilization. Our mills, our factories, our workshops of every kind are mainly engaged in supplying our people with the comforts and the luxuries of modern life, and in converting crudeness and barbarism into cultured civilization. Measured by this gauge, we are fifty percent more comfortable than in 1880, sixteen times as comfortable as were our parents in 1850, and our children, in 1900 to 1910, will have twice as many luxuries and live twice as easy and comfortable lives."[37] This message of uplift also echoed in churches, in Chautauqua meetings, in Washington, and in many publications.[38]

Thurston was confident that progress was directly caused by the continual increases in power available for manufacturing and transportation.[39] Indeed, "from the date of the perfection of Watt's steam engine and its application to mills and factories, and to steamboats and railroads, wealth has accumulated with a continually increasing *rate* of accumulation."[40] New inventions and new factories made people not only richer but also more able to "enjoy the advantages of leisure for thought and study and intellectual growth." Control of more energy made Americans richer, happier, wiser, and more educated. It gave them leisure and made them healthier. Thurston concluded that these trends would not merely continue, but accelerate.[41] Furthermore, in 1884, well before Edison predicted that material development would stimulate mental development, Thurston presented a similar argument to the American Association for the Advancement of Science, of which he was a vice-president. Science and engineering produced not only a better physical life, but also a more powerful mind. Thurston confidently expected that "greater frontal development," "increased mental and nerve power," "growing endurance," and "probably lengthening life" would be "products of this scientific development."[42]

Thurston's message was what most Americans wanted to hear, but a more pessimistic view was familiar among scientists. Like Edison's fellow inventor Nikola Tesla, many scientists firmly bracketed human evolution within limits set by physics. Tesla accepted Lord Kelvin's view that human life on earth was limited: "From an incandescent mass we have originated, and into a frozen mass we shall turn. Merciless is the law of nature, and rapidly and irresistibly we are drawn to our doom. Lord Kelvin, in his profound meditations, allows us only a short span of life, something like six million years, after which the time the sun's bright light will have ceased to shine, and its life-giving heat will have ebbed away, and our own earth will be a lump of ice, hurrying on through eternal night."[43] Tesla hoped that life might continue elsewhere in the universe, however, and believed that science and art could "illuminate our path." He urged readers of *Century Magazine* to "conceive . . . man as a mass urged on by a force."[44] The problem of survival was essentially "the problem of increasing human energy" so that mankind could move more rapidly forward, like an ocean wave through time. Man "is capable of increasing or diminishing his velocity of movement by the mysterious power he possesses of appropriating more or less energy from other substances, and turning it into motive energy."[45] Science could accelerate the movement of the human wave through history in three different ways. First, it might increase the sheer mass of mankind by improving the food supply and general health, so that more and healthier humans existed. Second, mankind's movement forward could be increased by reducing resistance, most obviously that caused by ignorance and warfare. Tesla doubted that war would end if airplanes were invented or more powerful guns produced, but he hoped that the number of people engaged in warfare would decline. He proposed that weapons become increasingly automated, until only "machines will meet in a contest without bloodshed, the nations being simply interested, ambitious spectators. When this happy condition is realized, peace will be assured."[46] But the third and most promising way to help mankind was to "increase the force accelerating the human mass." Tesla saw "this busy world about us" as "an immense clockwork driven by a spring":

In the morning, when we rise, we cannot fail to note that all the objects about us are manufactured by machinery: the water we use is lifted by steam-power; the trains bring our breakfast from distant localities; the elevators in our dwelling and in our office building, the cars that carry us there, are all driven by power; in all our daily errands, and in our very life-pursuit, we depend upon it; all the objects we see tell us of it; and when we return to our machine-made dwelling at night, lest we should forget it, all the material comforts of our home, our cheering stove and lamp, remind us how much we depend on power. And when there is an accidental stoppage of the machinery, when the city is snow-bound, or the life-sustaining movement otherwise temporarily arrested, we are affrighted to realize how impossible it would be for us to live the life we live without motive power. Motive power means work. To increase the force accelerating human movement means, therefore, to perform more work.[47]

Tesla expected that science could provide "food to increase the mass, peace to diminish the retarding force, and work to increase the force accelerating human movement."[48] He devoted the remainder of his long article to improved iron production, advantages of using aluminum, more efficient coal use, improved windmills, development of "solar engines," geothermal power, heat engines that might draw perpetual energy out of the earth or the air, and electrical transmitters that needed no wires. With enough energy, "men could settle down everywhere, fertilize and irrigate the soil with little effort, and convert barren deserts into gardens, and thus the entire globe could be transformed."[49] Tesla recognized an irreconcilable contradiction between human evolution and the law of entropy, but nevertheless he confidently wrote a foundation narrative. He expected that technology could speed human progress and prepare an eventual escape from the solar system before its heat death.

Whereas Edison, Thurston, and Tesla anticipated rapid technological development, Henry Adams feared it. He did not expect new inventions to transform human nature, to make life happier, or to facilitate escape to another part of the universe. Rather, he was filled with a sense of doom. He shared this foreboding with his younger brother Brooks, who sought history's underlying structures in economics. Brooks had been a student at

Harvard when Henry taught there, and later both searched for the laws that governed history. In the 1890s, Brooks Adams wrote *The Law of Civilization and Decay,* a survey from ancient times to the present that was influenced by Marx, Spencer, and Comte. Henry Adams read it in manuscript, offered suggestions for revision, helped pay for its publication, and supervised proofreading and printing. He also distributed copies to influential friends and praised the book in private letters. Yet Brooks's approach to history was quite distinct from Henry's. Brooks developed a cyclical theory in which humans alternate between fear and greed. "Fear," his biographer summarized, "stimulates the imagination to create a belief in an invisible world; and greed dissipates energy in trade."[50] If Henry developed considerable respect for the cult of the Virgin, for his brother religion was only superstition, whereas banking and trade were central to both progress and decline. Brooks Adams saw the 1890s as the close of an epoch; Western society was declining much as imperial Rome had, though there was hope for some reprieve should bimetalism be adopted, thereby increasing the money supply. Like his brother, he saw the prospects as bleak. Humans could not control their destiny; they were gripped by immutable laws.[51] Yet the brothers' theories were hardly identical. Ultimately, Brooks conceived of laws based on psychology and economics; Henry worked through analogies with the sciences. Despite common interests, they were fundamentally at odds, and their face-to-face meetings usually were stressful. "They excited, then irritated, and finally enraged each other during and after any long session of talk."[52]

Henry Adams saw history as a sequence of forces, not in terms of contingency and "roads not taken." In "The Problem," he begins by describing the mechanical theory of force, which science had taken as a given for 300 years: "Directly succeeding the theological scheme of a universe existing as a unity by the will of an infinite and eternal Creator, it affirmed or assumed the unity and indestructibility of Force or Energy, as a scientific dogma or Law, which was called the Law of the Conservation of Energy. Under this Law the quantity of matter in the universe remained invariable; the sum of movement remained constant; energy was indestructible; 'nothing was added; nothing lost'; nothing was created, nothing was destroyed."[53] This was the conception of mechanical force common among ordinary Americans until the end of the nineteenth century. But Adams knew that the "indestructibility of Force or Energy" had ceased to be an acceptable view among physicists. For fifty pages he discussed the new laws of thermodynamics, the widening circle of sciences that accepted them, and the apparent conflict between the pessimistic assertion that energy inevitably dissipated and the popular view, based on a simplified conception of evolution, that asserted a gradual but inexorable movement toward higher forms of life. Physicists of the day thought the solar system seemed doomed to shrinkage and heat death, and for some of them the length of time until the demise of all forms of life was measured in thousands of years, not millions. Indeed, as Adams explained the scientific conclusions of his day, the earth had already passed through its period of greatest warmth and fecundity, and it was on the downward slope toward a new ice age that might well extinguish humanity. In this view, humans had evolved near the end of terrestrial history. In striking contrast to the fears of global warming that concern scientists at the start of the twenty-first century, most of the experts of Adams's day agreed with Lord Kelvin, who saw no escape from the second law of thermodynamics, which he had formulated as follows: "Any restoration of mechanical energy, without more than an equivalent dissipation, is impossible in inanimate material processes, and is probably never effected by means of organized matter, either endowed with vegetable life or subjected to the will of an animated creature." This law, when applied to the earth itself, lead to the stark formulation of Kelvin's third law: "Within a finite period of time past, the earth must have been, and within a finite period of time to come, the earth must again be, unfit for habitation of man. . . ."[54]

Applied to human history, the second and third laws negated American beliefs in progress and increasing material abundance.[55] Instead, every conversion of energy only hastened the demise of life on earth. It seemed possible that "nothing remains for the historian to describe or develop except the history of a more or less mechanical dissolution."[56] Humans were latecomers. Their numbers increased, their prosperity advanced, and, taking the short view of conventional history books, it might seem that social evolution occurred. But taking the long views offered by geology, physical anthropology or physics, Adams could only conclude that the sense of progress was a shortsighted illusion. Geology told Adams that the forces of life had been far more powerful millions of years before, "in the wild exuberance of the carboniferous forests . . . which then developed into the wilder exuberance of the Eocene mammals."[57] Leading geologists also taught him that whole species had become extinct as the energy level dissipated, that more species had disappeared in recent times,[58] and that there was good reason to expect man soon to become extinct as well: "Already the anthropologists have admitted man to be specialized beyond the hope of further variation, so that, as an energy, he must be treated as a weakened Will—an enfeebled vitality—a degraded potential."[59] The psychologists Adams preferred to read concurred, finding the modern intellect to be a degenerated form of will.[60] He pondered the probability that "thought then appears in nature as an arrested—in other words, as a degraded—physical action."[61] He found another proof of general decline in Gustave Le Bon's studies of the crowd and political psychology, which proclaimed that mankind was becoming enfeebled. Underlying all these systems of knowledge were, of course, the laws of thermodynamics.

Mankind seemed "to have no function except that of dissipating or degrading energy,"[62] thereby hastening the workings of the second law: "Man dissipates every year all the coal which nature herself cannot now replace, and he does this only in order to convert some ten or fifteen per cent of it into mechanical energy immediately wasted on his transient and commonly purposeless objects."[63] Man did the same with natural gas, "digging out even the peat-bogs in order to consume them as heat." And "he has largely deforested the planet, and hastened its desiccation."[64] Overall, Adams felt that man "seizes all the zinc and whatever other minerals he can burn, or which he can convert into other forms of energy, and dissipate into space."[65] Since the universe was a vast machine whose fixed supply of energy would eventually become unavailable according to the law of entropy, it seemed mathematically certain that the ultimate result of exploiting these resources would be not progressive but apocalyptic. Adams cited Kelvin's warning that "if we burn up our fuel supplies so fast, the oxygen of the air may become exhausted, and that exhaustion might come about in four or five centuries." It seemed that mankind's "chief pleasures, so far as they are his own invention, consist in gratifying the same unintelligent passion for dissipating or degrading energy."[66] "Man," Adams could only conclude, "is a bottomless sink of waste unparalleled in the cosmos."[67]

However, if many scientists at the turn of the twentieth century concluded that entropy applied to physical, social, and psychological evolution as well as to physics, much of the general public, like Edison, Thurston, and Tesla, still believed in human progress and defined human thought as the highest form of energy.[68] In reply, Adams defined energy broadly to include not only steam engines and electricity but any force, whether of instinct, intellect, or art, that was capable of organizing and directing people. In "The Problem" he staged a dialogue in which an evolutionist asserts that the ability of the mind to reason is proof that humans are progressing. A physicist, who is a spokesman for Adams, replies: "The claim that Reason must be classed as an energy of the highest intensity is itself unreasonable. On the contrary, Reason is the last in time, and therefore the lowest in tension. According to our Western standards, the most intense phase of human Energy occurred in the form of religious and artistic emotion—perhaps in the Crusades and Gothic churches—but since then, though vastly increased in apparent mass, human energy has lost intensity and continued to lose it with accelerated rapidity."[69]

Like his physicist narrator, Adams saw religion as an embodiment of force in the medieval period. When he contrasted the Virgin's power with that of the modern electric dynamo, Adams was seeking to do more than to trace a pattern that led from the unity of the medieval world to the multiplicities of science and technology of 1900. He contrasted the French Gothic cathedrals with the dynamos and electrical machines displayed at the Paris Exposition to reach conclusions that operated in three temporal dimensions. Considered within the larger entropy of the universe, the movement was that of disintegration and loss of force. Here there was little for the historian to say; Lord Kelvin had already said it. Adams focused instead on the much shorter span of human history, within which he told the story of a geometrical acceleration of the forces in society, achieved at the price of a matching acceleration of entropy. Adams searched for a formula to express this second development, eventually settling on "the rule of phase."

Adams adapted this "rule" from Willard Gibbs, a Yale University physicist whose work had been influential in chemistry after c. 1870. Gibbs had elaborated a theory, born out by experimentation, that all substances pass through quite distinct phases depending on temperature and pressure.

H_2O, for example, existed as ice, water, and steam. Similarly, all else existed as solid, liquid, and vapor, and required only the proper temperature and pressure to make a sudden and radical shift from one phase to another. Adams seized on this element of Gibbs's theory, but as Ernest Samuels notes, he made no use of "the key element Gibbs's theory—the simultaneous coexistence of several phases when a chemical equilibrium was established."[70] This was unfortunate, since human societies likewise maintained multiple sets of institutions, evolved in different epochs. Adams might have advanced a far more complex argument that focused on the simultaneous coexistence of different energy systems and their corresponding economic systems and mentalités. But he concentrated instead on the notion of dramatic and sudden transformations, such as occur when water is heated from 99 to 100 degrees Celsius and suddenly becomes steam. From his scientific reading, Adams decided that solid, liquid, and gas might be augmented by four other states: the electron or electricity, the ether, space, and hyper-space.[71]

In adapting these ideas to history, Adams substituted attraction for pressure and acceleration for temperature: "Under the Rule of Phase, therefore, man's Thought, considered as a single substance passing through a series of historical phases, is assumed to follow the analogy of water, and to pass from one phase to another through a series of critical points, which are determined by the three factors of Attraction, Acceleration, and Volume. . . ."[72] He thus attempted to provide a scientific foundation for August Comte's three phases of history—theological, metaphysical, and practical. He further attempted to account for the periodization of each historical phase through the "law of squares," also adapted from Gibbs's chemistry.[73] Adams arrived at a conclusion that seemed to him inescapable: the theological phase had lasted 90,000 years until the Renaissance, and the metaphysical phase would therefore last only about the square root of 90,000 (that is, 300) years—roughly from 1600 to 1900. The rather astonishing result was that the third phase would be marked by extreme acceleration, lasting only about the square root of 300 (that is, about 171/2) years.[74] Implicit in these calculations was the possibility of further rapid acceleration during the four additional phases, which corresponded to electricity, the ether, space, and hyper-space. Each would last only as long as the square root of the previous phase, which I calculate to be, respectively, 1,521 days, 39 days, 6.25 days, and 2.25 days. The absurd result was that four epochs in human thought, each as important as the Renaissance, would occur in extremely rapid succession. This projection into the future remained unexpressed in the essay, however. Adams perhaps realized that few readers would believe such a result. Perhaps he could not believe it himself.

Adams not only flinched from extrapolating the rule of phase to its logical conclusion in the future; he never applied it with any rigor or detail to the nineteenth century, which he knew so well. He had composed nine volumes on its early decades, and his *Education* had traced events from 1838 until the early twentieth century and enfolded them within the story of a man with an eighteenth-century education seeking to grasp the politics and science of the twentieth century. Rather than examine the energy systems closely, however, Adams became querulous about the rapid transformation of society. He clearly was displeased that by 1860 "Northern society betrayed a preference for economists over diplomats or soldiers." Nor was he entirely satisfied to be in many ways an eighteenth-century figure, "nearer the type of his grandfather than to that of a railway superintendent."[75] Though he admired his presidential grandfather greatly, he realized that the mental makeup of railway superintendents was what society wanted.

Yet Adams's ancestors and their political allies had loosed the new technological forces upon the world. Adams was well aware of Albert Gallatin's proposals for canals, roads, and other improvements, which he described in detail in an admiring biography.[76] Both his grandfather, John Quincy Adams, and his uncle, Edward Everett, had been warm advocates of canals and railroads. Yet the antebellum narratives of the axe, the mill, and the canal already seemed so completely outmoded that, although they figured in his history of the early republic and in his life of Gallatin, Adams scarcely mentioned them in his late works. Each technology had once seemed to affirm the rational development of a geometrical spatial order within a laissez-faire economy. Each had also been questioned by counternarratives that challenged abstract geometry with personal experiences in specific locations. But Adams chose not to hear Native Americans, homesteaders, farmers, and workers who had experienced not foundation and progress but dislocation and hardship. Instead, he focused on the railroad system.

When Henry Adams wrote of the railroad in his autobiography, he had considered the subject carefully. "Adams," he wrote, "had been born with the railway system; had grown up with it; had been over pretty nearly every mile of it with curious eyes. . . ."[77] His elder brother, Charles Francis Adams, who had been a critic of the ruthless management methods of Erie Railroad and of the labor practices of railroads generally before the great strike of 1877, later served as president of the Union Pacific Railroad from 1884 until 1890. Yet the later writings never drew together this considerable stock of information on railroads and integrated it into a larger theory. Instead, in *The Education* the growth of the railroad system is a metaphor for the integration of national economies and an icon of irresistible change.[78]

Railroads and steam power dissolved American regionalism and further linked agricultural society to industrialization. When Henry Adams returned to the United States from London, he saw that the 1860s "had given to the great mechanical energies—coal, iron, steam—a distinct superiority in power over the old industrial elements—agriculture, handwork, and learning." During this shift, "society dropped every thought of dealing with anything more than the single faction called a railroad system. This relatively small part of its task was still so big as to need the energies of a generation, for it required all the new machinery to be created—capital, banks, mines, furnaces, shops, powerhouse, technical knowledge, mechanical population, together with a steady remodeling of social and political habits, ideas and institutions, to fit the new scale and suit the new conditions. The generation between 1865 and 1895 was already mortgaged to the railways, and no one knew it better than the generation itself."[79]

By the 1890s, Henry Adams recognized that American trains and boats often were far superior to their European equivalents. After freezing during a train trip from Paris to London, he wrote to John Hay: "The comforts of European travel in winter fill me with admiration for America."[80] Not only were American trains superior; the coal output of the United States was almost as great as that of Britain, and "one held one's breath at the nearness of what one had never expected to see, the crossing of courses, and the lead of American energies. The moment was deeply exciting to a historian."[81] The shift in the preponderance of energy toward the

United States spelled a shift in world power that was confirmed by the American victory over Spain in the war of 1898 and by the diplomatic successes of his friend John Hay as Secretary of State. Yet none of these examples and events from *The Education* can be found in the more formal theorizing written afterward.

For Adams, history was the result of impersonal forces and sequences. Nature acted on man, not the reverse. Mind followed the movement of matter, making a mockery of the creed of individualism.[82] If Adams sought to understand human history in terms of the acceleration of social forces as mankind controlled greater energy resources, and if the railway and the motorcar might serve as a convenient shorthand for this development, the statistical details were less important than the fact that humanity was caught within a larger, declining movement. The energy of the earth and that of the solar system were slowly evaporating; the heat death of the universe was not only unavoidable; in geological terms, it was near. The immediately obvious acceleration of energy in society was ultimately less important than the inexorable decline that enveloped it.

Yet the more science Adams read, the more fallible it appeared. Ronald Martin observed of his reading: "Down through the years his marginalia have two characteristic themes: a querulous bafflement in the face of science's refusal to claim universality and finality for its theories, and, almost conversely, an emphatic delight (almost as if he were vindicated) in any sign of uncertainty, confusion, or contradiction, admitted or manifested by a scientific writer."[83] This double movement in Adams's thought—a demand for certainty yoked to a delight in the discovery of inconsistency—reappears in his theories of history. Adams sought uniformity, but found multiplicity. He applied the rule of phase to human history; however, he never completely extrapolated it into the future, nor did he apply it to geologic time. The human control of energy increased, and events accelerated, yet human evolution failed, a microcosmic expression of the universal dissipation of energies. The law of entropy trumped all human striving. In their own ways, the universe and human society were each descending into chaos. But to put it this way suggests that the universe started out as coherent and then became disordered. Such was not Adams's argument. Rather, as he put it, "Chaos was the law of nature; Order was the dream of man."[84]

In contrast to technological foundation narratives of progress achieved through the mastery of ever-greater forces, Adams composed a counter-narrative that substituted irony for optimism, displacement for settlement, and determinism for individualism. His meta-narrative worked on three temporal levels: the human, the geologic, and the cosmic. It hurried human progress into destructive acceleration; it then framed this disaster within catastrophic geology, and then framed both within the inevitable heat death of the universe. Physics provided a framework that humans could never escape: ". . . the law of Entropy imposes a servitude on all energies, including the mental. The degree of freedom steadily and rapidly diminishes."[85] In Adams's system, technological creation stories became mere vanity. If the first creation was running down, the outlook for second creation could only be bleak.

During his lifetime, Adams refused to publish his autobiography. Printed privately, it circulated only among his friends. After his death at age 80 in 1918, however, *The Education of Henry Adams* was published, and it became the leading non-fiction work of 1919.[86] The First World War seemed to confirm his theory that technological forces were accelerating out of control. In the 1920s many intellectuals embraced his cultural pessimism, which seemed confirmed by Oswald Spengler's *Decline of the West* and by the many anti-war novels written by veterans from both sides.

Yet such narratives of inevitable decline still competed with the stories of progress. The belief in material advance expressed by Edison, Tesla, and Thurston proved durable. As the economy expanded in the 1920s, Henry Ford became one of the three most popular men in America.[87] His best-selling autobiography, *My Life and Work,* explained not entropy but the assembly line, and assured Americans that the future would offer them more abundance than the past. Mass production accelerated society, Ford assured his readers, not toward the entropic heat death of the universe, but toward more leisure time to enjoy more consumer goods. Indeed, Ford did not sense the imminent end of history; he found the world to be almost entirely "in the raw material stage," waiting for men to develop it.[88]

Conclusion

Second Creation, Conservation, and Wilderness

In 1909, the railroad magnate James J. Hill declared: "You might put a railroad in the Garden of Eden and if there was nobody there but Adam and Eve, it would be a failure."[1] His statement underlined the American aspiration to improve on God's handiwork. The second-creation story asserted that the original perfection of the world could be enhanced with new technologies.

Though Hill was particularly blunt, similar declarations had circulated for generations. Henry Adams's uncle Edward Everett was an early champion of second creation. One of the most famous orators of the early nineteenth century, Everett trained for the ministry, taught at Harvard, served in Congress, and later became governor of Massachusetts and president of Harvard. At the height of his powers, in 1831, the American Institute invited him to speak in New York on the occasion of its annual exhibition. He began by praising the "numerous specimens of the useful arts" on display as examples of American "mechanical ingenuity." The tools and labor-saving machinery had been made from "the lifeless elements that surround us."[2] Mechanics and inventors had transformed "dull, unorganized matter" into useful and beautiful things. In contrast, Native Americans, who were "possessed of all the powers which belong to the human race," had not developed "any of the improvements" on exhibit at the fair. This contrast framed the central question of Everett's address: "Whence are the power and skill that have produced this new creation?"[3] His answer was as follows: "It is these ingenious and useful arts—the product at once and the cause of civilization—acting upon society and themselves in turn, carried to new degrees of improvement and efficiency by social man, till they have been brought to the state in which we find them in these halls—it is these which form the difference between the savage in the woods and civilized, cultivated, moral, and religious man."[4]

To Everett and many others during the nineteenth century, it seemed clear that the practical arts improved humanity. In 1895 the U.S. Commissioner of Labor devoted the final chapter of a book on the industrial evolution of the nation to "the ethical influence of machinery on labor."[5] Like Everett, he concluded that better machinery not only gave humans more comfortable lives but also improved their morals. The second creation not only improved the "unorganized" material of nature; it improved Americans themselves.

Abraham Lincoln delivered a similar message in 1859, when he lectured on discoveries and inventions. It also seemed clear to Lincoln that the Garden of Eden was merely a starting point for progress: "Take, for instance, the first of the old fogies, father Adam. There he stood, a very perfect physical man, as poets and painters inform us; but he must have been very ignorant and simple in his habits. He had had no sufficient time to learn much by observation; and he had no near neighbors to teach him anything. No part of his breakfast had been brought from the other side of the world; and it is quite probable, he had no conception of the world having any other side. In all of these things, it is very plain, he was no equal of Young America."[6] Rather than see Adam as perfect before the fall, Lincoln suggested that he lacked the advantages of education and rapid transportation. "In the way of land and live stock," Lincoln admitted, "Adam was quite in the ascendant. He had dominion over all the earth, and all the living things upon, and round about it. The land has been sadly divided out since; but never fret, Young America will re-annex it."[7] Indeed, in Lincoln's own lifetime "Young America" had settled the Midwest, annexed Texas and the Southwest, and moved rapidly into California and the Oregon territory. Lincoln supported the transcontinental railway, and as president in 1862 he signed the bill that authorized it. Like Everett in 1830 and Hill in 1909, Lincoln believed that the world was unfinished and waiting to be improved. All three embraced the technological creation story of improving any Eden whose inhabitants were few or ignorant or lacked a railroad.

During the twentieth century, technological creation stories continued to provide a historical road map for most Americans. Indeed, until 1970 or so the common pattern of American history textbooks was to follow technological expansion from the East, just after the Revolution, into the empty grid of the West. For most Americans, the meaning of technologies such as the axe and the mill was contained in and inseparable from the foundation stories. They were taught and they believed that their second creation did not exploit nature but improved it. The new order seemed essentially in harmony with God's first creation because the improvements seemed latent within the land. Far from violating an original perfection, the settlers were seen as working in partnership with nature. The clearing in the forest, the mill in the valley, the canal and railroad heading west, the dam in the canyon, and the irrigation ditch in the desert enhanced the bounty of the earth while helping mankind follow the biblical injunction to "be fruitful and multiply." Each improvement justified the taking of land that had "lain idle." As James Fenimore Cooper put it: "There is a pleasure in diving into a virgin forest and commencing the labors of civilization, that has no exact parallel in any other human occupation. . . . [It] approaches nearer to the feeling of creating."[8]

Technological creation stories all had a common structure. The steel "plow that broke the plains" and the steamboat that opened western rivers to upstream travel became the bases for narratives similar to those about the axe and the railroad. The steamboat brought the "unused" land of the Ohio Valley into the national market after 1810. Already in 1787, promoters had tried to establish New Athens "at the confluence of those majestic rivers, the Mississippi and Missouri" with the claim that it was "perhaps the most desirable spot in the known world"[9]; however, with no reliable upstream transportation, this particular boomtown never materialized. After the steamboat, in contrast, Wheeling, Marietta, Cincinnati, Louisville, St. Louis, Memphis, and many other cities mushroomed. Steamboats spread land speculation and the platting of new towns through the heart of the continent.[10] Streets were laid out in a rectilinear pattern, and as early as 1800 travelers in the Ohio Valley found the grid to be the typical form of town organization.[11] Land sales in such communities relied on retailing the steamboat narrative of settlement and expansion.

As early as 1819, journalists parodied claims about the future greatness of planned towns. A mock advertisement for the metropolis of Skunksburgh included "a reservation for seventeen banks, to each of which may be attached a lunatic Hospital." The city was sure to prosper "as soon as a canal shall be cut through the Rocky Mountains." The prospectus was signed by "Andrew Aircastle," "Theory M'Vision," and "L. Moonlight, Jr."[12]

New towns, particularly those "located" on major rivers, continued to attract investors throughout the antebellum period. Speculation led to exaggeration and sometimes to fraud. Some settlers, for example, were persuaded to buy lots in the nonexistent community of Nininger, Minnesota, on the banks of the Mississippi:

The city of Nininger, as delineated on the large and beautifully engraved and printed maps . . . was a well-built metropolis capable of containing ten thousand people. As delineated, it had a magnificent courthouse, this city being the country seat of Dakota County, Minnesota. Four or five church spires sprang a hundred feet each into the atmosphere. It had stores and warehouses, crowded with merchandise, and scores of drays and draymen were working with feverish energy to keep the levee clear of the freight being landed from half a dozen well-known steamboats.[13]

The speculators who invented Nininger made it "real" by means of engravings and newspaper stories. They published a *Nininger Daily Bugle* and filled it with advertising for local shops of every kind. The hapless investors who "bought" Nininger believed in the same story of instantaneous creation as their many compatriots who were moving west.

Once steamboat stories proliferated, counter-narratives followed. If Cincinnati, Louisville, Memphis, and St. Louis testified to the rapid growth that steamboats made possible, other sites never flourished. Floods destroyed some; others proved uncompetitive. There were appalling accidents, and the average boat did not last even 5 years.[14] Nor were steamboats always in service. They reached some localities only during high water in the spring and the fall.[15] Even when steamboat service was available, some travelers preferred other forms of transportation, including rafts, keelboats, and flatboats.

Alluring counter-narratives praising non-steamboat travel appeared as early as the 1820s. James Hall, for example, published a series of *Letters from the West* recounting a trip down the Ohio River. Hall disdained the steamboat, preferring the slower movement of a keelboat, which was quieter and stopped more frequently. When he felt like it, he could walk along the bank and talk to local inhabitants, while keeping pace with the boat. And when the keelboat got stranded on a sandbar, he simply took the occasion to go hunting in the woods. Thousands of others floated down the rivers on rafts and other homemade craft, but like Huck Finn they were in constant danger of being run over by steamboats. Many travelers considered steamboats too expensive, too noisy, or too fast. The steamboat passenger was confined to the deck most of the time,[16] could not hear the birds, the plash of the water, or the other river sounds, and often felt "a deep longing to penetrate the mysteries of the dark woods that accompany him like a living panorama along either shore."[17]

Thus, half a century before Mark Twain's *Life on the Mississippi* the steamboat was already the center of a technological creation story and a series of counter-narratives. Twain drew on these traditions and put them into dynamic tension with a nostalgic vision about a lost world that was fading away as railroads and barges took over the traffic.[18]

The steel plow also stood at the center of a system of stories. This was a distinctively American technology, developed to cut through the thickly matted sod of the prairie and turn over the moist and sticky soil beneath it. Just as anonymous eastern blacksmiths evolved the American axe in the eighteenth century, during the 1830s midwestern blacksmiths created the strong new plow, with its polished steel mouldboard. John Deere played a central role in this development, and by 1857 his factory alone was producing 10,000 plows a year.[19] Much as the solitary pioneer of the axe narrative entered a "primeval" forest, the settler on the western prairie entered an "empty" space with his team and his steel plow. Pioneers used the plow not only to prepare the land for agriculture but also to produce building material for their sod houses. The plow entered folk tradition in song and story, and it became a part of the national epic of self-creation. In response, some early counter-narratives emphasized Native Americans' rights to the land. Later, environmental critiques focused on the overuse of the plow, which had led to massive erosion and helped create the "dust bowl." One of the most famous documentary films of the 1930s celebrated "the plow that broke the plains" but also evoked the devastation caused by overfarming.[20]

Each foundation story—whether about the axe, the mill, the canal, the steamboat, the railway, the steel plow, or irrigation—accepted the imposition of a grid on an empty landscape. Each rejected English notions of scarcity, as expressed by Ricardo and Malthus, and instead embraced an expansive belief in resource abundance. Each rejected price controls and government regulation in favor of the free market. Finally, each assumed a world in which access to force and efficiency in using it improved constantly. How could anything but progress result from using natural forces to develop the immense resources of an empty continent in a free-market economy?

The technological creation story has by no means disappeared. On television, pioneers still enter the empty space of the American West. Children play computer games, such as Sim City, that invite them to create new communities from scratch in an empty virtual landscape where a grid defines the contours of roads and the arrangement of houses, factories, and commercial districts. Nor are such visions limited to children's games. The most popular exhibit at the New York World's Fair of 1964, General Motors' Futurama, depicted a series of idealized second creations.[21] With its communities under the ocean, its air-conditioned utopias in the jungle, and its lunar colonies, Futurama told 29 million visitors that Americans could transform any location and make it a comfortable second creation.

Many Americans think about ways to inhabit outer space. In March 1998, newspapers carried excited stories about the discovery of ice on the moon, in craters near its north and south poles. Scientists estimated that there were between 10 million and 100 million tons of frozen water, and this prospect set off a wave of speculation. The mainstream media exhibited little environmental concern.[22] They assumed that the water was a resource there for the taking, and that it might be used to make rocket fuel, irrigate greenhouses, and supply tourist hotels. Preliminary market studies suggested that upmarket lunar travelers would want condominiums, tennis courts, and even golf courses. This speculation continued the long-standing tradition in which American entrepreneurs exploit natural resources in order to found new communities that will lure settlers and tourists. A private American corporation, having discovered that the international treaty currently in force prohibits nations but not individuals from owning land on the moon, began selling lunar lots.[23]

Such designs for the moon recapitulate the foundation narrative. Once again Americans expect to use technology to transform the raw material of an unsettled space. As was the case with western irrigation, lunar speculators plan to build communities based on water. Colonizing the moon would require a far more complex technological infrastructure and greater amounts of energy than the dams on the Colorado River. Habitable buildings would require materials capable of withstanding temperature swings from –200 to +200 degrees Celsius. Lunar plumbing, heating, and air conditioning would have to function in weak gravity, and, in view of the distance to the nearest hardware store, inhabitants would require reliable backup systems. Moon colonies also would need regular shuttles to and from the earth to carry supplies and spare parts. Even this short list suggests the enormous costs and technological demands such a second creation would entail.

Americans in 1998 did not see water on the moon in quite the same way as their predecessors had viewed irrigation in 1898. There was no sense that God had put water on the moon in order to challenge men to make a second creation. No latter-day William Smythe proclaimed the gospel of lunar irrigation before Congress. Many nineteenth-century Americans had seen technology as a means of carrying out divine intentions "manifested" in the landscape. Few twenty-first-century Americans see providential designs in nature, though Americans still have the urge to transform new environments. A popular science fiction trilogy describes the settlement and biotic transformation of Mars, although it does take care to insert running environmental and other counter-narratives along the way.[24] A Mars Society advocates building greenhouses in the carbon-dioxide-rich Martian environment and recommends melting the ice caps, thickening the atmosphere, and "terraforming" the planet.[25] Space colonies "on the frontier" are a staple in science fiction, and Robert Zubrin's book *The Case for Mars* rather predictably compares the colonization of the planet to the settling of the American West. (The magazine *Wired* compared it to the settling of Antarctica.[26]) Second creation on the moon, on Mars, or further out in space demands more complex technology, but the underlying story remains the classic nineteenth-century narrative.

Because some Americans continue to tell these stories, it is important to underscore that the assumptions underpinning the second-creation narrative had lost much of their credibility by 1910. Historical experience had undermined all four of the enabling assumptions. The grid, for example, dictated that roads follow abstract geometry rather than topography. This worked reasonably well in the flat Midwest but became less and less serviceable as settlers moved into the arid and mountainous West. And the grid dispersed farmers into solitary homesteads, in effect declaring the community less important than the individual. This choice worked reasonably well on 160-acre farms; farther west, however, where scant rainfall dictated ever-larger farms, it increased isolation.

Furthermore, there was a fundamental contradiction between the grid and the very premise of second creation. How could Americans discover land's intended purpose if land was uniform? The latent potential of a forest, a river, or a watershed lay in its distinctive features. A cabin, a mill, a rail line, or an irrigation ditch was best situated not by relying on an abstract mathematical system but by studying the topography of specific locations. No homesteader or land speculator believed that land was everywhere the same. Under the Homestead Act, one undeveloped tract had the same value as all other tracts of the same size; however, settlers knew better and looked hard for the best land. Some tracts were wooded and had good soil; others were sandy or rock-strewn wastes. West of the 100th meridian, few homesteaders could afford the high initial investment that irrigation usually required. Irrigation created relatively few family farms but a good deal of agribusiness employing landless laborers. As the population moved west, a system of land sales intended to produce Jeffersonian democracy increasingly produced oligarchy instead. The geometrical division of space was egalitarian in principle, but in practice it denied the rights of first inhabitants, imposed inconvenient boundaries, created social isolation, and often helped investors more than homesteaders.

Then there was the assumption of free markets. In theory, the founders of each new homestead, mill village, or town along a river, a canal, or a railway were free from the crippling restrictions of local monopolies or legislative edicts. In practice, though, the market was not uncontrolled. Government laid out the ground rules for land division and settlement. The courts gave special rights to mills, canals, railroads, and irrigation districts. Railroads dictated the cost of transportation. Water companies controlled the flow of precious water. In theory, Americans had abandoned the "just price" in favor of competition; in practice, a few parties often controlled limited resources, and national markets in agricultural products favored middlemen and traders in grain futures over small farmers. At the time of its founding, a family farm, a mill village, or a railroad town seemed to embody the laissez-faire ideal, but each farming region and each new community soon manifested economic inequalities.

The third underlying assumption—resource abundance—also proved illusory. Americans argued that technology would allow them to avoid the downward resource spiral that Malthus and Ricardo predicted. Certainly the earliest settlers farmed the most easily developed tracts, but to farm many of the best lands required that one master sophisticated technologies. The dams, canals, and ditches of Utah, Colorado, Arizona, and California seemed evidence that the most productive lands were the last to be developed. As John Wesley Powell declared, "the arid lands of the West, last to be redeemed" were "the best agricultural lands of the continent."[27]

Irrigation required state and ultimately federal assistance, however, and guaranteeing abundance through government programs ultimately contradicted the ideology of free markets. Similar contradictions underlay earlier versions of the foundation story. In theory, an entire continent lay open to homesteaders; in practice, Congress and state legislatures handed over much valuable land to railroads, and more came under the control of speculators. When Frederick Jackson Turner pointed to the "closing" of the frontier, much was wrong with his theory, but he was right to recognize that the land itself was not limitless. Settlers hoping for homesteads in California had discovered this limit a generation before Turner. Furthermore, a generation of plains farmers learned to their sorrow that rain does not follow the plow, and that even massive federal assistance cannot ensure an abundance of irrigation water.

The fourth assumption of the foundation story was a Newtonian conception of force in which every human action had an equal and opposite reaction, with no loss. From this point of view, cutting down millions of trees or plowing millions of acres of sod merely gave sunlight to crops rather than to trees and wild grasses. By the middle of the nineteenth century, however, early conservationists such as George Perkins Marsh were warning that deforestation could lead to erosion and drought. After about 1870, contrary to the assumption of abundance, it began to seem possible that all the best farmland had been taken and that natural resources such as wood and pure water were running out. In the equilibrium of a Newtonian system, such facts were only alarming; in an entropic universe, they were catastrophic. After about 1860, as scientists developed the theory of thermodynamics, the idea that nothing was lost in the exploitation of resources became untenable. Leading scientists such as Lord Kelvin saw no escape from entropy, which made long-term progress impossible. The forces and resources available to humankind would steadily diminish. The late writings of Henry Adams emphasized these drastic implications of thermodynamics, which would inevitably lead to the heat death of the universe.

By 1910, then, no mathematician could justify the grid; no historian could vouch for the actuality of the free market during the canal era, the railroad boom, or the era of irrigation; few economists could argue for perpetual agricultural abundance; and no physicist could argue away the laws of thermodynamics. With their underlying assumptions in tatters, second-creation stories began to lose their centrality. Americans devised other narratives based on new technologies: the automobile, the assembly line, the motion picture, the airplane, radio. The new stories focused not on strict geometries but on fluidity. They were still about a free market, but they were less about the exchange of commodities than about movement, communication, replication, and simulation. None were about settling down in one place. All were disconnected from the land. All suggested new kinds of equality, new constructions of the self, and new ways to obliterate the past and start anew. All emphasized not only the conquest of space but also the manipulation of time.

By the 1920s, second-creation stories no longer described the future or even the present; they primarily referred to the past. Yet the trope of second creation by no means disappeared. Like captivity narratives and Puritan jeremiads, technological foundation narratives continue to be rewritten and to circulate in the repertoire of stories many Americans tell to make sense of their existence in the New World. Even if the beliefs implicit in the stories have become intellectually indefensible, Americans remain loath to abandon the vision of second creation. The narrative has become so deeply embedded in American thinking that it has ceased to be merely a story. It has become a national myth of origin.

Some, however, rejected the technological creation stories that began to appear in the late eighteenth century from the start. Because most of the counter-narratives were based mainly on detailed local knowledge, they formed a less unified tradition than the foundation stories they attacked. Yet it can be said that many a counter-narrative began by denying the proposition that America was empty, undeveloped space waiting for the fructifying touch of a new people. Many recalled a meaningful community

and a particular landscape that existed before "development," and chron-
icled the losses alongside the gains that resulted from using new technologies. They doubted the ultimate advantages of technological creation. As
Nathaniel Hawthorne put it: "We who are born into the world's artificial
system can never adequately know how little in our present state and circumstances is natural, and how much is merely the interpolation of the
perverted mind and heart of man. Art has become a second and stronger
nature."[28] In such formulations, the technological arts that Edward Everett
praised as uplifting became perverted distractions. More recent counter-
narratives have attacked the expansionist logic of second-creation stories,
which taken as a whole moved from the East in colonial times toward
national development in the West during the nineteenth century. They
reconceive the frontier as a borderland where different cultures clash and
coalesce.[29] They retell the national story from the viewpoints of Native
Americans, Hispanic Americans, failed farmers, factory workers, migrants,
and others who did not control the technologies of development.[30]

Some who protested the second-creation narrative of North America
recast the story in terms of conquest. Mary Austin, Frank Waters, and
Louise Erdrich focused on earlier inhabitants who lost out to the newcomers. Another counter-narrative tradition dealt with the hardships
and disappointments of white settlers. A host of nineteenth-century journalists and travelers described the suffering of pioneers who believed in
second creation but who had the misfortune to attempt it in the wrong
places and who failed because of floods, drought, fires, or other disasters.
Henry George, Frank Norris, Sherwood Anderson, and John Steinbeck
attacked monopolies that made second creation difficult or impossible to
realize because they controlled land, water, or transportation. Yet another
counter-narrative tradition focused on damage to the environment;
examples may be found in the work of Henry David Thoreau, John
Muir, Henry James, Aldo Leopold, and William Faulkner. In "Big Two-
Hearted River" Ernest Hemingway described a logged-over district in
northern Michigan that had suffered repeated fires and become a wasteland.[31] A solitary angler, recently returned from World War I, finds no
solace in the ruined land.

Despite the counter-narratives, however, the technological creation story
long remained dominant. Giving it up was far more difficult than trading

in an old automobile for a new one; it was comparable to admitting that owning and driving automobiles was environmentally destructive and switching to mass transit and denser housing. Rejecting the foundation story required recognizing that the geometrical order of the grid was profoundly unnatural and that its egalitarian erasure of difference was an illusion. Giving it up meant recognizing historical injustices to the first inhabitants, accepting environmental limits, and acknowledging the ideological nature of the free market. Rejecting the foundation story implied the loss of white entitlement to the continent. Discarding second-creation stories required acknowledging cultural conflicts and listening to counternarratives. Giving up technological foundation narratives proved so difficult, in fact, that it required most of the twentieth century even to begin the process.

Both the narratives of settling new lands and creating new communities and the counter-narratives that emphasized injustice, misuse, and environmental destruction remain important parts of the discourse of what it meant to inhabit the land. But these are not the only stories that shaped public debate about resources, wilderness, and land use. Rather than reject the technological creation story outright, many twentieth-century Americans embraced another story: the recovery narrative. Essentially about remaking a despoiled landscape, the recovery narrative begins not with empty space waiting to be improved by new settlers, but with a place corrupted and degraded by human misuse. At such sites, the free market has unleashed selfish individualists who have exploited the land for short-term gain. To restore natural beauty and environmental harmony, a countervailing social force enters the area in the form of a nonprofit organization or government agency. After taking control away from short-sighted private interests, this institution redevelops the area. It cleans up pollution, halts erosion, plants new trees and shrubs, restocks rivers and lakes with native species, protects wildlife, and gives back to the public a restored version of the natural world. This restoration is not identical with first creation, or the world as it was before the environment was tampered with. Nor is it simply a "correction" of abuses that provides the public with an idealized second creation that "should" have emerged through free enterprise. Instead, revising the assumptions of the second-creation story, the result is a managed site that seeks to recoup the virtues of the first landscape while making it accessible to tourists and profitable for private enterprise. This was the underlying narrative of the conservation movement.

The recognition that second creation had failed led many to embrace new initiatives aligned with the recovery narrative. The process began with the preservation of Yosemite in 1864 and Yellowstone in 1871 and continued with the establishment of state and national forests.[32] Well before 1900, Congress deemed the protection of natural sites to have national significance and, by preserving them, recognized that the free market in land had a destructive side. The history of Niagara Falls provides an excellent example. During much of the nineteenth century the area around the falls had been privately owned and developed for commercial purposes. High fences hid the falls from visitors until they paid an admission fee. Souvenir stands proliferated, and a commercial atmosphere prevailed. In 1886, responding to rising public protest against these conditions, the Canadian and American governments began to expropriate the many private properties that bordered the falls and hired Frederick Law Olmsted to landscape the area. Olmsted removed fences and other commercial obstructions and created a "natural" frame.[33] Niagara Falls became an international park with no admission fee.

De-commercializing Niagara Falls and transforming its shores into a park exemplified the recovery narrative that would become dominant in the twentieth century. This narrative was inscribed in laws that established state and federal agencies to clean up, restore, and protect the environment. Like the second-creation story, this was also a technological narrative. If Americans had misused machines to log too many trees, plow too intensively, irrigate too assiduously, or otherwise damage the land, the wiser use of technologies would make a region beautiful and productive again. Indeed, in the early decades of the twentieth century the new professional foresters, game managers, and park rangers used science to maximize productivity. Aldo Leopold wrote the standard book on game management, which began with this declaration: "Game management is the art of making land produce sustained annual crops of wild game for recreation use."[34] A forester from the same years would have written much the same thing, substituting trees for game and log production for recreation. The conservation ethic treated nature as raw material that could be better managed than in the past, but it did not conceive of animals or trees as having rights, nor did it give first priority to the protection of endangered species. Nor did conservation oppose all forms of development. Niagara Falls had been recovered as a tourist site, but water was still diverted from it to drive factories and to make electric power.

During the New Deal, the recovery narrative encompassed entire landscapes that earlier generations had ravaged. Franklin D. Roosevelt put the Civilian Conservation Corps to work replanting millions of acres of cut-over forest land, and he took marginal lands in the dust bowl out of production. Likewise, the Tennessee Valley Authority showed how government dams could exercise rational control in a river system that drained an area larger than England. Not only would the TVA end ravaging floods; it would also improve navigation, generate electricity, show farmers how to diversify and rotate their crops, produce fertilizer for them, and speed the modernization of rural areas.

The incorporation of a counter-narrative in the recovery narrative is exemplified by the 1937 documentary film *The River*, which begins by dramatizing the destruction of entire watersheds during the nineteenth century. Directed by Pare Lorentz, *The River* emphasizes how earlier Americans had used technologies to build the nation but in the process had abused the land. Accompanied by images of falling rain that washes away the soil, the narrator intones:

We built new machinery and cleared new land in the West. We built a new continent. We built a hundred cities and a thousand towns—But at what a cost! We cut the top off the Alleghenies and sent it down the river. We cut the top off Minnesota and sent it down the river. We left the mountains and the hills slashed and burned, and moved on. We played with a continent for fifty years.

After invoking this counter-narrative, the film describes a process of restoration that will remake the Mississippi Valley. The misuse of earlier technologies (excessive lumbering and wasteful farming techniques) will be remedied by more modern technologies. Trees will be replaced, levees improved, dams built, and farmers reeducated. The narration continues:

Today a million acres of land in the Tennessee Valley are being tilled scientifically. Power for the farmers of the Valley, power for the villages and cities and factories of the Valley, West Virginia, North Carolina, Tennessee, Mississippi, Georgia and Alabama, power to give a new Tennessee Valley to a new generation, power enough to make the river work!

Flood Control? Of the Mississippi? Control from Denver to Helena, from Itasca to Paducah; from Pittsburgh to Cairo—control of the wheat, the corn and the cotton land; control enough to put back a thousand forests; control enough to put the river together again before it is too late; before it has picked up the heart of a continent and shoved it into the Gulf of Mexico.[35]

In such technological recovery narratives, humankind still plays the dominant role, as conservation replaces individualistic waste. The repetition of the word 'control' emphasizes that, in this story, nature remains subservient. Gifford Pinchot and other foresters generally shared this anthropocentric view, according to which natural environments could be recovered and managed to serve human needs.

A countervailing movement slowly emerged that demanded preservation of the wilderness, with as little intervention by humans as possible. John Muir, an early spokesman for this movement, called for certain areas to be set aside from all forms of development in perpetuity. Such ideas, which had seldom prevailed before 1900, became increasingly popular in the twentieth century. As Roderick Nash has pointed out, the increasing popularity of primitive and pristine nature in the first years of the twentieth century can be seen in the creation of the Boy Scouts and in the speeches of Theodore Roosevelt, who extolled the harsh but invigorating conditions of the West and denounced overcivilized urban life.[36] Organizations such as the Sierra Club gradually became advocates of preserving sublime landscape for its own sake. At first, the wilderness ethic may seem to contain no narrative at all, but simply to deny the essential value of second creation. From this perspective, primeval forests should not be cut down, dams and railways should not be allowed to violate national parks and forests, and irrigation projects should not be allowed to destroy delicate desert ecological systems. If the conservation movement rewrote second creation into a recovery narrative, the wilderness movement wanted to keep some portion of nature pure and prevent it from becoming a part of any human story.

But wilderness was also a part of the narrative of American culture that Frederick Jackson Turner had told so convincingly—a narrative in which the primeval landscape strips European culture away, creating Americans in the process. In this narrative, it was necessary to preserve wilderness so that men and women would always be able to relive the confrontation with nature as a powerful "other." Wilderness became the essential foil to civilization. It was needed in order to reveal the artificiality of culture. If the second-creation story was about using new technologies to build a new society from the raw material of nature, the wilderness story was about creating a new self by leaving civilization behind and stripping life to its bare essentials. Taken to its logical extreme, the wilderness story was about ecologists who tried to think like a mountain.

Cultivating such perspectives, one could imagine the settlement of the United States from the viewpoint of animals. Jim O'Brien's often-reprinted essay "A Beaver's Perspective on North American History" begins before any humans had arrived. Beavers dammed streams, created small canals and ponds, and reshaped the landscape in ways that affected the larger biotic community. The beaver population, which may have reached 60 million at its height, went into decline with the arrival of the European fur trade. O'Brien's narrative radically reconceives American history as a catastrophe. From the beaver's point of view, there was "a wave of destruction fanning out from the St. Lawrence River, from the Dutch and English colonies on the Atlantic seaboard, later from Hudson's Bay, and still later from John Jacob Astor's fistful of the Oregon coast." This story assumes that the land was not empty in 1492 or even in 30,000 BC. In it, the intrusion of new technologies (the steel trap and the gun) destroys the beavers' world.

During the 1960s, preservation of wilderness was written into federal law, and for the first time areas were set aside not for game management, forestry, or recreation but for themselves. Giving legal sanction to the idea of wilderness signaled a public willingness to see nature as a biotic community in which humans had to respect the rights of other species. Such legislation gave sanction to the time before the second-creation story could begin. It sought to return to the historical moment before the national grid had been imposed, or even before Native Americans could lay claim to the land. The idea of wilderness erased all humans and posited a landscape outside history, with no development—a landscape where the only possible narratives included neither roads nor technologies nor the linear sense of time. In the wilderness narrative there was no grid; each space was unique and precious, and places that had never figured in a second creation became the most valuable of all.

Of the two major political parties, Republicans have been less willing than Democrats to suspend second creation by establishing wilderness areas; however, both Republicans and Democrats have added land to the untouchable public domain, using much the same rhetoric. In 1984 President Ronald Reagan signed a law that designated thousands of acres as wilderness in North Carolina, New Hampshire, Vermont, and Wisconsin.[37] "Each of these areas," Reagan proclaimed, "is intended to be completely natural—no housing developments, no power lines, just forest, rock, wind, and sky." "These wilderness areas," he said, "will remain just as they are, places of beauty and serenity for hikers, campers, and fishermen. Generations hence, parents will take their children to these woods to show them how the land must have looked to the first Pilgrims and pioneers. And as Americans wander through these forests, climb these mountains, they will sense the love and majesty of the Creator of all of that."[38] Not only would visitors be able to imagine what F. Scott Fitzgerald once called the "fresh green breast of the new world"[39]; Reagan expected that they would also recover Puritan piety at the same time. These were muddled expectations, for the continent did not seem to the first settlers to be a "green breast" so much as a howling wilderness. The attitude that Reagan attributed to the Pilgrims was a nineteenth-century view of the land, one that was already a nostalgic vision in 1830. This point of view, developed by the Hudson River School of painters, was expressed as an ethic by Thomas Cole, Henry David Thoreau, John Muir, and those they influenced in the twentieth century.

By the 1980s, "wilderness" had become an integral part of the national landscape. It was expected that in these places future generations would be able to recover the Europeans' first ecstatic encounter with the land. Wilderness areas, in Reagan's formulation, reproduced the magical first moment of America's discovery without the historical inconvenience of a pre-existing Native population. However, as Fitzgerald noted in the famous passage at the end of *The Great Gatsby* that evokes the response of the Dutch sailors to the New World, the vision of a wilderness "pandered" to their imaginations. They were seduced by an illusion of virginal perfection that awaited them.[40]

As Americans entered the third millennium, they embraced three interlocking stories that together defined a sequence: the wilderness tale, the second-creation story, and the recovery narrative. The imagination and preservation of a pristine America did not exclude but rather prefaced the second-creation story. If one part of the government was charged with preserving the wilderness in its pre-narrative state, it did so in order to provide a point of departure for the technological foundation story, which still served as a basic account of entitlement to the national land mass. The recovery narrative of conservationism was institutionalized in the Forest Service, the Environmental Protection Agency, and a host of state and local institutions, all of which worked to ensure that the second-creation narrative did not become a story of rapacious land exploitation. Outside government, a panoply of organizations and movements repeated these three stories. The Sierra Club, Friends of the Earth, and Earth First! demanded more wilderness areas. In opposition, private enterprises lobbied for more technological creation as they sought land for mining, oil exploration, ranching, farming, and logging. On January 12, 2001, the *Wall Street Journal* editorialized on their behalf: "Radicals like to frame the environmental debate by dividing people into those for nature and those against it. In fact, the divide is much more practical, between those who would manage our land and those who would not. Because we necessarily intrude upon nature, we must take it upon ourselves to manage it— through road building, fire policies and, most importantly, logging. . . . To pretend that forests are pristine ecosystems that should be left to grow wild is absurd." Thus the narratives of wilderness, second creation, and conservation were woven into public debate.

In the 2000 presidential campaign, George W. Bush stood primarily within the second-creation tradition. He doubted the existence of global warming, called for oil drilling in the Arctic National Wildlife Refuge, and promised to appoint a secretary of the interior acceptable to loggers and paper manufacturers. Al Gore had less utilitarian policies and rhetorically embraced the wilderness ethic. Likewise, just before leaving office, President Bill Clinton banned road building on 58 million acres of national forest land (an executive order that his successor promptly countermanded). Yet the differences between the two national parties were less decisive than the election rhetoric suggested. The Clinton presidency disappointed many environmentalists. Bush, after he reached the White House, eventually admitted that global warming existed and found it prudent to back down on some environmental issues, adopting instead the recovery narrative of conservation. And like Reagan, any Republican inside a designated wilderness area could speak reverently and sincerely about the sublime lessons to be learned from untouched nature.

Once not only the Sierra Club but also Ronald Reagan and Barry Goldwater embraced the three narratives of wilderness, second creation, and conservation, critics began to understand that each of these stories was in its own way about human domination of the natural world. In response, critics began to develop another counter-narrative, which some called eco-feminism. Carolyn Merchant called for "a partnership ethic would bring humans and nonhuman nature into a dynamically balanced, more nearly equal relationship." Significantly, she recognized that a partnership ethic required a new narrative. This new story "would not accept the idea of subduing the earth, or even dressing and keeping the garden, since both entail total domestication and control by human beings. Instead, each earthly place would be a home, or community, to be shared with other living and nonliving things."[41]

The new counter-narratives consider "nature" to be a complex set of human ideas projected onto an even more complex set of phenomena. Rather than seeing people as standing outside the environment and manipulating it—a fundamental proposition in the ideologies of wilderness, second creation, and conservation—these counter-narratives recognize that people are inseparable from the spaces they inhabit. William Cronon put it as follows: "Nature is far less natural than we think. . . . So many modern ways of thinking about nature too easily accept the false dualism between nature and culture, positing an inescapably fallen humanity that cannot help being unnatural—whereas the paired naturalness and unnaturalness of humanity is the very thing we most need to linger over and understand."[42] In contrast to fantasies of escape into pure nature, Cronon recognizes that humans can never be outside nature. They remain inextricably a part of it, and they need a partnership ethic in order to rethink the use and abuse of land and resources.

Wilderness, and the narrative of escape from culture that it implies, is not necessarily a rejection of the second-creation story. Rather, wilderness becomes a prehistoric baseline against which Americans can measure the achievements of second creation. Wilderness areas naturalize the division into two forms of space: an untouchable remnant of original nature (or first creation) and a generic, blank space waiting to be surveyed into squares and inserted into a national narrative of inevitable progress. Setting aside wilderness areas makes it easier to hold onto the constellation of stories that justify white Americans' entitlement to the land.

Yet, as the many counter-narratives have suggested, there are other possibilities. A multi-cultural America could move beyond stories of entitlement to stories of partnership. A people that recognizes environmental limits and co-dependency could imagine narratives based on stewardship. A creative society could use new technologies not as tools of control but as means of achieving sustainable development. Such changes will only become acceptable choices, however, to the extent that Americans embrace new stories that move beyond second creation.

Notes

Introduction

1. For a discussion of the experience of new technologies, see Nye, *American Technological Sublime.*

2. Silko, "Landscape, History and the Pueblo Imagination," in Glotfelty and Fromm, eds., *The Ecocriticism Reader,* pp. 271–273.

3. Kelley and Francis, *Navajo Sacred Places,* pp. 38–39.

4. On the Hopi, see Silko, "Landscape, History and the Pueblo Imagination," pp. 269–271. On the Chippewa, see Graham, *A Face in the Rock,* pp. 15–17, 29, 45.

5. Slotkin, *Regeneration through Violence.*

6. Rowlandson, *A Narrative of the Captivity, Sufferings and Removes, of Mrs. Mary Rowlandson.*

7. White, "Frederick Jackson Turner and Buffalo Bill," in Grossman, ed., *The Frontier in American Culture,* p. 9.

8. See the early chapters of Tichi, *New World, New Earth.*

9. These narrative conceptions of the settlement of America have been analyzed often. See, e.g., Jones, *O Strange New World;* Marx, *The Machine in the Garden.*

10. On the Spanish in the New World, see Savelle, *Empires into Nations;* Meinig, *The Shaping of America,* pp. 3–76.

11. Miller, *Errand into the Wilderness.*

12. See, e.g., Bercovitch, *Puritan Origins of the American Self.*

Chapter 1

1. Tocqueville, "A Fortnight in the Wilds," in Mayer, ed., *Journey to America,* p. 329.

2. Whittier, "Pawtucket Falls," in Whittier, ed., *Prose Works,* volume 2, pp. 302–303.

3. Emerson, "Wealth," in *Selected Writings,* p. 698.

4. Ibid.

5. I will not be concerned here with the writings of inventors and engineers, who provide technical expositions about how machines work. These are narratives, to be sure, but usually they are not foundation stories.

6. See Nye, *Consuming Power,* pp. 234–239.

7. McDermott, ed., *Before Mark Twain.*

8. The steamboat as the center of a system of foundation narratives is discussed in chapter 12.

9. In another common sense of the word, no narrative is original; one can usually find an earlier variant.

10. When technologies are quite new, they will often be described as wonders in terms of the sublime. Some writers believed, as Charles Caldwell put it in 1832, that the railroad had a morally uplifting influence, and nineteenth-century Americans celebrated in turn their canals, railroads, bridges, and skyscrapers in these terms. See Nye, *American Technological Sublime.*

11. Emerson, "Wealth," p. 698.

12. Adas, *Machines as the Measure of Men,* p. 15.

13. White, *Tropics of Discourse,* p. 96.

14. See Harden, *A River Lost,* pp. 49–50, 221, 239–245.

15. Neihardt, *Black Elk Speaks.*

16. For a typology of energy narratives that focuses on the twentieth century, see Nye, "Energy Narratives," in Nye, *Narratives and Spaces,* pp. 75–91.

17. Etzler, *The Paradise within the Reach of All Men, without Labour, by Powers of Nature and Machinery,* p. 1. All citations from the British edition, courtesy of the British Library.

18. Etzler proposed harnessing as much as possible of what he estimated to be the 20,750 horsepower in the wind that blew over every square mile of the earth's surface (ibid., p. 13).

19. Ibid., pp. 64–65.

20. Whittier, *A Stranger in Lowell* , reprinted in Folsom and Lubar, eds., *The Philosophy of Manufactures,* p. 424. Thoreau was also sharply critical of Etzler; see Thoreau, "Paradise (to Be) Regained," reprinted in ibid., pp. 413–414.

21. Webster, reprinted in Rozwenc, ed., *Ideology and Power in the Age of Jackson,* p. 33.

22. Ibid., p. 36.

23. Greeley, *Art and Industry as Represented at the Exhibition at the Crystal Palace, 1853–4,* p. 52.

24. See Robinson, *American Apocalypses.*

Chapter 2

1. Macherey, *A Theory of Literary Production,* p. 155.

2. Hall, *The Hidden Dimension,* pp. 146–147. The classic work on the layout of American cities, showing the spread of the rectilinear pattern, has been done by John V. Reps; see, e.g., *Cities of the American West.*

3. Onuf, "Liberty, Development, and Union," p. 208.

4. "The Accessible Landscape," in Jackson, *A Sense of Place, a Sense of Time,* p. 4.

5. Hart, *The Rural Landscape,* p. 155.

6. See Stilgoe, *Common Landscape of America,* p. 103.

7. See Johnson, "Towards a National Landscape," pp. 127–132.

8. The classic work is Smith, *Virgin Land.*

9. Humphreys, "Address to the People of the United States," reprinted in Folsom and Lubar, *The Philosophy of Manufactures,* p. 132.

10. Congress adopted the land units of the section and the township in order to sell farms in advance of settlement, but it rejected purely geometrical state boundaries and left the definition and naming of new states to the future. For an overview, see Meinig, *The Shaping of America,* volume 1, pp. 341–343, or Boorstin, *The National Experience,* pp. 241–248. Boorstin points out that part of Georgia was laid out into square-mile sections as early as 1717.

11. This inscription is beautifully visualized in Corner and MacLean, *Taking Measures across the American Landscape.*

12. Cosgrove, "The Measures of America," p. 9.

13. Philip Fisher, "Democratic Social Space," in Fisher, *Still the New World,* p. 47.

14. Ibid., p. 48.

15. Ibid., p. 50.

16. Chevalier, *Society, Manners, and Politics in the United States,* p. 236.

17. Hall, *Travels in North America,* volume 1, p. 144.

18. Richardson, *Beyond the Mississippi,* p. 30.

19. Jackson, *A Sense of Place, A Sense of Time,* p. 154.

20. Cather, *My Antonia,* p. 7.

21. Rajchman, *Constructions,* p. 91.

22. Ibid.

23. "Farming," in Emerson, *Selected Writings,* p. 756.

24. Ibid. The most important of Henry Carey's many works was his three-volume *Principles of Political Economy.* See Dawson, "Reassessing Henry Carey."

25. Emerson, "Farming," pp. 756–757.

26. Ibid., p. 757.

27. Allen, *The Practical Tourist,* volume 1, pp. 234–235.

28. *Pioneers: Narratives of Noah Harris Letts and Thomas Allen Banning,* p. 39.

29. Lincoln, "Second Lecture on Discoveries and Inventions," pp. 356–357.

30. Ibid., p. 357.

31. Dempsey, "Just Price in a Functional Economy," pp. 471–476.

32. Smith, "Food Rioters and the American Revolution," p. 3.

33. Ibid., p. 23.

34. Ibid., p. 24.

35. Ibid., p. 25.

36. See Henretta, "The Transition to Capitalism in America," pp. 218–237.

37. Inexpensive power to propel transportation networks made the idea of the free market into a reality. When energy is abundant and inexpensive, the "right" to have it remains implicit and is seldom stated. But when this "right" is threatened at the gasoline pump, as it was in the 1970s or again in the spring of 2000, Americans become angry. They feel that threats to automotive mobility, heating, air conditioning, and economic well-being threaten the free market itself. Because Americans took for granted the relationship between energy and the free market, during the 1970s oil crisis they were surprised to see how much inflation was caused by the sudden rise in oil prices. It had become "normal" that transportation represented only a small part of total costs, and the ability to ship anything just about anywhere was taken for granted.

38. Thomas Jefferson had read and known many of the Physiocrats when he served as ambassador to France. Smith devotes only a chapter to the agrarian system they proposed. He emphasized that it had never been tried in any nation, and he suggested it was unlikely to be put into practice. In the same year that Smith published his book, however, Jefferson was writing the Declaration of Independence, which would give such ideas an opportunity to be tried. See Smith, *The Wealth of Nations,* book IV, chapter 9.

39. Smith, *The Wealth of Nations,* book V.

40. Smith could take British mastery of sail power for granted. Fleets of merchant vessels circling an island with many good ports and driven by cost-free wind made the idea of a national free market a reality for Britain long before many other nations. For the United States, a different kind of free market of continental dimensions emerged, as North America was transformed into a single system of surveyed land, served by canals, railroads, and steamboats.

41. Daniel Webster, reprinted in Rozwenc, ed., *Ideology and Power in the Age of Jackson,* p. 38.

42. Ibid., pp. 34–35.

43. Certeau, *The Practice of Everyday Life,* p. 37.

44. Ibid., p. 36. Indeed, Certeau's analysis draws upon Clausewitz's *On War.*

45. Bigelow, *The Useful Arts,* volume II, p. 81.

46. Ibid. Bigelow also asserts: "It is the office of machines, to receive and distribute motion, derived from an external agent, since no machine is capable of generating motion, or moving power, within itself. The sources from which the moving power, applied to machinery, is obtained, are various, according to the nature of the object, and the amount of force, which is required."

47. Jammer, *Concepts of Force,* p. 53.

48. Ibid., p. 86

49. Ibid., pp. 103–104.

50. Martin, *American Literature and the Universe of Force,* pp. 10–11.

51. There were difficulties here, however, as Newton realized. See Jammer, *Concepts of Force,* pp. 124–139. See also Westfall, *Force in Newton's Physics.*

52. In view of the absence of a single term equivalent to what today is commonly meant by 'energy', I adopt the term 'force', understood in the rather mechanistic sense in which it was variously applied to gravity, heat, and mechanical movement. If 'force' presented insuperable problems of definition to the philosopher and scientist, in colloquial speech it ultimately referred to physical effort, expanding steam, and falling water.

53. Rather than being concerned with philosophical or scientific problems, the nineteenth-century American was intensely interested in questions of competition between forces: How many men were required to perform as much work as a horse? Could a new railroad train outrun a horse? Could the fastest sailing ship, the clipper, defeat a steamboat? Which steamboat was fastest? Could the champion steel-driving man, John Henry, best a steam drill? Such questions all concerned increasing speed or gaining access to more force, and they were widely studied in the nineteenth century. See Ferguson, "The Measurement of the 'Man Day'"; Hunter and Bryant, *A History of Industrial Power in the United States, 1780–1930,* volume 3, pp. 26–28.

54. On the simultaneous development of the water turbine in the United States and Europe, see Hunter, *A History of Industrial Power in the United States,* volume 1, pp. 292–318.

55. Macherey, *A Theory of Literary Production,* p. 155.

Chapter 3

1. Hall, *Travels in North America,* volume 1, p. 146. Hall continues, "In the course of the first year he raises a little Indian corn, and other things, which keep him alive and enable him to supply various wants. Next season, he makes a fresh start with improved means, and a few less discomforts."

2. Cooper, *The Chainbearer,* p. 97.

3. John Locke, *Second Treatise on Civil Government,* book V, p. 27, discussed in Sibley, *Political Ideas and Ideologies,* pp. 376–377.

4. Hough, *The Way to the West*, p. 7.

5. I thank Bruce Clunies Ross for pointing out to me that Native Americans had developed expertise in working with hickory before Europeans arrived. This suggests the delicious irony that the symbol of the white settler was itself a product of interaction with the previous inhabitants.

6. Pursell, *The Machine in America*, pp. 14–15.

7. For a brief summary of the elimination of woodlands, see Nye, *Consuming Power,* pp. 19–21.

8. Cited in Kouwenhoven, *Made in America,* p. 16. On the history of the American axe, see Kauffman, *American Axes;* Russell, *Firearms, Traps, and Tools of the Mountain Men,* pp. 232–310.

9. Testimony of William Vickers before British Parliament, *Second Report . . . Exportation of Machinery* (1841), p. 73, cited in Pursell, *The Machine in America,* p. 15.

10. Richard Frame, "Short Description of Pennsylvania," cited in Glacken, *Traces on the Rhodian Shore,* p. 697.

11. Kulik, "American Difference Revisited," p. 30.

12. Ibid., p. 26.

13. One of these was on display in the Cornell University Library just before the 2000 election.

14. Turner, *Rise of the New West,* p. 332.

15. Hough, *The Way to the West,* p. 7.

16. Ibid., pp. 8–9.

17. Martin, *Profits in the Wilderness,* p. 9.

18. Ibid., p. 38.

19. Shurtleff, *The Log Cabin Myth,* p. 211.

20. Ibid., p. 168.

21. Weslager, *The Log Cabin in America,* p. 200–201. An early investigation that makes a good case for the precedence of the Swedish log house is Mercer, "The Origin of Log Houses in the United States."

22. Weslager, *The Log Cabin in America,* pp. 169–170

23. Ibid., pp. 85–92.

24. Ibid., p. 231.

25. Ibid., pp. 238–240.

26. Powell, *Historic Towns of New England,* pp. 84–88.

27. Byrd, *Histories of the Dividing Line betwixt Virginia and North Carolina,* p. 37.

28. Byrd, cited in Weslager, *The Log Cabin,* p. 244.

29. Cited in Nobles, *Divisions Throughout the Whole,* p. 116.

30. Nobles, *Divisions Throughout the Whole,* p. 119.

31. Ibid., pp. 120–121.

32. White, "Frederick Jackson Turner and Buffalo Bill," p. 17. In April 2001, the New York Public Library mounted a large map exhibition that made the same point. Until the Revolution, maps of the American West consistently indicated the presence of Native American nations.

33. Wright, *The Prose Works of William Byrd of Westover,* p. 251.

34. Benjamin Rush, "Letter to Thomas Percival," October 26 1786, in Butterfield, *Letters of Benjamin Rush,* volume I, p. 400. This essay was composed in the form of a long letter to an English physician, who read it to the Manchester Literary and Philosophical Society and then published it in *Memoires of the Manchester Literary and Philosophical Society* (III, 1790: 183–197). In the United States, Rush published the essay in the *Columbian Magazine* (1: 117–122). Rush later published it again in his *Essays* (1798).

35. Rush, "Letter to Thomas Percival," p. 401. Many of Rush's generalizations are confirmed by others. For example, a German doctor traveling through western Pennsylvania in the 1780s met a man who lived primarily by hunting, which he said was spoiled by new neighbors although no one lived within 7 miles. Such hunters, he recorded, "live very much like Indians and acquire similar ways of thinking. They shun everything which appears to demand of them law and order, dread anything which breaths constraint." ("Travels of Johann David Schoepf," in Harpster, *Crossroads,* pp. 133–134)

36. Rush, "Letter to Thomas Percival," p. 402.

37. Ibid.

38. Ibid., p. 403.

39. Ibid., p. 404.

40. Franklin, interview with Gottfried Achewall, in Labaree, *Papers of Benjamin Franklin,* volume 13, pp. 353–354.

41. There are some interesting differences. Whereas Crèvecoeur described a general intermarriage among descendants of Europeans, Rush, in a section that he later removed from the published essay, classified his three figures according to ethnic stereotypes: the Scots-Irish and the old-stock Americans are frontiersmen; the Yankees, the Irish, and émigré gentlemen are transient squires; the Germans and the English are fully civilized farmers. Rush qualifies these remarks, admitting that he has seen some frontiersmen stick to one spot during subsequent transitions and that some people from every ethnic group may be found in each of the categories. Franklin saw the frontiersmen as older-stock white Americans, the second wave as Scots-Irish (who improved the land somewhat before selling out), and the third and final wave as made up primarily of Germans. The idea of an ethnic succession survived for a long time as a description of immigrant experience, but this particular version did not. Crèvecoeur's conception of a new race created by intermarrying among the settlers became more prevalent.

42. Dwight, *Travels, in New England and New York*, volume 2, p. 321, letter 13.

43. In contrast, the captivity narrative put the armed conflict with Native Americans at the center, but this dramatic tale is only half the story. The other, ultimately inseparable tale is that of the transformation of the land. Indeed, Native Americans often appear as willing assistants in the story of settlement. In Byrd's *History of the Dividing Line*, for example, the Native Americans seem to offer no resistance to the surveying party; indeed, several of them work as hunters and scouts.

44. Isaac Weld, *Travels through the States of North America*, cited in Williams, *Americans and Their Forests*, pp. 124–125.

45. Franklin, "To ——: Information to Europeans Who Are Disposed to Migrate to the United States," in Labaree, *Papers of Benjamin Franklin*, volume 13, p. 550.

46. Zelinsky, *Exploring the Beloved Country*, p. 193.

47. "Diary of David McLure" in Harpster, *Crossroads*, p. 119.

48. "Journal of Arthur Lee," in Harpster, *Crossroads*, p. 157.

49. Birkbeck, *Notes on a Journey in America*, pp. 123–124.

50. See Slotkin, *Regeneration through Violence*, pp. 268–312.

51. Ibid., p. 305.

52. Ibid., p. 311. A surveyor who traveled from Connecticut into the Ohio Valley in 1819 found that settlers did not seek out the open prairie. They assumed that forested land was more fertile than open land, and they needed the trees for construction. Therefore, "the new settler builds his cabin in the edges of the timbered land, and fences in the prairie ground, sufficient for his tillage, which he has no trouble of clearing. But the great distance between the timbered land in many places, it being from twelve to twenty, thirty, or forty miles, will leave it thinly settled in places, for some time." (Benjamin Harding, *A Tour Through the Western Country* , p. 9) This surveyor noted the possibility that trees could grow in these open prairies if the Native American practice of setting fire to them was halted.

53. Kulik, "American Difference Revisited," p. 35.

54. Cited in ibid., p. 35.

55. Ward, *Andrew Jackson*, pp. 92–97.

56. Ibid., pp. 30–45.

57. Dwight, *Travels*, letter XIII, p. 462

58. Ibid., p. 463.

59. Allen, *The Practical Tourist*, volume 1, p. 264.

60. Ibid., p. 263.

61. Ibid.

62. Ibid.

63. Ibid., p. 264.

64. Tocqueville, *Journey to America*, p. 322.

65. Bachelard, *The Poetics of Space,* p. 36.

66. The importance of proximity to transportation was visible to any observant traveler. When Basil Hall traveled through upstate New York in 1827, he observed that the land near the roads and canals was extensively cleared, but that further from the arteries of transport the forest remained intact. "Most of the houses are built of rough unbarked logs," Hall observed, "nicked at the ends so as to fit closely and firmly; and roofed with planks." These houses, "being scattered about without order, look more like a collection of great packing boxes, than the human residences which the eye is accustomed to see in old countries." Yet Hall went on to note that such houses were eventually replaced by wooden frame houses with "rows of wooden columns in front." (Hall, *Travels in North America,* pp. 130–135)

67. After 1848, Root would serve as superintendant in Samuel Colt's revolver factory in Hartford. See Hoke, *Ingenious Yankees,* pp. 102–121.

68. Edward Everett, "The Western Railroad," in *Speeches of Edward Everett,* p. 145.

69. Ibid. p. 278.

70. Daniel Drake, *Discourse,* p. 9.

71. Ibid., pp. 9–10.

72. Weslager, *The Log Cabin in America,* pp. 261–266.

73. Ibid., p. 267. See also Norton, *Reminiscences of the Log Cabin and Hard Cider Campaign.*

74. Ibid., p. 268.

75. Ibid.

76. Emerson, *Journals and Miscellaneous Notebooks,* volume VII, p. 379.

77. *Yeoman's Gazette,* Saturday, July 11, 1840; cited in Emerson, *Journals and Miscellaneous Notebooks,* volume 7, p. 379.

78. Greeley, *Recollections of a Busy Life,* pp. 133–134.

79. Gunderson, *The Log Cabin Campaign,* p. 16.

80. Cited in ibid., p. 181.

81. However, when all his paintings are considered, Cole's attitude is ambiguous, balanced between nature and civilization. Cole did not simply celebrate the conquest of nature; indeed, he wrote a poem about the axe titled "The Complaint of the Forest."

82. These four images were published in 1849 by Orsamus Turner in his *Pioneer History of the Holland Purchase.* They are reproduced on pp. 247 and 267 of Meinig, *Continental America.*

83. See Stilgoe, *Common Landscape of America,* pp. 144–145.

84. Weslager, *The Log Cabin in America,* p. 99.

85. First given the title "Broad-Axe Poem" in 1856, it was revised slightly and retitled "Chants Democratic No. 2" for the 1860 edition. Whitman gave it the present title in 1867. All quotations here are from the 1973 Norton edition of *Leaves of Grass.*

86. For example, Alfred B. Street had published "Song of the Axe" in the April 1855 issue of *Graham's Magazine.*

87. Whitman, *Leaves of Grass,* p. 185.

88. Ibid.

89. Ibid., p. 186.

90. Ibid., p. 187.

91. Ibid., p. 191.

92. Ibid., p. 192.

93. Ibid.

94. Ibid., p. 193.

95. Ibid., p. 195.

Chapter 4

1. Cited in Williams, *Americans and Their Forests,* p. 373.

2. Abbey, *The Monkey Wrench Gang,* p. 188.

3. Cited in Novak, *Nature and Culture,* p. 160.

4. Olmsted, *The Cotton Kingdom,* p. 28. See also p. 31.

5. Cited in Weslager, *The Log Cabin in America,* p. 245.

6. Olmsted, *The Cotton Kingdom,* pp. 160–161.

7. Ibid., p. 11.

8. Ibid.

9. Ibid., p. 16.

10. Ibid., p. 17.

11. Dickens, *American Notes,* pp. 133–135.

12. "Travels of Johann David Schoepf," in Harpster, *Crossroads,* p. 136.

13. Here Crèvecoeur was propounding the beginnings of a narrative about the beneficence of technology. Without a navigable river or a passable road, these people were wretched. Once better connected to the marketplace, they would acquire industrious virtues. In this narrative, lack of inexpensive transportation prevented development and made life slow and difficult. Better roads, by implication, would make them more careful, diligent, and active. The virtues advocated by Benjamin Franklin were thus wedded to the transportation infrastructure, of which more in subsequent chapters. See Moore, ed., *More Letters from the American Farmer,* p. 67.

14. Thoreau, *The Maine Woods,* p. 320.

15. Novak, *Nature and Culture,* pp. 160–165.

16. Ibid., pp. 157–165.

17. Novak, *American Painting of the Nineteenth Century,* pp. 61–76. On how the Hudson Valley became a "sanctified landscape," see Schuyler, "The Sanctified Landscape," pp. 93–110.

18. Novak, *Nature and Culture,* pp. 161–162.

19. Bureau of the Census, *Agriculture, 1860,* p. clxix.

20. Marsh, *Man and Nature,* p. viii.

21. Ibid., p. 43.

22. Williams, *Americans and Their Forests,* pp. 371–372.

23. Ibid., p. 373.

24. Greeley, "Mr Greeley's Letters from Texas and the Lower Mississippi . . . ," p. 20.

25. Ibid., p. 21.

26. Williams, *Americans and Their Forests,* p. 354.

27. Ibid., p. 355.

28. Cited in Jackson, *American Space,* p. 91.

29. Parkman, "The Forests and the Census," p. 828.

30. Ibid., p. 839.

31. See Terrie, *Contested Terrain,* pp. 86–96.

32. Ibid., pp. 96–105.

33. Muir, "The American Forests," p. 157.

34. See Nash, *Wilderness and the American Mind;* Sears, *Sacred Places.*

35. Turner, "The Hunter Type," p. 153.

36. Wicks, *Log Cabins and Cottages,* p. 7.

37. Ibid., p. 8.

38. Cited in Hyde, *An American Vision,* p. 257.

39. Weslager, *The Log Cabin in America,* p. xxiii.

40. Ibid., pp. 289–287.

41. Ibid., p. 307.

42. Ibid., pp. 298–299.

43. Ibid., p. xxiv.

44. Johnson, *A Home in the Woods,* p. 3.

45. Tenner, *Why Things Bite Back,* pp. 101–104.

46. See Bealer and Ellis, *The Log Cabin,* pp. 178, 185.

47. Dorson, *Folklore and Fakelore*, p. 291.

48. Ibid., pp. 292–298.

49. Benjamin Britten and W. H. Auden wrote a "chamber opera" about Paul Bunyan; it premiered at Columbia University in 1941. Whatever its artistic merits, the American audience rejected the work, perhaps because its Paul Bunyan was a leader of the masses who imposed his will on other lumbermen. Reviewers found this proletarian giant a misinterpretation. Such attempts to invest him with socialistic ideological content were uncharacteristic, however.

50. Dorson, *Folklore and Fakelore*, p. 335.

51. Hyde, *An American Vision*, pp. 255–258.

52. Hough, *The Way to the West*, p. 7.

53. Muir, "The American Forests," p. 157.

54. Harrison, *Forests*, p. 122.

55. Kates, *Planning a Wilderness*, pp. 6–11.

56. Cited in ibid., p. 6.

57. Butler, "Henry Ford's Forest," pp. 725–731.

58. Ronald Reagan, Proclamation 4874, National Forest Products Week, October 9, 1981. The next year, Reagan again declared a National Forest Products Week, and said: "Under careful management, our forests can produce more than twice the volume of timber now being grown, without damaging our environment. This means that we can meet our own increasing demands and still export wood products, thus strengthening both our economy and our independence." (Proclamation 4975, September 23, 1982).

59. Abbey, *The Monkey Wrench Gang*, p. 188.

60. Callenbach, *Ecotopia*, p. 63.

61. Ibid., pp. 54–55.

62. Friedenberg, *Life, Liberty, and the Pursuit of Land*, p. 307.

63. Ibid.

64. Erdrich, *Tracks*, pp. 219–220.

65. See Neihardt, *Black Elk Speaks*, pp. 199–200.

66. Erdrich, *Tracks*, p. 220.

67. Ibid., p. 223.

68. Ibid., p. 225.

Chapter 5

1. Allen, *The Practical Tourist*, pp. 154–155.

2. Thoreau, "Paradise (to Be) Regained," in Folsom and Lubar, *The Philosophy of Manufactures*, pp. 413–414.

3. Kendall, *Travels through the Northern Parts of the United States in the Years 1808 and 1809,* volume 3, pp. 33–34.

4. On colonial lumber mills, see Bishop, *A History of American Manufacturers, from 1608 to 1860,* pp. 93–115.

5. Daniels, "Economic Development in Colonial and Revolutionary Connecticut," p. 439.

6. Stilgoe, *Common Landscape of America,* p. 308.

7. Clark, *The Eastern Frontier,* p. 203.

8. Moore, ed., *More Letters from the American Farmer,* p. 64.

9. Johnson, *A Home in the Woods,* pp. 65–66. These reminiscences are confirmed by Morris Birkbeck's contemporary account of settlers in the Ohio Valley (*Notes on a Journey in America,* p. 120): "These lonely settlers are poorly off—their bread corn must be ground thirty miles off, requiring three days to carry to the mill, and bring back, the small horse load of three bushels."

10. Allen, *Practical Tourist,* volume 1, p. 159.

11. Stilgoe, *Common Landscape,* pp. 310–311.

12. Ibid., p. 314.

13. Moore, ed., *More Letters,* p. 170. Crèvecoeur also noted that flour and other mill products had to be transported to market when the roads were good. As soon as snows fell, sleighs and sleds were used to haul goods to "the edges of great Rivers," where they would wait to be transported in the spring. If snow did not come or if the grain was not shipped to the water's edge in good time, wagons would have to be used to move it through the spring mud. That was slow, teams could not pull a full wagon, and planting time was lost.

14. Ibid., p. 179. See also p. 190.

15. Ibid., p. 179.

16. Ibid., p. 13.

17. Ferguson, "Industrial Power Sources of the Nineteenth Century."

18. Marquis de Chastellux, *Travels in North-America,* pp. 343–344.

19. Armstrong, *Factory under the Elms,* pp. 2–3.

20. William Wycoff, *The Developer's Frontier,* p. 2.

21. Ibid., p. 128.

22. Walsh, *The Manufacturing Frontier,* pp. 14–15.

23. Cited in Hunter, *History of Industrial Power in the United States,* volume I, p. 34.

24. Smith, *The Power Policy of Maine,* p. 27. At this time Maine was a part of Massachusetts.

25. Hunter, *History of Industrial Power in the United States,* volume I, p. 33.

26. Ibid., p. 34.

27. Smith, *Power Policy of Maine*, p. 30.

28. Ibid., p. 35.

29. U.S. Census Office, Sixth Census, 1840, pp. 360–364. Hunter (*History of Industrial Power in the United States*, volume I, pp. 37–38) estimated that the number of mills was closer to 71,000.

30. Hunter, *Steam Power*, pp. 73–75.

31. Hunter, *History of Industrial Power in the United States*, volume I, pp. 37–38.

32. "Ollapondiana," *Knickerbocker* 8 (September, 1836), p. 348.

33. Whitworth and Wallis, *The Industry of the United States*, p. 9. In 1818, Morris Birkbeck, visiting the Ohio Valley, found that the location of a town was usually determined by "the goodness of the soil or vicinity to a mill." Usually, "some enterprising proprietor finds in his section what he deems a good scite for a town: he has it surveyed and laid out in lots, which he sells, or offers for sale by auction." If the sale is successful, "the new town then assumes the name of its founder: a store-keeper builds a little framed store, and sends for a few cases of goods; and then a tavern starts up, which becomes the residence of a doctor and a lawyer, and the boarding-house of the store-keeper, as well as the resort of the weary traveler: soon follow a blacksmith and other handicraftsmen in useful succession. . . . Thus the town proceeds, if it proceeds at all, with accumulating force, until it becomes the metropolis of the neighbourhood." He was well aware that founding a community was a matter of financial speculation, for "Hundreds of these speculations have failed." But yet "hundreds prosper; and thus trade begins and thrives, as population grows around these lucky spots." (Birkbeck, *Notes on a Journey in America*, pp. 103–104)

34. Thoreau, "Paradise (to Be) Regained," pp. 413–414.

35. "Samuel Slater," in Hunt, *Lives of Famous American Merchants*, volume 1, p. 453.

36. On the comparative use of mill power and muscle power in the North and the South, see Nye, *Consuming Power*, pp. 54–60.

37. See Marmor, "Anti-Industrialism and the Old South," pp. 377–406.

38. Gregg, *Essays on Domestic Industry*, reprinted in Folsom and Lubar, *The Philosophy of Manufactures*, p. 434.

39. Benjamin Franklin, cited in Licht, *Industrializing America*, p. 15.

40. Logan, "Letters Addressed to the Yeomanry of the United States," reprinted in Folsom and Lubar, *The Philosophy of Manufactures*, p. 107.

41. See Kasson, *Civilizing the Machine*.

42. George W. P. Custis, "An Address on the Importance of Encouraging Agriculture and Domestic Manufactures" (1808). Reprinted in Folsom and Lubar, *The Philosophy of Manufactures*, p. 155.

43. Ibid.

44. Logan, "Letters Addressed to the Yeomanry of the United States," reprinted in Folsom and Lubar, *The Philosophy of Manufactures*, p. 110.

45. Licht, *Industrializing America*, pp. 17–18.

46. Jefferson, *Notes on the State of Virginia*, p. 157.

47. Kasson, *Civilizing the Machine*, p. 24.

48. Thomas Jefferson to Colonel David Humphreys, January 20, 1809, reprinted in Folsom and Lubar, *The Philosophy of Manufactures*, p. 27.

49. U.S. Department of State, *Sixth Census or Enumeration of the United States* (Washington: Blair and Rives, 1841), p. 55. Maryland had four times as many people as Rhode Island but had the same number (21,325) working in all forms of manufacturing. Of South Carolina's nearly 600,000 people (55% of them slaves), only 10,325 (less than 2%) worked in trades or manufacturing (ibid., pp. 201).

50. Ibid., p. 47.

51. Martineau, *Society in America*, p. 173.

52. Gregg, reprinted in Folsom and Lubar, *The Philosophy of Manufactures*, p. 432.

53. Ibid, p. 433.

54. Cited in Olmsted, *The Cotton Kingdom*, p. 586.

55. See Boorstin, *The Americans*, pp. 125–131.

56. Birkbeck, *Notes on a Journey in America*, p. 21.

57. See Parker, *Lowell*, pp. 61–69.

58. U.S. Department of State, *Sixth Census*, p. 45.

59. Hall, *Travels in North America*, volume 2, p. 135.

60. See Theodore Steinberg, *Nature Incorporated*, pp. 99–112.

61. Emerson, journal, 1847, cited in Folsom and Lubar, *The Philosophy of Manufactures*, p. 447.

62. Yates, ed., *Harriet Martineau on Women*, p. 161.

63. McKelvey, *Rochester*, pp. 35–37.

64. Ibid., p. 38.

65. Ibid., pp. 95–96.

66. Pope-Hennessy, ed., *Aristocratic Journey*, p. 238.

67. Ibid., p. 240.

68. American newspaper article, cited in Arfwedson, *The United States and Canada in 1832, 1833, and 1834*, volume 2, pp. 2–3.

69. Willoughby, *Flowing through Time*, p. 68.

70. Ibid.

71. Olmsted, *The Cotton Kingdom*, p. 213.

72. McKelvey, *Rochester*, p. 71.

73. Ibid., p. 91.

74. Cited in Boorstin, *The Americans*, p. 131.

75. Cited in Marx, *The Machine in the Garden*, pp. 182–183.

76. Ibid., p. 185.

77. White, *Memoir of Samuel Slater*, p. 121.

78. Cooper, *Notions of the Americans*, volume 1, p. 249.

79. Reprinted in *Orations and Speeches of Edward Everett*, volume 2, pp. 67–68.

80. Everett, "Address on Fourth of July at Lowell," in *Orations and Speeches*, volume 2, pp. 52–53.

81. Ibid., p. 63.

82. Allen, *The Practical Tourist*, volume 1, p. 153.

83. Jeremy, "Cotton Mills in Developing Regions, 1820–1840," p. 191.

84. Allen, *The Practical Tourist*, pp. 154–155.

85. Ibid., p. 155.

86. Wright, "The Village of Little Falls," p. 244.

87. White, *Tropics of Discourse*, p. 96.

Chapter 6

1. Faulkner, *Light in August*, p. 4.

2. Melville, "A Tartarus of Maids," in Lewis, ed., *Herman Melville*, p. 192.

3. On the distribution of water power in industry, see Nye, *Consuming Power*, pp. 44–55.

4. Hunter, *Water Power*, p. 144.

5. Cowley, *History of Lowell*, pp. 129–130.

6. Plumb, *History of Hanover Township*, pp. 231–232.

7. Hunter, *Water Power*, pp. 144–147.

8. Ibid., p. 145.

9. Cited in Merchant, *Ecological Revolutions*, p. 240.

10. See Steinberg, *Nature Incorporated*, p. 189.

11. Audubon, cited in Simpson, *Visions of Paradise*, p. 30.

12. Thoreau, *Journal*, volume III, p. 193 (February 28, 1856).

13. Ibid., volume IV, p. 151 (June 26, 1852).

14. Ibid., volume VIII, p. 348 (May 19, 1856).

15. Ibid., volume IV, p. 23.

16. Thoreau, *The Maine Woods*, pp. 329–330.

17. Williams, *Americans and Their Forests,* p. 280.

18. Ibid., p. 282.

19. Faulkner, "Delta Autumn," p. 311.

20. Faulkner, *Light in August,* p. 4.

21. Ibid.

22. See Ware, *The Industrial Worker, 1840–1860,* pp. 135–148.

23. Thoreau, *Walden,* p. 20.

24. Leighton, *Life at Puget Sound,* pp. 25–26.

25. Indeed, the potential for such a development was contained within even Kendall's narrative, which recognized that many people in the town formed around a mill became indebted to the miller and store owners, and noted the early need for a lawyer to collect unpaid debts.

26. Whittier, "The City of a Day," p. 288.

27. Ibid.

28. Ibid., p. 289.

29. Ibid., pp. 290–291.

30. On Lowell's labor unrest, see Robinson, *Loom and Spindle,* pp. 135–149. The best general account is Dublin, *Women at Work.*

31. Whittier, "The City of a Day," p. 290.

32. Hughes, *A Treatise on Sociology,* pp. 286–291.

33. George Fitzhugh, *Sociology for the South,* p. 226.

34. Melville, "A Tartarus of Maids," p. 187.

35. Ibid., pp. 187–188.

36. Ibid., p. 188.

37. Ibid., p. 190.

38. Ibid., p. 192.

39. Denning, *Mechanic Accents,* p. 186.

40. Walkowitz, *Worker City, Company Town,* pp. 52–54.

41. Hine, "The High Cost of Child Labor," reproduced in Stange, *Symbols of Ideal Life,* p. 69.

42. See Stange, *Symbols of Ideal Life,* p. 70.

43. Reprinted in Hurley, *Industry and the Photographic Image,* p. 66. Original in George Eastman House, Rochester (negative 3786).

44. Proceedings of the National Consumer's League, Second Annual Meeting, 1901, p. 7; Proceedings of the National Consumer's League, Seventh Annual Meeting, 1901, p. 15.

45. Walkowitz, *Worker City,* pp. 48–51, 56, 249–250.

46. On the differences between paternalism in the antebellum North and the industrial New South, see Philip Scranton, "Varieties of Paternalism."

47. Neill, *Report on Strike of Textile Workers in Lawrence, Mass, in 1912*, p. 9.

48. Ibid., p. 27.

49. Ibid., pp. 10–11. The chief exceptions were the 300 people (mostly Belgian and English immigrants) who had joined the Industrial Workers of the World. The "Wobblies" emphasized that the "historic mission of the working class [is] to do away with capitalism" and abolish the wage system" (ibid., p. 17). The leaders of the strike committee were not members of the labor union, which only actively joined the walkout a month after it began.

50. Fewer than 300 would be arrested in the course of the 63-day strike. Only 54 went to jail.

51. Ibid., pp. 18–19. This was still only one-third of the lost wages. The average per capita income during a sample week late in 1911 was $8.76, amounting to about $192,000 for 21,922 employees.

52. For a full and sympathetic account, see Steve Golin, *The Fragile Bridge*. Also useful is Green, *New York 1913*.

53. Vorse, *The Passaic Textile Strike, 1926–27*, p. 4. Another important strike occurred at New Bedford, Massachusetts, in 1928; see Georgianna and Aaronson, *The Strike of '28*. As had been the case in New Jersey the year before, the strike was provoked by management's attempt to impose a 10% wage cut. The union won back 5% of the cut after a protracted strike.

54. Cited in Vorse, *Passaic Textile Strike*, pp. 4–5.

55. Ibid., pp. 121–123.

56. Cited in Ross, "Struggles for the Screen," p. 355.

57. Ibid., p. 356.

58. Ibid.

59. Anderson, *Poor White*, p. 327.

60. Ibid., p. 329–330.

61. Ibid., pp. 343–344.

62. Anderson, "A Great Factory," p. 52.

63. Muir continued: "A small stone, that a man might carry under his arm, is fastened to the vertical shaft of a little home-made, boyish-looking, back-action water-wheel, which, with a hopper and a box to receive the meal, is the whole affair. . . . No dam is built. The water is conveyed along some hillside until sufficient fall is obtained, a thing easily done in the mountains. On Sundays you may see wild, unshorn, uncombed men coming out of the woods, each with a bag of corn on his back. From a peck to a bushel is a common grist. They go to the mill along verdant footpaths, winding up and down over hill and valley, and crossing many a rhododendron glen. The flowers and shining leaves brush against their shoulders and knees, occasionally knocking off their coon-skin caps. The first

arrived throws his corn into the hopper, turns on the water, and goes to the house." He reported that such small mills could grind between 10 and 20 bushels a day. (*A Thousand-Mile Walk to the Gulf*, pp. 35–36)

64. Ibid., p. 37

65. See Beatty, "Lowells of the South."

66. Cash, *The Mind of the South*, p. 205.

67. Ibid.

68. Beatty, "Lowells of the South," passim.

69. The most persuasive work in this vein is *The Political Economy of Slavery*, in which Eugene Genovese argues that the South was a pre-capitalist region with feudal tendencies.

70. See Fitzhugh, *Cannibals All*, pp. 50–51.

71. Mitchell and Mitchel, *The Industrial Revolution in the South*, pp. 1–13.

72. Lockwood, Greene & Company, cited in ibid., p. 38.

73. Mitchell and Mitchel, *The Industrial Revolution in the South*, p. 15.

74. Galenson, *The Migration of the Cotton Textile Industry from New England to the South*, chapter 1.

75. New Southern mills commonly adopted electric motors to drive their machinery. Electric motors ran at higher speeds and produced goods of more uniform quality than water-driven machinery. An electrified mill no longer had to be close to a river, with its uncertain flow and flood danger. On electrification of textile mills, see Nye, *Electrifying America*, pp. 196–200.

76. Lewis, "Cheap and Contested Labor."

77. Cowley, *The Dream of the Golden Mountains*, p. 250.

78. Salmond, *Gastonia 1929*, pp. 10–11. See also Draper, "Gastonia Revisited."

79. Salmond, *Gastonia 1929*, p. 13.

80. Aronowitz, *False Promises*, p. 191.

81. Page, *Gathering Storm*, p. 21. (The first edition of this work was printed in the Soviet Union.)

82. Ibid.

83. Ibid., p. 371.

84. Anderson, *Beyond Desire*, pp. 49–50.

85. Ibid., p. 50.

86. Ibid., pp. 53–54.

87. Ibid., p. 285.

88. Ibid., p. 286.

89. Ibid., p. 336.

90. Ibid., p. 340.

91. Ibid., p. 32.

92. Ibid., pp. 33–34.

93. Ibid., p. 58.

94. All three perspectives can be found at the Lowell complex of mills, which have become a national historic site and center for research on the history of textile manufacturing. The visitor can see the engineering needed to create Lowell, and can follow its power system from the dam through the canal system and the water turbines to the gears and shafting that drive the machines. The lives of workers, from the boarding house to the workspace and from paternalism to unionization, can also be imagined. More recently, environmental history has been introduced as an interpretive component of the site, though it relies less on artifacts than on displays.

95. Gutman, *Work, Culture, and Society in Industrializing America*, p. 236.

96. Wallace, *Rockdale,* p. 374.

97. Scranton, *Figured Tapestry,* p. 452. See also pp. 502–503.

98. Ibid., p. 437.

99 Hareven and Langenbach, *Amoskeag:,* p. 71.

100. Ibid., p. 315.

101. Lucic, *Charles Sheeler and the Cult of the Machine,* p. 108.

102. George Van Wagenen and Monroe Rosenfeld, "By The Dear Old Village Mill Down In The Valley" (New York: Seminary Music Company, 1908). A copy can be found in MIT's Lewis Music Library.

Chapter 7

1. *Annals of Congress,* 14th Congress, 2nd Session, 851–960, cited in Shaw, *Canals for a Nation,* p. 26. For the early scholarship about internal improvements, see Goodrich, "Internal Improvements Reconsidered."

2. Emerson, *Journals* VIII, p. 434, cited in Kasson, *Civilizing the Machine,* p. 133.

3. Chudacoff and Smith, *The Evolution of American Urban Society,* pp. 87–88.

4. Emerson, *Journals and Miscellaneous Notebooks,* ed. Gilman et al., volume VII, p. 268. Note, however, that here Emerson rails against trade, not the machine. In another passage (ibid., p. 436), he allows that "the activity of the Engineer, of the railroad builder, & the manufacturer is real & inventive, & deserves regard."

5. Madison et al., *The Federalist,* no. 14 (New American Library, 1960), pp. 102–103. Alexander Hamilton, in no. 22, also argued that the Constitution would improve trade.

6. Stilgoe, *Common Landscape of America,* pp. 108–109.

7. Taylor, *The Transportation Revolution,* p. 21. Under Adams the average annual appropriation had been $700,000.

8. Edmund Dana, "Geographical Sketches on the Western Country Designed for Emigrants and Settlers" (1819)," in Lindley, ed., *Indiana as Seen by Early Travelers,* p. 207.

9. Taylor, *Transportation Revolution,* p. 47. See also Flint, *A Condensed Geography and History of the Western States,* volume II, pp. 185–190.

10. Sheriff, *The Artificial River,* p. 34.

11. Ibid., p. 35.

12. There were no railroads in North America at this time. Hall must have been thinking of shipments by the newly available canal or through Lake Ontario.

13. Hall, *Travels in North America,* volume 1, p. 160.

14. Ibid., p. 161.

15. Ibid., p. 162.

16. Ibid., p. 161.

17. Stilgoe, *Common Landscape,* p. 120.

18. Hawthorne, "Sketches from Memory," in *Complete Works,* volume 2.

19. "Ollapondiana," *Knickerbocker* 8 (September, 1836), p. 353.

20. Anonymous review, *The New England Magazine* 3 (1826), p. 17.

21. Edward Everett, "The Western Railroad," in *Orations and Speeches of Edward Everett,* volume II, p. 147.

22. Ibid.

23. Washington, cited in Everett, "The Western Railroad," p. 153.

24. Smith, *The Book of the Great Railway Celebrations of 1857,* p. 1.

25. Ibid., pp. 2–3.

26. Everett, "Opening of the Railroad to Springfield," in *Orations and Speeches of Edward Everett,* volume II, p. 367.

27. Smith, *Book of the Great Railway Celebrations of 1857,* p. 215.

28. Everett, "Opening of the Railroad to Springfield," p. 369.

29. White, *Memoir of Samuel Slater,* p. 121.

30. On railroads and the sublime, see Nye, *American Technological Sublime,* pp. 45–75.

31. Lyell, *A Second Visit to the United States,* volume 2, p. 357.

32. Emerson, "The Young American" in *Complete Works,* volume 1, p. 364.

33. Greeley et. al., *The Great Industries of the United States,* volume 2, p. 1032.

34. See Kasson, *Civilizing the Machine,* pp. 178–179.

35. "Verses on the Prospect of Planting Arts and Learning in America," in Berkeley, *The Works of George Berkeley, D. D.,* ed. Fraser, volume IV, p. 364. This declaration dates from the 1720s.

36. "English and American Railways," reprinted in *American Railroad Journal* 33 (1860), p. 926.

37. Ibid.

38. Richardson, *Beyond the Mississippi,* p. 567. Also cited in Danly and Marx, *The Railroad in American Art,* p. 22.

39. Bell, *New Tracks in North America,* pp. 507–510.

40. Thomas Prichard Rossiter, "Opening of the Wilderness" (1858), Boston Museum of Fine Arts, M. and M. Karolik Collection.

41. Reps, *Cities of the American West,* pp. 3–10.

42. Willis Drummond Jr. et al., "Land Grants in Aid of Internal Improvements," in Powell, *Report on the Lands of the Arid Region of the United States,* pp. 178–196.

43. Stover, *The Routeledge Historical Atlas of the American Railroads,* p. 33.

44. Emmons, *Garden in the Grasslands,* p. 28.

45. See, e.g., Santa Fe Railroad, *Kansas in 1875.*

46. The railroads were only granted every other lot. This was done for the same reasons: so that large tracts could not be easily assembled, and so that some of the value created in lands by its propinquity to railroads would accrue to the government. See Sakolski, *The Great American Land Bubble,* pp. 277–279.

47. Gates, *The Illinois Central Railroad and Its Colonization Work,* pp. 211–231.

48. Dicey, *Six Months in the Federal States,* pp. 132–139.

49. Cited in Reps, *Cities of the American West,* p. 529.

50. Clampitt, *Echoes from the Rocky Mountains,* pp. 133–134.

51. Reps, *Panoramas of Progress,* passim.

52. Strong, *Our Country,* p. 18.

53. Ibid., p. 198.

54. Richardson, *Beyond the Mississippi,* p. 571.

55. Strong, *Our Country,* pp. 198–199. The book sold 175,000 copies.

56. Ibid., p. 255.

57. See Strong, *Expansion under New World Conditions,* p. 6. This book began where the previous one had ended. The first chapter examined the "exhaustion of our arable public lands" and advocated imperialist expansion as the destiny of the "Anglo-Saxon race."

58. Stilgoe, *Metropolitan Corridor,* p. 218.

59. Stromquist, *A Generation of Boomers,* pp. 161–163.

60. Cist, *Cincinnati in 1851,* p. 132.

61. *Western Monthly Review,* May 1827, cited in Turner, *The Rise of the New West,* pp. 104–105.

62. Richardson, *Beyond the Mississippi,* p. i.

63. Ibid., p. ii.

64. Ibid., p. 18.

65. Ibid., p. 25.

66. Ibid., p. 59.

67. Ibid.

68. Ibid., p. 564.

69. Ibid.

70. Ibid., p. 571. A certain playfulness with the truth was common among settlers, many of whom, Richardson found, filed homestead claims for land they never lived on, thereby purchasing that land for speculation. Richardson even reported that a small cabin was put on wheels and moved from one claim to another to establish the appearance of residence (ibid., p. 141).

71. Rusling, *The Great West and Pacific Coast*, p. 22.

72. Porter, *The West, from the Census of 1880*, p. 555.

73. Ibid., p. 576.

74. Ibid.

75. Beale, *Picturesque Sketches of American Progress*, passim.

76. Ibid., p. 597.

77. Stover, *Routeledge Historical Atlas of the American Railroads*, pp. 40–41, 52–53.

78. Hough, *The Way to the West*, p. 362.

79. Ibid., p. 364.

80. Daniel Webster, reprinted in Rozwenc, ed., *Ideology and Power in the Age of Jackson*, pp. 38–39. On the early enthusiasm for the railroad, see Nye, *American Technological Sublime*, pp. 45–79.

81. Whitman, "To a Locomotive in Winter," in *Leaves of Grass*, p. 472.

82. Whitman, "Passage to India," in *Leaves of Grass*, pp. 413, 420.

83. Turner, *Rise of the New West, 1819–1829*, p. 224.

84. Stilgoe, *Metropolitan Corridor*, pp. 73–104.

85. On the industrial sublime, see Nye, *American Technological Sublime*, pp. 112–118, 125–127, and 131–132.

86. James, *The American Scene*, pp. 44 and 462–465.

87. Marx, *The Machine in the Garden*, pp. 352–353.

Chapter 8

1. Beadle, *The Undeveloped West*, p. 62.

2. Muir, "The American Forests," p. 154.

3. Hough, *The Way to the West*, p. 427.

4. Kasson, *Civilizing the Machine,* p. 131.

5. Cited in ibid., p. 130.

6. Emerson, *Journals* VIII, p. 434 (cited in Kasson, *Civilizing the Machine,* p. 133).

7. Cited in Marx, *The Machine in the Garden,* p. 13.

8. Emerson, *Journals,* volume VI, p. 322.

9. Marx, *The Machine in the Garden,* pp. 32–33.

10. Ibid., pp. 260–261.

11. Cited in ibid., p. 261.

12. See Nye, *American Technological Sublime,* pp. 34–36, 47–51.

13. "Ollapondiana," *Knickerbocker* 8 (September 1836), p. 346.

14. "Ollapondiana," *Knickerbocker* 8 (July 1836), p. 74.

15. "Travel in the United States," *The Atlantic,* April 1867, p. 479.

16. George, *Progress and Poverty,* p. 197.

17. Decker, "The Railroads and the Land Office."

18. *Oregon City Enterprise,* February 27, 1874, cited in Robbins, *Our Landed Heritage,* p. 257.

19. Robbins, *Our Landed Heritage,* pp. 257–260.

20. *Turner vs. Atchison, Topeka and Santa Fe Railroad Company,* 1879.

21. See House Miscellaneous Document 132 (43rd Congress, first session) for the resolution from the legislature of Kansas, and Document 185 for that from Minnesota.

22. For a brief summary of the case, see McWilliams, *Factories in the Field,* pp. 16–17.

23. George, *Progress and Poverty,* p. 197.

24. Marcy, *The Prairie and Overland Traveler,* p. 1.

25. Beadle, *The Undeveloped West,* p. 62.

26. Hazen, "The Great Middle Region of the United States."

27. Beadle, *The Undeveloped West,* p. 130.

28. Ibid., p. 131.

29. Meinig, *Transcontinental America,* p. 63.

30. Fishkin, *Lighting Out,* p. 179.

31. Hodgskin, "The Truth about Land Grants," p. 417.

32. Safire, ed., *Lend Me Your Ears,* p. 833.

33. George, *Our Land and Land Policy,* p. 9.

34. Ibid., p. 10.

35. Ibid.

36. Ibid., p. 13.

37. Robbins, *Our Landed Heritage*, p. 277.

38. George, *Our Land and Land Policy*, p. 45.

39. Way, *Workers and the Digging of North American Canals*, pp. 181–187.

40. Drucker, *Men of the Steel Rails*, p. 59.

41. Reps, *Cities of the American West*, p. 529.

42. See Scheriff, *The Artificial River*, pp. 138–155.

43. Cited in ibid., p. 170.

44. Kennedy, *Agriculture of the United States in 1860*, p. clxix.

45. Stromquist, *A Generation of Boomers*, p. 45.

46. Ibid., p. 271.

47. Licht, *Industrializing America*, p. 167.

48. Stowell, *Streets, Railroads, and the Great Strike of 1877*, p. 88.

49. Ibid., p. 144.

50. Holt, *Forging a Majority*, p. 228.

51. Adams, "The State and the Railroads," pp. 691, 695.

52. Greeley, *The Great Industries of the United States*, volume 2, p. 1032.

53. Ibid., p. 1035.

54. Ibid., p 1036.

55. McWilliams, *Factories in the Field*, p. 17.

56. Greeley, *Great Industries of the United States*, volume 2, p. 1036.

57. Greeley, *An Overland Journey from New York to San Francisco in the Summer of 1859*, p. 14.

58. Ibid., p. 15.

59. Dumke, *The Boom of the Eighties in Southern California*, pp. 20–21.

60. Gates, *Fifty Million Acres*, pp. 171–175.

61. For a contemporary explanation as to why the rates varied so much, see Grosvenor, "The Railroads and the Farms," pp. 608–609.

62. Miller, *Railroads and the Granger Laws*, p. 19.

63. Ibid., p. 127.

64. Stover, *The Routeledge Historical Atlas of the American Railroads*, pp. 46–47.

65. Miller, *Railroads and the Granger Laws*, p. 132.

66. See "Petition for Return of Railroad Lands," reprinted in Barnes, ed., *The People's Land*. See also *Oregon and California Railway Company v. United States*, 243 U.S., 549 (1917).

67. Mark Aldrich, *Safety First*, pp. 11, 17.

68. Ibid., p. 16.

69. O'Malley, *Keeping Watch,* pp. 100–144.

70. Aldrich, *Safety First,* pp. 17, 23.

71. Cease, "The Death and Disability Toll of Our Railways Employees," p. 1217.

72. Ibid., p. 1218–1219

73. Hone, *Diary of Philip Hone,* June 28, 1838 volume 1, p. 321.

74. Aldrich, *Safety First,* p. 25.

75. Adams, "Of Some Railroad Accidents, II."

76. Cohen, *Long Steel Rail,* p. 171.

77. Schivelbusch, *The Railway Journey,* pp. 142–144.

78. Ibid., pp. 139–140.

79. Cohen, pp. 169–171.

80. Miner, *The Corporation and the Indian,* pp. 24–25.

81. Both cited in ibid., p. 108.

82. See Stradling, *Smokestacks and Progressives,* pp. 50–51.

83. For a masterful overview of coal smoke in the urban landscape, see Tarr, *The Search for the Ultimate Sink.*

84. Stradling, *Smokestacks and Progressives,* p. 78.

85. Williams, *Americans and Their Forests,* pp. 345–436.

86. Cited in ibid., p. 346.

87. Ibid, p. 352.

88. Cited in Olson, *The Depletion Myth,* p. 8.

89. Roosevelt, "The Forest in the Life of a Nation," p. 8.

90. Muir, "The American Forests," p. 155.

91. Ibid., p. 154.

92. Norris, *The Octopus,* p. 369.

93. Fletcher, "American Railways and American Cities," p. 805. Another book that attacked the railroads was Hudson, *The Railways and the Republic.* The inability of state legislatures to regulate them was a common subject; see e.g. Warner, "Railroad Problems in a Western State."

94. Bercovitch, *The American Jeremiad.*

95. Hough, *The Way to the West,* p. 385.

96. Ibid., p. 393.

97. Ibid. p. 427.

98. Dreiser, "The Railroad and the People," p. 479.

99. James, *The American Scene,* p. 463.

100. Ibid.

101. Ibid., p. 464.

102. Ibid., pp. 464–465.

103. Schumpeter, *Business Cycles,* volume I, p. 303.

104. Fishlow, *The American Railroads and the Ante-Bellum Economy,* p. 166.

105. Ibid., p. 165.

106. Ibid., p. 176.

107. The complex relationship between railroads and the separate corporations they often created to plan and sell towns complicates the argument further, though not to Schumpeter's advantage. Such activities by railroads clearly stimulated anticipatory demand.

108. Fogel, *The Union Pacific Railroad,* pp. 96–97.

109. Fogel, *Railroads and American Economic Growth,* p. 92.

110. Ibid. pp. 100–105.

111. Ibid., p. 110.

112. Martin, *Railroads Triumphant,* p. 401. For more on the critique of Fogel, see Chandler, *The Visible Hand,* p. 531.

113. Stilgoe, *Metropolitan Corridor,* p. 339.

114. Loree, "Address of L. F. Loree at the Hotel Astor."

115. Ibid.

116. *Official Guide,* A Century of Progress, 1933, pp. 46–48.

117. *Building the World of Tomorrow—Official Guide Book* (New York: Exposition Publications, 1939), pp. 163–164.

118. Ibid., p. 167. On this fair, see also Nye, *Narratives and Spaces,* pp. 104–108, 129–145; Nye, *American Technological Sublime,* pp. 199–224.

119. Updike, *Rabbit at Rest,* p. 187.

120. See e.g., "Chair Car" (1965), reproduced in Goodrich, *Edward Hopper.*

121. Hopper, "Hotel by a Railroad" (1952), reproduced in Renner, *Edward Hopper.*

122. Stover, *Routeledge Historical Atlas of the American Railroads,* pp. 116, 124.

Chapter 9

1. Powell, "The Irrigable Lands of the Arid Region,"p. 767.

2. Smythe, *The Conquest of Arid America,* p. 331.

3. As late as 1911 some were still trying to farm the arid west without the benefit of irrigation. Homesteaders in the Dakotas and in eastern Montana were drawn west by glossy brochures of railroad companies. They were the last true believers in the doctrine that "rain follows the plow" in an updated version that included modern dry farming. See Raban, *Bad Land.*

4. Smythe, *Conquest*, p. 327.

5. Cited in Worster, *Rivers of Empire*, p. xi.

6. Cy Warman, in *McClure's Magazine*, September 1894, quoted in "The Opening of Empire," *Literary Digest* 19 (1894), no. 20 (September 15).

7. Cited in Opie, *Ogallala*, p. 50.

8. See Smith, *Virgin Land*.

9. See Kolodny, *The Land Before Her*.

10. Cited in Emmons, *Garden in the Grasslands*, pp. 35–36.

11. Hinton, *Irrigation in the United States*, p. 5.

12. Ibid., p. 43

13. Ibid.

14. Ibid., p. 46.

15. Smythe, *Conquest*, p. 107.

16. Greeley, *Mr. Greeley's Letters from Texas and the Lower Mississippi*, p. 14. Greeley advocated irrigation not only in arid lands but also in New England. There, he believed, mountain streams could easily be diverted into meadows to increase the hay harvest, and even rainy places such as Connecticut could benefit.

17. "Mr. Greeley's Letters," p. 14.

18. Hinton, *Irrigation*, p. 39.

19. Worster, *Rivers of Empire*, pp. 83–88.

20. Hinton, *Irrigation*, p. 39.

21. Reprinted in Rusling, *The Great West and Pacific Coast*, p, 493.

22. Ibid., p. 495.

23. Whitman, *Collected Writings of Walt Whitman, Prose Works*, volume 2, ed. Stovall, p. 220.

24. Rusling, *The Great West and Pacific Coast*, p. 199.

25. Ibid., p. 202. Nevertheless, a century later only 2.1 percent of Utah could be irrigated, with another 1.1 percent developed as dry farms using techniques developed in the twentieth century (Nelson, *Utah's Economic Patterns*, pp. 22–24).

26. *Census of 1910* (Government Printing Office, 1912), p. 837.

27. Hunter, *A History of Industrial Power in the United States, 1780–1930*, volume 1: *Water Power*, p. 57.

28. Sherow, *Watering the Valley*, p. 83. The first irrigator had had experience in California.

29. Hinton, "Irrigation in the Arid West," p. 729.

30. Sherow, *Watering the Valley*, pp. 84–87.

31. Ibid., p. 88.

32. Opie, *Ogallala,* p. 118. Such early irrigation was not part of a regional planning process, however. It often brought landowners into conflict, because the water available was insufficient for all parties.

33. Smythe, *Conquest,* p. 111.

34. Ibid., pp. 111–112.

35. Ibid., p. 111.

36. Barbour, "Home-Made Windmills in Nebraska," pp. 116–117.

37. Opie, *Ogallala,* pp. 119–121.

38. Smythe, *Conquest,* p. 115.

39. Ibid., p. 118.

40. Cited in ibid., p. 119.

41. Cy Warman, in *McClure's Magazine,* Sept. 1894, quoted in "The Opening of Empire," *Literary Digest* 19 (1894), no. 20 (September 15).

42. Quoted in "The Opening of Empire," *Literary Digest* 19 (1894), no. 20 (September 15).

43. John Downey, in Kennedy, *Census of 1860,* p. clxxi.

44. Ibid. On pp. 328–331 of *An Overland Journey from New York to San Francisco* (1859), Greeley noted that California's best crop was winter wheat. Artesian wells and windmills served smaller gardens, and fruits did well in the mild climate, but Greeley foresaw no vast irrigation projects. In 1862 an English visitor noted that the Spanish missions had used irrigation to cultivate oranges, olives, walnuts, and other fruits. Yet he also found the Central Valley arid, even though "Hundreds of windmills" pumped water "from the wells for the cattle and for irrigating the lands." In summer, the streams were completely dry. See William Brewe, *Up and Down California in 1860–64,* pp. 22, 121.

45. Hittel, *The Resources of California,* p. 162.

46. Ibid., pp. 151–152, 160–161.

47. Ibid., p. 161.

48. Ibid., p. 160.

49. Ibid., p. 153.

50. Porter, *The West,* p. 460.

51. Hinton, *Irrigation,* p. 28. Hinton was intrigued by the eastern parts of Washington and Oregon. He described "12,000,000 acres, now almost rainless" near the Snake River, where "the opening of the Northern Pacific railroad, since 1880" had caused "a large increase of land occupation and population" (p. 35).

52. Porter, *The West.* p. 331.

53. "Nowhere else have pumpkins been seen to reach two hundred and fifty pounds in weight each, beets one hundred and twenty pounds, white turnips twenty-six pounds. Some cabbages and beets have spontaneously become perennials here"; indeed, "the abundance, excellence, and variety of our fruit astonish

the stranger, though he may have come from the markets of London or New York" (Hittel, *The Resources of California,* pp. 434–435). Greeley (*Recollections of a Busy Life,* p. 386) was enthusiastic about Californian grapes and noted that peach and apple trees yielded so well that much of their fruit rotted in the fields because it could not be harvested fast enough. He described mustard fields "self sown and growing wild from year to year" yet producing a full crop.

54. Porter, *The West,* pp. 505–508.

55. Starr, *Material Dreams,* p. 5.

56. Pisani, *From the Family Farm to Agribusiness,* p. 82.

57. Roske, *Everyman's Eden,* p. 409.

58. The Spanish had introduced oranges into California in the late eighteenth century (ibid., pp. 398–399).

59. *Appleton's Guide to the United States,* p. 380.

60. Porter, *The West,* p. 88. By the 1870s, one tourist speculated, "if husbanded properly, with the same care exercised at Salt Lake, [the Los Angeles River] might be made to irrigate many times the present breadth of land" (Rusling, *The Great West and Pacific Coast,* p. 334).

61. Richardson, *Beyond the Mississippi,* p. 368.

62. Worster, *Rivers of Empire,* p. 133.

63. Newell, *Irrigation in the United States,* p. 23.

64. Powell, *Report on the Lands of the Arid Region of the United States,* pp. 77–80.

65. Ibid., p. 85.

66. Cited ibid.

67. Ibid., p. 106.

68. Ibid., p. 53.

69. Ibid., p. 54.

70. Ibid.

71. Ibid., p. 55.

72. Ibid.

73. Ibid., p. 51.

74. Powell, "The Irrigable Lands of the Arid Region," p. 767.

75. Ibid.

76. Ibid., p. 768.

77. Ibid.

78. Ibid.

79. Starr, *Material Dreams,* pp. 10–13.

80. Ibid., 15–16.

81. "Irrigating Arid Lands in the West," *Scientific American* 65 (1891), July 25, p. 49.

82. *Abstract, Twelfth Census of the United States, 1900* (Government Printing Office, 1904), table 150, p. 296.

83. In 1893, e.g., the South Gila Canal Company was constructing a dam on the Gila River and a 125-mile canal, the Sonora Canal Company was surveying the route for its works in California, and a second dam on the Gila was planned ("Gigantic Irrigation Project, Gila River Dam," *Scientific American* 69, July 25, 1891, p. 24).

84. *Census of 1910* (Government Printing Office, 1912), p. 834.

85. Ibid.

86. Smythe, *Conquest*, p. 327.

87. Ibid., p. 331

88. Ibid., p. 328.

89. Ibid., p. 329.

90. Ibid., p. 331.

91. Starr, *Material Dreams*, p. 3.

92. Reprinted in Smythe, *Conquest*, p. 55.

93. Nason, *The Vision of Elijah Berl*, pp. 5–6.

94. Smythe, *Conquest*, p. 101.

95. Nason, *The Vision of Elijah Berl*, pp. 210–211.

96. Ibid., p. 136.

97. Smythe, *Conquest*, p. 95.

98. Austin, *The Ford*, p. 34.

99. Ibid.

100. Ibid., p. 294.

101. Ibid., p. 35.

102. Ibid., p. 273.

103. Austin, *The Land of Little Rain*, p. 85.

104. Ibid., p. 103.

105. Ibid., p. 107.

106. Ibid., p. 102.

107. In the 1870s, two leaders of federal expeditions into the Colorado Basin, George Wheeler and John Wesley Powell, also saw the possibilities for large-scale dam building, but they focused their attention further upstream (Worster, "Comment," p. 218). On Wheeler, see Dawdy, *George Montague Wheeler*.

108. Cited in Dawdy, *Wheeler*, p. 74.

109. For a fuller account of these events, see Starr, *Material Dreams*, pp. 22–44.

110. Ibid., p. 43.

111. Wright, *The Winning of Barbara Worth*, p. 136.

112. Ibid., p. 137.

113. Ibid., pp. 139–140.

114. Ibid., p. 144.

115. Ibid., p. 153.

116. Ibid., p. 154.

117. Ibid., p. 312.

Chapter 10

1. U.S. Department of the Interior, *The Colorado River*, p. 25.

2. Steinbeck, *The Grapes of Wrath*, pp. 324–235.

3. "Irrigating Arid Lands in the West," *Scientific American* 65 (1891), July 25, p. 49.

4. Cited in "Results of the Irrigation Congress," *Literary Digest* 9 (1894), September 29, p. 637.

5. *Chicago Times,* cited in *Literary Digest* 9 (1894), September 29, p. 637.

6. Bokovoy, "Inventing Agriculture in Southern California," pp. 3–4.

7. Truman, *History of the World's Fair,* p. 505.

8. For more on the Wright Act, see Pisani, *From the Family Farm to Agribusiness,* pp. 250–282.

9. Cited in "Irrigation Law of Eminent Domain," *Literary Digest,* December 5, 1896, p. 134.

10. Ibid., p. 133.

11. See Pisani, *To Reclaim a Divided West,* pp. 102–103.

12. Cited in "Irrigation Law of Eminent Domain," *Literary Digest,* December 5, 1896, p. 134.

13. Pisani, *To Reclaim a Divided West,* p. 103.

14. Ibid., 289–291.

15. Ibid., pp. 291–292.

16. Roosevelt, cited in Newell, *Irrigation in the United States,* pp. 393–394.

17. Language of original law, quoted in U.S. Department of the Interior, *The Colorado River,* p. 2.

18. "The Gardens of the West," *National Geographic* 16 (1905), March, p. 118.

19. Ibid., p. 123.

20. Ibid.

21. Davis, *Irrigation Works Constructed by the United States Government,* pp. 6–7.

22. On construction and early use, see ibid., pp. 6–33.

23. Newell, *Irrigation in the United States,* pp. 11–12.

24. U.S. Department of the Interior, *The Colorado River,* pp. 138–139.

25. Nye, *Electrifying America,* p. 300.

26. Newell, *Irrigation in the United States,* p. 1.

27. Ibid.

28. Ibid., p. 3.

29. *Census of 1920* (Government Printing Office, 1923), volume 7, *Irrigation,* pp. 41.

30. Ibid., p. 46.

31. Ibid., p. 32.

32. *Census of 1910* (Government Printing Office, 1912), chapter XI, "Irrigation," p. 829.

33. Teele, *Irrigation in the United States,* p. 234.

34. Davis, *Irrigation Works,* p. vii.

35. Teele, *Irrigation,* p. 235.

36. Census of 1910, chapter XI, "Irrigation," p. 831.

37. Pisani, *From the Family Farm to Agribusiness,* pp. 443–445.

38. Ibid., p. 449.

39. Ibid., p. 451.

40. In 1920, 91% of the capacity for pumping water from wells was in private hands, distributed among 30,000 individuals and partnerships. The Bureau of Reclamation held less than 1% of such equipment. In California alone, 1,126,000 acres were supplied by private pumping. Groundwater levels fell drastically, forcing farmers to drill ever-deeper wells. Most such pumps were concentrated in just three states: California, Texas, and Louisiana. Source: *Census of 1920.* (Government Printing Office, 1923), volume 7, *Irrigation,* p. 37.

41. McWilliams, *Factories in the Field,* pp. 19–21.

42. George, *Our Land and Land Policy,* pp. 9, 24.

43. Ibid., p. 25

44. Davis, *Irrigation Works,* p. 393.

45. Teele, *Irrigation,* p. 240.

46. Ibid., p. 241.

47. Such an argument meant, of course, that the government subsidized western farmers to compete with farmers in the East. During the 1890s, when there had

been a glut of commodities and when prices had been low, that kind of subsidy had been difficult to sell to representatives from the Midwest, Pennsylvania, or New York. With the generally higher farm prices after c. 1900 and the rapid growth of urban markets, a federal stimulus for western production seemed more reasonable. Source: ibid., pp. 241–242.

48. Kluger, *Turning on Water with a Shovel,* pp. 118–120.

49. Ibid., pp. 128, 130,

50. Roosevelt, "Address at Dedication of Boulder Dam," p. 397.

51. Ibid., p. 398.

52. Ibid., p. 402.

53. Ibid., p. 401.

54. U.S. Department of the Interior, *The Colorado River,* p. 25.

55. Ibid.

56. Ibid.

57. McWilliams, *Factories in the Field,* pp. 66–92.

58. Starr, *Endangered Dreams,* pp. 223–224.

59. Steinbeck, *The Grapes of Wrath,* p. 319.

60. Ibid., p. 325.

61. Ibid., p. 324.

62. Waters, *People of the Valley,* p. 179.

63. Ibid., pp. 173–174.

64. Ibid., p. 141.

65. Ibid., p. 143.

66. Reisner, *Cadillac Desert,* p. 61.

67. U.S. Supreme Court, *Winters v. United States,* 207 U.S. 564 (1908).

68. U.S. National Water Commission, *Water Policies for the Future,* cited in Reisner and Bates, *Overtapped Oasis,* p. 93.

69. Reisner, *Cadillac Desert,* pp. 194–199.

70. See U.S. Supreme Court, *Arizona v. San Carlos Apache Tribe,* 463 U.S. 545 (1983). See also Arizona v. San Carlos Apache tribe et al., certiorari to U.S. Court of Appeals for the Ninth Circuit, no. 81–2147.

71. S. Rep. No. 1507, 89th Cong., 2d Sess., 2 (1966). [463 U.S. 545, 578]. Justice Marshall cited the same source in his separate dissent: "These cases thus implicate the strong congressional policy, embodied in 28 U.S.C. 1362, of affording Indian tribes a federal forum. Since 1362 reflects a congressional recognition of the 'great hesitancy on the part of tribes to use State courts,' S. Rep. No. 1507, 89th Cong., 2d Sess., 2 (1966), tribes which have sued under that provision should not lightly be remitted to asserting their rights in a state forum."

72. Reisner and Bates, *Overtapped Oasis,* p. 43.

73. "In some cases also the rise of ground water, resulting from over-irrigation or seepage from canals, has destroyed the fertility of lands otherwise fertile. The latter difficulty has been remedied largely by the construction of drainage works, but not until after some hardship had been suffered by some of the settlers." Davis, *Irrigation Works Constructed by the United States Government*, p. 396.

74. Berger, *There Was a River*, pp. 60–61.

75. Dawdy, *Congress in Its Wisdom*, p. 116.

76. Ibid., p. 119.

77. Ibid., p. 120.

78. Ibid., pp. 123–125.

79. Cited in ibid., p. 125.

80. See *Carson-Truckee Water Conservancy District v. Clark*, 741 F.2d 257, 261 (9th Circuit Court. 1984).

81. Reisner and Bates, *Overtapped Oasis*, pp. 42–43.

82. A single wildlife refuge cannot, in practice, take care of its wildlife without dealing with other government agencies and private organizations. It cannot be managed or understood on its own. The area shares water with local irrigators, sets quotas for hunters on public land, and coordinates its work with owners of private wetlands and hunting preserves. See Lopez, *Crossing Open Ground*, pp. 29–36.

83. Reisner, *Cadillac Desert*, pp. 148–149.

84. Ibid., p. 153.

85. Ibid., pp. 189–194.

86. Ibid., pp. 503–504.

87. Ibid., p. 502.

88. For widely read statements marking this change in environmental history, see the essays in Cronon, ed. *Uncommon Ground*; also see White, "Discovering Nature in North America."

89. Pisani, *From the Family Farm to Agribusiness*, pp. 291–292.

Chapter 11

1. Turner first presented his thesis at the meeting of the American Historical Association at the Chicago World's Fair of 1893. It was published in the *Annual Report of the American Historical Association for the Year 1983* (Turner, "The Significance of the Frontier," pp. 199–227). By the 1950s, a summary of the defects of the Turner thesis had become standard in textbooks—for an overview, see Taylor, ed., *The Turner Thesis*, and for more recent discussions see Malone, "The Historiography of the American West," Noble, "Frederick Jackson Turner and Henry Nash Smith Revisited, 1890–1950–1990." For recent reflections on Turner, see White,

"Frederick Jackson Turner and Buffalo Bill" and Limerick, "The Adventures of the Frontier in the Twentieth Century."

2. Cited in Boller, *American Thought in Transition*, p. 212.

3. Macherey, *A Theory of Literary Production*, p. 155.

4. Recent readings of Henry Adams have explored the private life that is largely absent from his published writings. They have castigated him for his racial views, his relations with women, and his support for the emergence of the United States as an imperialist power. They have shown that Adams, for all his erudition, was in some ways not an admirable figure in his private life and personal opinions. (See e.g. Rowe, ed., *New Essays on The Education of Henry Adams;* Kaledin, *The Education of Mrs. Henry Adams.*) Rather than focus on these personal failings, I will focus on Adams's attempt to reconcile evolutionary theory and thermodynamics in his last works.

5. Letter cited in Samuels, *Henry Adams*, p. 320.

6. Adams, *The Education of Henry Adams*, p. 60.

7. Martin, *American Literature and the Universe of Force*, p. 99. Adams's late essays, published posthumously in *The Degradation of the Democratic Dogma*, contain references to scientific works from as late as 1909, so it appears that Adams continued this intellectual interest until the end of his writing life. See Adams, *The Degradation of the Democratic Dogma*, p. 239. Adams was still revising the manuscript and incorporating material published in 1910; see ibid., p. 252.

8. At first the debate was between Newtonians and those who followed Leibniz's dynamic theory, which argued that moving objects were inherently imbued with force. According to Max Jammer (*Concepts of Force*, pp. 161–162, 180–183), "relativity of space and absoluteness of motion led Leibniz to the existence of force, in contrast to Newton, for whom absoluteness of force proved the absoluteness of motion and, consequently, the absoluteness of space." Kant attempted to overcome the difference between these arguments in a theory that presupposed the existence of two kinds of force, but his approach failed to convince most scientists.

9. Adams also recognized that new energies would likely become available through the discoveries of science and the work of inventors, seen variously in the discoveries of radium and x rays and in the transformations in all aspects of life made possible by railways, electrification, the gasoline engine, and air travel.

10. See Martin, *American Literature and the Universe of Force*, pp. 13–26.

11. Spencer, *First Principles*, p. 169.

12. Ultimately both the controversy over force and the theories of thermodynamics were superseded by the development of the theory of relativity. The definition of force was but one of the problems of the Newtonian synthesis. For the classic discussion, see Thomas Kuhn, *The Structure of Scientific Revolutions*, second edition (University of Chicago Press, 1970).

13. A prolific author, Adams was also an avid reader. Having begun with his father's history library, Charles Dickens, and Ralph Waldo Emerson, he continued

reading not only in English but also in French. Adams also knew many writers, notably James Russell Lowell, Robert Browning, Algernon Swinburn, Rudyard Kipling, Robert Louis Stevenson, and Henry James. If Adams's own novels were not notably experimental, they were well above the average production of his day, and he certainly was aware of late-nineteenth-century experiments with point of view. He wrote his autobiography in the third person, as though it were the story of another person; this device allowed him to omit most of his private affairs.

14. Adams, *The Education*, p. 335.

15. Samuels, *Henry Adams*, p. 351.

16. Ibid., p. 361–362.

17. There were other reasons for Adams to become distant from himself, of course, not least his beloved sister's death, his wife's suicide, the passing of his parents, the death of friends such as Clarence King, and other personal losses. *The Education* omits all reference to his marriage. The entire decade of the 1880s disappears, and the book changes tone and mood in the sections dealing with the 1890s and after. In the longer, earlier portion, Adams searches for a career. In the second part, he is old; the mood is more retrospective and philosophical, but his wife's suicide is not permitted to function as a part of his education.

18. Adams, *The Degradation of the Democratic Dogma*, p. 233.

19. Spencer was widely read and quoted, and his ideas were seminal for many realist and naturalist writers. See Martin, *American Literature and the Universe of Force*. Spencer was, if anything, even more widely influential outside literature; see e.g. Wright, *The Industrial Evolution of the United States*, p. 343.

20. Adams, *The Education*, p. 239

21. The sequence of force is a striking conception that undoubtedly influenced my book *Consuming Power*, though I did not consciously recall it while writing.

22. See Tichi, *Shifting Gears*, pp. 137–164.

23. Lears, *No Place of Grace*, pp. 286–297.

24. Marx, *The Machine in the Garden*, p. 347.

25. Adams, *The Education*, p. 382.

26. Ibid., p. 380.

27. Ibid.

28. Ibid., p. 381.

29. My concern here is not to determine precisely what Edison thought about these matters and what was added by journalists; it is the articulation of a certain popular narrative that many Americans found attractive and persuasive.

30. "Edison's Prophecy," *Literary Digest*, November 14, 1914, pp. 966–968.

31. "The Woman of the Future," *Good Housekeeping Magazine*, October 1913, p. 436.

32. See Marshall, "Thomas A. Edison on Immortality," p. 610; Uzzell, "The Future of Electricity," p. 8.

33. Edison's ideas resembled less Darwinian natural selection than the Lamarckian idea that human beings acquire new characteristics from changes in their environment.

34. "America's Future in Light of the Past," *Literary Digest* XI (1895), no. 23, p. 668.

35. See Mulhall, "The Increase of Wealth," pp. 78–85.

36. Thurston, "Our Progress in Mechanical Engineering," pp. 7–8. This address is replete with such facts and figures for the major industries of the day. Thurston's argument was common at the time; see e.g. Wright, *The Industrial Evolution of the United States,* pp. 345–351. I thank Merritt Roe Smith for drawing my attention to this example.

37. Thurston, "The Trend of National Progress," p. 310.

38. See Wright, *The Industrial Evolution of the United States,* pp. 346–349.

39. Thurston, "The Trend of National Progress," p. 311.

40. Ibid., p. 306

41. "Of the Trend of Modern Progress, in direction and rate of movement, there is no reasonable doubt." Ibid., p. 312.

42. Thurston, "The Mission of Science," p. 28. See also Thurston, "The Borderland of Science," p. 73, where he predicts the evolution of a far more intelligent race.

43. Tesla, "The Problem of Increasing Human Energy," p. 175.

44. Ibid., p. 176.

45. Ibid., p. 177.

46. Ibid., p. 183. To this end, Tesla did experiments with "controlling the movements and operations of distant automatons" (ibid., p. 186).

47. Ibid., pp. 191–192.

48. Ibid., p. 193.

49. Ibid., p. 210. To Tesla it seemed "highly probable that if there are highly intelligent beings on Mars they had long ago realized this very idea, which would explain the changes on its surface noted by astronomers."

50. Beringause, *Brooks Adams,* p. 117.

51. Ibid., p. 122.

52. Ibid., p. 111.

53. Henry Adams, "The Problem," in Adams, *The Degradation of the Democratic Dogma,* p. 140.

54. Cited in ibid., pp. 141–142.

55. For a discussion, see Nye, *Narratives and Spaces,* pp. 75–91.

56. Adams, "The Problem," p. 206

57. Ibid., p. 224.

58. Ibid., pp. 26–228.

59. Ibid., p. 195. See also p. 225. Adams also cited German scientists who argued that early man had a vigorous intellect (ibid., p. 240).

60. Ibid., pp. 205–206.

61. Ibid., p. 243.

62. Ibid., p. 216.

63. Ibid.

64. Ibid.

65. Ibid.

66. Ibid., p. 217.

67. Ibid., p. 218.

68. Ibid., p. 220.

69. Ibid., p. 229.

70. Samuels, *Henry Adams*, p. 412.

71. Henry Adams, "The Rule of Phase Applied to History," in *The Degradation of the Democratic Dogma*, pp. 274–276.

72. Ibid., p. 281.

73. Ibid., pp. 285–286.

74. Ibid., pp. 290–292, 308.

75. Adams, *The Education*, p. 59.

76. Adams, *The Life of Albert Gallatin*, pp. 350–352.

77. Adams, *The Education*, p. 330.

78. What Adams loved about the Germany of his youth was its still-fragmented eighteenth-century culture, which even at the time of his first visit was giving way to mechanization. Until coal power and railways were created, Germany "was mediaeval by nature and geography." Afterward, it was welded into a single nation. When confronting European politics, Adams found that "his morals were the highest, and he clung to them to preserve his self-respect; but steam and electricity had brought about new political and social concentrations, or were making themselves necessary in the line of his moral principles—freedom, education, economic development and so forth—which required association with allies as doubtful as Napoleon III, and robberies with violence on a very extensive scale." The railroad and the telegraph thus presented themselves not simply as forms of economic development, but as political instruments of unification (Germany) and war (France and Austria). In Central Europe, technologies transformed the old order, a lesson reiterated by the American Civil War (Adams, *The Education*, pp. 83–84, 72).

79. Adams, *The Education*, p. 240.

80. After describing a painful journey, where nothing functioned any better than it had 30 years before, Adams concluded: "Actually, in Europe I see no progress—none! . . . The people are stupid." (Henry Adams to John Hay, January 9, 1892, in Levenson et al., eds., *The Letters of Henry Adams*, pp. 599–600) Adams wrote in the same vein to Elizabeth Cameron two days later: "The Frenchmen and the Englishman are just where they were 30 years ago, with a certain halo of vulgarity and commonness added to their stupidity." (ibid., p. 601).

81. Adams, *The Education*, p. 330.

82. Ibid., pp. 478, 484–485.

83. Martin, *American Literature and the Universe of Force*, p. 99.

84. Adams, *The Education*, p. 451.

85. Adams, "The Problem," in *The Degradation of the Democratic Dogma*, p. 251.

86. See Miller, *Pretty Bubbles in the Air*, pp. 211–215.

87. *Colliers Weekly Magazine*, July 14, 1923, p. 5; *Literary Digest*, June 30, 1923.

88. See Nye, *Henry Ford*, p. 83.

Conclusion

1. Hill, "Development of the West," p. 35.

2. Edward Everett, "American Manufacturers," in *Orations and Speeches of Edward Everett*, volume 2, pp. 69–70.

3. Ibid., p. 70.

4. Ibid., p. 71.

5. Wright, *The Industrial Evolution of the United States*, pp. 347–352.

6. Lincoln, "Second Lecture on Discoveries and Inventions," pp. 357–358.

7. Ibid., p. 358.

8. Cited in Tichi, *New World, New Earth*, p. 172. I thank Cecelia Tichi for drawing my attention to this passage.

9. Cited in Wade, *The Urban Frontier*, p. 31. After the steamboat arrived, many projected towns also failed, however. In 1809 a city was laid out and advertised at the junction of the Ohio and Mississippi Rivers with the name "Town of America," and potential investors were asked to consider why this location should not support one of the greatest cities of the United States (*Pittsburgh Gazette*, October 15, 1809).

10. On the rich antebellum literature on steamboats and the Mississippi, see McDermott, ed., *Before Mark Twain*.

11. Reps, *Cities of the American West,* pp. 3–10.

12. *Liberty Hall* (Cincinnati), October 1, 1819, cited in Wade, *The Urban Frontier,* p. 33.

13. Merrick, *Old Times on the Upper Mississippi,* p. 180. Similarly, in the spring of 1852, several hundred people from the East, many from New York, arrived on the upper Mississippi looking for an imaginary colony with the unlikely name of Rolling Stone. They had "beautiful maps and bird's eye views of the place, showing a large greenhouse, lecture hall, and library." Each had purchased a house in town and a farm outside town. Like everything else on their maps, the houses and the farms did not exist. The unfortunate people made sod houses for themselves and tried to eke out a living as best they could (ibid., p. 182).

14. Hunter, *Steamboats on Western Rivers,* pp. 100–101.

15. Prentice, *A Tour of the United States,* p. 48. See also Hunter, *Studies in the Economic History of the Ohio Valley,* pp. 15–17.

16. Another traveler complained that "for days and days together" he had heard "no sound but the equable working of the engine, the periodically recurring rattle of the table preparations, and the deafening noise of the dinner bell." Furthermore, steamboats were too unwieldy to tackle rapids, such as those at Louisville, which some adventurous travelers enjoyed. See Mollhausen, *Diary of a Journey from the Mississippi to the Coasts of the Pacific with a United States Government Expedition,* p. 1–2.

17. Ibid., p. 2

18. Hunter, *Steamboats on Western Rivers,* pp. 585–589.

19. Pursell, *The Machine in America,* pp. 111–112.

20. Pare Lorentz, *The Plow That Broke the Plains* (film, 1936).

21. *Official Guide, New York World's Fair, 1964/1965,* pp. 201–202.

22. Carlin, "Fly Me to the Moon," p. 20.

23. Real estate on Mars, Venus, and the Moon, divided on the grid system, is being offered on the World Wide Web (http://www.lunarembassy.com/lunar/index_e.shtml). Those offering it are exploiting a loophole in the 1967 United Nations Outer Space Treaty.

24. Kim Stanley Robinson, *Red Mars* (Bantam, 1993), *Green Mars* (Bantam, 1995), *Blue Mars* (Bantam, 1997).

25. McNichol, "The New Red Menace," p. 146.

26. Ibid.

27. Powell, "Irrigable Lands of the Arid Region," p. 768.

28. "The New Adam and Eve," in Hawthorne, *Mosses from an Old Manse,* p. 247.

29. See e.g. Gloria Anzaldúa, *Borderlands/La Frontera: The New Mestiza* (San Francisco: Spinsters/Aunt Lute, 1987); David G. Gutierrez, "Significant to Whom? Mexican Americans and the History of the American West," *Western Historical Quarterly* 24 (1993), November: 519–539.

30. African-Americans, in good part because most were slaves until 1865 and sharecroppers during the rest of the nineteenth century, were not given much of a role in these narrative or counter-narratives. They were subalterns to the machine.

31. "Big Two-Hearted River," in Hemingway, *In Our Time.*

32. On the creation of national parks, see Nash, *Wilderness and the American Mind*, pp. 96–140.

33. On this and subsequent projects to reframe Niagara Falls, see Anne Whiston Spirn, "Constructing Nature: The Legacy of Frederick Law Olmsted," in Cronon, ed., *Uncommon Ground*, pp. 95–96.

34. Worster, *An Unsettled Country*, p. 81.

35. Narration to *The River* (Pare Lorentz, 1937), reprinted in Filler, ed., *The Anxious Years*, pp. 364–375.

36. Nash, *Wilderness and the American Mind*, pp. 146–151.

37. The Wisconsin Wilderness Act of 1984 became Public Law 98–321. The Vermont Wilderness Act of 1984 became Public Law 98–322. The New Hampshire Wilderness Act of 1984 became Public Law 98–323. The North Carolina Wilderness Act of 1984 became Public Law 98–324.

38. Ronald Reagan, "Remarks on Signing Four Bills Designating Wilderness Areas," June 19, 1984.

39. Fitzgerald, *The Great Gatsby*, p. 171.

40. I am indebted to Leo Marx, who pointed out the importance of Fitzgerald's word choice in a lecture titled "The Pandering Landscape" (Syddansk University, September 30, 2001).

41. Merchant, "Reinventing Eden," in Cronon, ed., *Uncommon Ground*, pp. 157–158.

42. Cronon, "Toward a Conclusion," in Cronon, ed., *Uncommon Ground*, p. 459.

Bibliography

Abbey, Edward. *The Monkey Wrench Gang.* Avon, 1975.

Adams, Brooks. *The Law of Civilization and Decay.* London: Swan Sonnenschein, 1895.

Adams, Charles Francis. "Of Some Railroad Accidents, II." *The Atlantic* 36 (1875), December: 736–748.

Adams, Charles Francis. "The State and the Railroads." *The Atlantic* 37, June, 1876: 691–699.

Adams, Henry. *The Life of Albert Gallatin.* Lippincott, 1879.

Adams, Henry. *The Degradation of the Democratic Dogma,* ed. B. Adams. Macmillan, 1919.

Adams, Henry. *The Education of Henry Adams.* Modern Library, 1931.

Adas, Michael. *Machines as the Measure of Men: Science, Technology, and Ideologies of Western Dominance.* Cornell University Press, 1989.

Aldrich, Mark. *Safety First: Technology, Labor, and Business in the Building of American Work Safety.* Johns Hopkins University Press, 1997.

Allen, Zachariah. *The Practical Tourist.* Providence: A. S. Beckwith, 1832.

Anderson, Sherwood. "A Great Factory: Problems and Attitudes in Life and Industry Considered from a Workman's Angle." *Vanity Fair* 27 (1926), November: 49–56.

Anderson, Sherwood. *Beyond Desire.* New York: Liveright, 1932.

Anderson, Sherwood. *Poor White.* Viking, 1966 (reprint of 1920 edition).

Appleton's Guide to the United States. Appleton, 1883.

Arfwedson, Carl David. *The United States and Canada in 1832, 1833, and 1834.* Johnson Reprint, 1969.

Armstrong, John B. *Factory under the Elms: A History of Harrisville, New Hampshire, 1774–1969.* MIT Press, 1969.

Aronowitz, Stanley. *False Promises: The Shaping of American Working Class Consciousness.* McGraw-Hill, 1973.

Austin, Mary. *The Land of Little Rain.* Houghton Mifflin, 1903.

Austin, Mary. *The Ford.* University of California Press, 1997 (reprint of 1917 edition).

Bachelard, Gaston. *The Poetics of Space.* Beacon, 1969.

Barbour, Edwin Hinckley. "Home-Made Windmills in Nebraska." *Literary Digest,* January 27, 1900: 116–117 (reprinted from *Scientific American*).

Barnes, Peter, ed. *The People's Land: Reader on Land Reform in the United States.* Rodale, 1975.

Barthes, Roland. "Introduction to the Structural Analysis of Narratives." *Image, Music, Text.* Hill and Wang, 1977.

Beadle, J. H. *The Undeveloped West; or Five Years in the Territories.* Philadelphia: National Publishing, 1873.

Beale, J. H. *Picturesque Sketches of American Progress.* New York: Empire Co-operative Association, 1889.

Bealer, Alex W., and John O. Ellis. *The Log Cabin: Homes of the North American Wilderness.* Barre, 1978.

Beatty, Bess. "Lowells of the South: Northern Influences on the Nineteenth-Century North Carolina Textile Industry." *Journal of Southern History* 53 (1987), no. 1: 37–62.

Bell, William A. *New Tracks in North America.* London: Chapman and Hall, 1870.

Bercovitch, Sacvan. *The American Jeremiad.* University of Wisconsin Press, 1978.

Bercovitch, Sacvan. *Puritan Origins of the American Self.* Yale University Press, 1986.

Berger, Bruce. *There Was a River.* University of Arizona Press, 1994.

Beringause, Arthur. *Brooks Adams: A Biography.* Knopf, 1955.

Berkeley, George. *The Works of George Berkeley, D. D.,* ed. A. Fraser. Oxford University Press, 1901.

Bigelow, Jacob. *The Useful Arts.* Boston: Thomas Webb, 1840.

Birkbeck, Morris. *Notes on a Journey in America.* London: James Ridgeway, 1818.

Bishop, J. Leander. *A History of American Manufacturers, from 1608 to 1860,* third edition 1868, ed. L. Hacker. Reprint: Augustus M. Kelley, 1966.

Bokovoy, Matthew F. "Inventing Agriculture in Southern California." *Journal of San Diego History* 45 (1999), no. 2.

Boller, Paul F. *American Thought in Transition: The Impact of Evolutionary Naturalism, 1865–1900.* Rand McNallly, 1969.

Boorstin, Daniel. *The Americans: The National Experience.* Random House, 1965.

Brewe, William. *Up and Down California in 1860–64.* University of California Press, 1966.

Bureau of the Census. *Agriculture, 1860.* Government Printing Office, 1864.

Butler, Ovid M. "Henry Ford's Forest." *American Forestry* 28 (1922), no. 348: 725–731.

Butterfield, L. H., ed. *Letters of Benjamin Rush.* Princeton University Press, 1951.

Byrd, William. *Histories of the Dividing Line betwixt Virginia and North Carolina.* Dover, 1967 (reprint of 1929 edition).

Callenbach, Ernest. *Ecotopia: A Novel.* Bantam, 1975.

Carey, Henry. *Principles of Political Economy.* A. M. Kelley, 1965 (reprint of 1837–1840 edition).

Carlin, John. "Fly Me to the Moon." *Independent on Sunday,* March 8, 1998, Focus section, p. 20.

Cash, W. J. *The Mind of the South.* Knopf, 1941.

Cather, Willa. *My Antonia.* Houghton Mifflin, 1918.

Cease, D. L. "The Death and Disability Toll of Our Railway Employees." *Charities and the Commons* 19 (1908): 1216–1222.

Certeau, Michel de. *The Practice of Everyday Life.* University of California Press, 1984.

Chandler, Alfred D. *The Visible Hand: The Managerial Revolution in American Business.* Harvard University Press, 1977.

Chastellux, Marquis de. *Travels in North-America, in the Years 1780–81–82.* New York: White, Gallagher, & White, 1827; reprint: Augustus K. Kelley, 1970.

Chatman, Seymour. "Reply to Barbara Herrnstein Smith." *Critical Inquiry* 7 (1980): 802–807.

Chevalier, Michael. *Society, Manners, and Politics in the United States.* Doubleday Anchor, 1961.

Chudacoff, Howard, and Judith Smith. *The Evolution of American Urban Society.* Prentice-Hall, 2000.

Cist, Charles. *Cincinnati in 1851.* Cincinnati: W. H. Moore, 1851.

Clampitt, John W. *Echoes from the Rocky Mountains.* Chicago: Belford, Clarke, 1889.

Clark, Charles E. *The Eastern Frontier: The Settlement of Northern New England.* Knopf, 1970.

Cohen, Norm. *Long Steel Rail: The Railroad in American Folksong.* University of Illinois Press, 2000.

Conzen, Michael P. *The Making of the American Landscape.* Routledge, 1994.

Cooper, James Fenimore. *The Chainbearer.* New York: James G. Gregory, 1864.

Cooper, James Fenimore. *Notions of the Americans.* Frederick Ungar, 1963.

Corner, James, and Alex S. MacLean. *Taking Measures across the American Landscape.* Yale University Press, 1996.

Cosgrove, Denis. "The Measures of America." In *Taking Measures across the American Landscape,* ed. J. Corner and A. MacLean. Yale University Press, 1996.

Cowley, Charles. *History of Lowell,* second edition, revised. Boston: Lee and Shepard, 1868.

Cowley, Malcolm. *The Dream of the Golden Mountains.* Viking, 1980.

Cronon, William, ed. *Uncommon Ground: Rethinking the Human Place in Nature.* Norton, 1996.

Custis, George Washington P. "An Address on the Importance of Encouraging Agriculture and Domestic Manufactures" (1808). Reprinted in *The Philosophy of Manufactures,* ed. M. Folsom and S. Lubar (MIT Press, 1982).

Daniels, Bruce C. "Economic Development in Colonial and Revolutionary Connecticut." *William and Mary Quarterly,* third series, 37 (1980), no. 3: 429–450.

Danly, Susan, and Leo Marx. *The Railroad in American Art: Representations of Technological Change.* MIT Press, 1988.

Dargan, Olive. *Call Home the Heart, by Fielding Burke [pseud.].* Longmans, Green, 1932.

Dargan, Olive. *A Stone Came Rolling.* Longmans, Green, 1935.

Davis, Arthur Powell. *Irrigation Works Constructed by the United States Government.* Wiley, 1917.

Dawdy, Doris Ostrander. *Congress in Its Wisdom: The Bureau of Reclamation and the Public Interest.* Westview, 1989.

Dawdy, Doris Ostrander. *George Montague Wheeler: The Man and the Myth.* Swallow/Ohio University Press, 1993.

Dawson, Andrew. "Reassessing Henry Carey (1793–1879): The Problem of Writing Political Economy in Nineteenth Century America." *Journal of American Studies* 34 (2000), December: 465–485.

Decker, Leslie E. "The Railroads and the Land Office: Administrative Policy and the Land Patent Controversy, 1864–1896." *Mississippi Valley Historical Review* 46 (1960), no. 4: 679–699.

Dempsey, Bernard W. "Just Price in a Functional Economy." *American Economic Review* 25 (1935), no. 3: 471–476.

Denning, Michael. *Mechanic Accents: Dime Novels and Working-Class Culture in America.* Verso, 1998.

Dicey, Edward. *Six Months in the Federal States.* Macmillan, 1863.

Dickens, Charles. *American Notes.* Fromm International, 1985.

Dorson, Richard. *Folklore and Fakelore.* Harvard University Press, 1976.

Drake, Daniel. *Discourse on the History, Character and Prospects of the West.* Gainesville: Scholars Facsimiles & Reprints, 1955.

Draper, Theodore. "Gastonia Revisited." *Social Research* 38 (1971): 3–29.

Dreiser, Theodore. "The Railroad and the People." *Harper's New Monthly Magazine* 29 (1900): 479–484.

Drucker, James H. *Men of the Steel Rails: Workers on the Atchison, Topeka & Santa Fe Railroad, 1869–1900.* University of Nebraska Press, 1983.

Dublin, Thomas. *Women at Work: The Transformation of Work and Community in Lowell, Massachusetts, 1826–1860.* Columbia University Press, 1979.

Dumke, Glenn S. *The Boom of the Eighties in Southern California.* San Marino: Huntington Library, 1944.

Dwight, Timothy. *Travels, in New England and New York,* ed. B. Solomon. Harvard University Press, 1969.

Emerson, Ralph Waldo. *Complete Works.* Houghton Mifflin, 1903.

Emerson, Ralph Waldo. *Journals of Ralph Waldo Emerson.* Houghton Mifflin, 1909–1914.

Emerson, Ralph Waldo. *Selected Writings of Emerson.* Modern Library, 1950.

Emerson, Ralph Waldo. *The Journals and Miscellaneous Notebooks of Ralph Waldo Emerson.* Harvard University Press, 1969.

Emmons, David M. *Garden in the Grasslands: Boomer Literature of the Central Great Plains.* University of Nebraska Press, 1971.

Erdrich, Louise. *Tracks.* Harper & Row, 1988.

Etzler, J. A. *The Paradise within the Reach of All Men, without Labour, by Powers of Nature and Machinery.* London: John Brooks, 1836.

Everett, Edward. *Orations and Speeches of Edward Everett.* Little, Brown, 1850, 1861, 1869.

Faulkner, William. *Light in August.* Modern Library, 1950.

Faulkner, William. "Delta Autumn." In *The Portable Faulkner.* Viking, 1967.

Ferguson, Eugene S. "The Measurement of the 'Man Day.'" *Scientific American* 225 (1971), October: 96–103.

Ferguson, Eugene S. Industrial Power Sources of the Nineteenth Century: The Hagley Example. Typescript, 1981.

Filler, Louis, ed. *The Anxious Years: America in the Nineteen Thirties.* Putnam, 1963.

Fisher, Philip. *Still the New World: American Literature in a Culture of Creative Destruction.* Harvard University Press, 1999.

Fishkin, Shelley Fisher. *Lighting Out for the Territory.* Oxford University Press, 1996.

Fishlow, Albert. *The American Railroads and the Ante-Bellum Economy.* Harvard University Press, 1965.

Fitzgerald, F. Scott. *The Great Gatsby.* Scribner, 1925.

Fitzhugh, George. *Cannibals All! or, Slaves without Masters.* Harvard University Press, 1960 (reprint of 1857 edition).

Fitzhugh, George. *Sociology for the South.* New York: Franklin, 1965 (reprint of 1854 edition).

Fletcher, Henry J. "American Railways and American Cities." *The Atlantic* 73 (1894), June: 803–811.

Flint, Timothy. *A Condensed Geography and History of the Western States, or the Mississippi Valley.* Scholars' Facsimiles and Reprints, 1970 (reprint of 1828 edition).

Fogel, Robert William. *The Union Pacific Railroad: A Case in Premature Enterprise.* Johns Hopkins University Press, 1960.

Fogel, Robert William. *Railroads and American Economic Growth.* Johns Hopkins University Press, 1964.

Folsom, Michael Brewster, and Steven D. Lubar, eds. *The Philosophy of Manufactures: Early Debates over Industrialization in the United States.* MIT Press, 1982.

Ford, Henry. *My Life and Work.* Doubleday and Page, 1922.

Franklin, Benjamin. "To ——: Information to Europeans Who Are Disposed to Migrate to the United States." In *The Papers of Benjamin Franklin,* volume 13, ed. L. Labaree. Yale University Press, 1972.

Franklin, Benjamin. "Some Observations on North America from 'Oral Information by Dr. Franklin.'" (interview with Gottfried Achewall). Translated and reprinted in *The Papers of Benjamin Franklin,* volume 13, ed. L. Labaree. Yale University Press, 1972.

Friedenberg, Daniel M. *Life, Liberty, and the Pursuit of Land.* Buffalo: Prometheus, 1992.

Galenson, Alice. *The Migration of the Cotton Textile Industry from New England to the South, 1880–1930.* Garland, 1985.

Gates, Paul Wallace. *The Illinois Central Railroad and Its Colonization Work.* Harvard University Press, 1934.

Gates, Paul Wallace. *Fifty Million Acres: Conflicts over Kansas Land Policy, 1854–1890.* Cornell University Press, 1954.

Genovese, Eugene. *The Political Economy of Slavery.* Pantheon, 1965.

George, Henry. *Our Land and Land Policy.* San Francisco: White and Bauer, 1871.

George, Henry. *Progress and Poverty.* Dutton, 1976 (reprint of 1880 edition).

Georgianna, Daniel, and Roberta Hazen Aaronson. *The Strike of '28.* New Bedford, Mass.: Spinner, 1993.

Glacken, Clarence J. *Traces on the Rhodian Shore.* University of California Press, 1967.

Golin, Steve. *The Fragile Bridge: The Paterson Silk Strike, 1913.* Temple University Press, 1988.

Goodrich, Carter. "Internal Improvements Reconsidered." *Journal of Economic History* 30 (1970): 289–311.

Goodrich, Lloyd. *Edward Hopper.* Abrams, 1983.

Graham, Loren. *A Face in the Rock: The Tale of a Grand Island Chippewa.* University of California Press, 1998.

Greeley, Horace, et. al. *The Great Industries of the United States.* Garland, 1974 (reprint of 1872 edition).

Greeley, Horace. *Art and Industry as Represented at the Exhibition at Crystal Palace, 1853–4.* New York: Redfield, 1853.

Greeley, Horace. *An Overland Journey from New York to San Francisco in the Summer of 1859.* New York: Saxton, Barker, 1860.

Greeley, Horace. *Recollections of a Busy Life.* New York: J. B. Ford, 1868.

Greeley, Horace. *Mr. Greeley's Letters from Texas and the Lower Mississippi, Address to the Farmers of Texas, and Speech on his return to New York.* New York: Tribune Office, 1871.

Green, Martin. *New York 1913: The Armory Show and the Paterson Strike Pageant.* Scribner, 1988.

Gregg, William. *Essays on Domestic Industry* (1845). Portions reprinted in *The Philosophy of Manufactures,* ed. M. Folsom and S. Lubar (MIT Press, 1982).

Grossman, James R., ed. *The Frontier in American Culture.* University of California Press, 1994.

Grosvenor, W. M. "The Railroads and the Farms." *The Atlantic,* November 1873: 591–610.

Gunderson, Robert Gray. *The Log Cabin Campaign.* University of Kentucky Press, 1957.

Gutman, Herbert G. *Work, Culture, and Society in Industrializing America.* Vintage, 1977.

Hall, Basil. *Travels in North America.* London: Simpkin and Marshall, 1830.

Hall, Edward. *The Hidden Dimension.* Doubleday, 1966.

Hall, James. *Letters from the West.* London: Henry Colburn, 1828.

Harden, Blaine. *A River Lost: The Life and Death of the Columbia.* Norton, 1996.

Harding, Benjamin. *A Tour through the Western Country.* New London: Samuel Green, 1819.

Hareven, Tamara, and Randolph Langenbach. *Amoskeag: Life and Work in an American Factory City.* Pantheon, 1978.

Harpster, John W., ed. *Crossroads: Descriptions of Western Pennsylvania, 1720–1829.* University of Pittsburgh Press, 1938.

Harrison, Robert Pogue. *Forests: The Shadow of Civilization.* University of Chicago Press, 1992.

Hart, John Fraser. *The Rural Landscape.* Johns Hopkins University Press, 1998.

Hawthorne, Nathaniel. *Complete Works.* Houghton Mifflin, 1887–88.

Hawthorne, Nathaniel. *Mosses from an Old Manse*. Ohio State University Press, 1974,

Hazen, William Babcock. "The Great Middle Region of the United States, and Its Limited Space of Arable Land." *North American Review* 120 (1875), January: 1–34.

Hemingway, Ernest. *In Our Time*. New York: Boni & Liveright, 1925.

Henretta, James A. "The Transition to Capitalism in America." In *The Transformation of Early American History*, ed. J. Henretta et al. Knopf, 1991.

Hill, James J. "Development of the West." In *Proceedings of the Fourth Dry Farming Congress, Billings, Montana, October 26–28, 1909*. Billings, 1910.

Hine, Lewis. "The High Cost of Child Labor." *Child Labor Bulletin* 3 (1914–15): 63–65.

Hinton, Richard J. "Irrigation in the Arid West." *Harper's Weekly* 32 (1888), September 22: 729–730.

Hinton, Richard J. *Irrigation in the United States*. Government Printing Office, 1887.

Hittel, John S. *The Resources of California*. San Francisco: A. Roman, 1863.

Hodgskin, J. B. "The Truth about Land Grants." *The Nation* 11 (1870), December: 417–418.

Hoke, Donald R. *Ingenious Yankees: The Rise of the American System of Manufactures in the Private Sector*. Columbia University Press, 1990.

Holt, Michael Fitzgibbon. *Forging a Majority: The Formation of the Republican Party in Pittsburgh, 1848–1860*. University of Pittsburgh Press, 1969.

Hone, Philip. *Diary of Philip Hone*. Arno, 1970.

Hough, Emerson. *The Way to the West, and the Lives of Three Early Americans, Boone, Crockett, Carson*. Grosset & Dunlap, 1903.

Hudson, James F. *The Railways and the Republic*. Harper, 1886.

Hughes, Henry. *A Treatise on Sociology, Theoretical and Practical*. Lippincott, 1854.

Humphreys, David. "Address to the People of the United States." In *Miscellaneous Works*. New York: T. and J. Swords, 1804. Reprinted in *The Philosophy of Manufactures*, ed. M. Folsom and S. Lubar (MIT Press, 1982).

Hunt, Freeman. *Lives of Famous American Merchants*. New York: Office of *Hunt's Merchants' Magazine*, 1856.

Hunter, Louis C. *Studies in the Economic History of the Ohio Valley*. Johnson Reprint, 1970.

Hunter, Louis C. *A History of Industrial Power in the United States, 1780–1930*, volume 1: *Water Power*. University Press of Virginia, 1979.

Hunter, Louis C. *A History of Industrial Power in the United States, 1780–1930*, volume 2: *Steam Power*. University Press of Virginia, 1985.

Hunter, Louis C., and Lynwood Bryant. *A History of Industrial Power in the United States, 1780–1930,* volume 3: *The Transmission of Power.* MIT Press, 1991.

Hunter, Louis C. *Steamboats on Western Rivers: An Economic and Technological History.* Harvard University Press, 1949; reprint: Dover, 1993.

Hurley, F. Jack. *Industry and the Photographic Image.* Dover, 1980.

Hyde, Anne Farrar. *An American Vision: The Far Western Landscape and National Culture, 1820–1920.* New York University Press, 1990.

Jackson, John Brinckerhoff. *American Space: The Centennial Years.* Norton, 1972.

Jackson, John Brinckerhoff. *A Sense of Place, a Sense of Time.* Yale University Press, 1994.

James, Henry. *The American Scene.* Indiana University Press, 1968.

Jammer, Max. *Concepts of Force.* Harper, 1962.

Jefferson, Thomas. *Notes on the State of Virginia.* Harper Torchbooks, 1964.

Jeremy, David J. "Cotton Mills in Developing Regions, 1820–1840." In *Géographie du capital marchand aux Amériques, 1760–1860,* ed. J. Chase. Paris: Editions de l'Ecole des hautes études en sciences sociales, 1987.

Johnson, Hildegard Binder. "Towards a National Landscape." In *The Making of the American Landscape,* ed. M. Conzon. HarperCollins, 1994.

Johnson, Howard. *A Home in the Woods: Pioneer Life in Indiana, Oliver Johnson's Reminiscences of Early Marion County as Related by Howard Johnson.* Indiana University Press, 1951, 1978.

Jones, Howard Mumford. *O Strange New World.* Viking, 1964.

Kahn, Joseph. "Cheney Promotes Increasing Supply as Energy Policy." *New York Times,* May 1, 2001.

Kaledin, Eugenia. *The Education of Mrs. Henry Adams.* Temple University Press, 1981.

Kasson, John. *Civilizing the Machine.* Penguin, 1977.

Kates, James. *Planning a Wilderness: Regenerating the Great Lakes Cutover Region.* University of Minnesota Press, 2001.

Kauffman, Henry J. *American Axes: A Survey of Their Development and Their Makers.* Stephen Greene, 1972.

Kelley, Klara Bonsack, and Harris Francis. *Navajo Sacred Places.* Indiana University Press, 1994.

Kemp, Oliver. *Wilderness Homes: A Book of the Log Cabin.* New York: Outing Publishing, 1908; second edition, 1911.

Kendall, Edward A. *Travels through the Northern Parts of the United States in the Years 1808 and 1809.* New York: I. Riley, 1809.

Kennedy, Joseph C. G. *Agriculture of the United States in 1860.* Government Printing Office, 1864.

Kluger, James R. *Turning on Water with a Shovel: The Career of Elwood Mead.* University of New Mexico Press, 1992.

Kolodny, Annette. *The Land Before Her.* University of North Carolina Press, 1984.

Kouwenhoven, John. *Made in America.* Doubleday, 1962 (reprint of 1948 edition).

Kulik, Gary. "American Difference Revisited: The Case of the American Axe." In *American Material Culture,* ed. A. Martin and J. Garrison. Henry Francis du Pont Winterthur Museum, 1997.

Lassen, Henrik R. The Idea of Narrative: The Theory and Practice of Analyzing Narrative Types, and Legends of Suppressed Inventions. Ph.D. thesis, Odense University, 1998.

Lears, Jackson. *No Place of Grace.* Knopf, 1981.

Leighton, Caroline C. *Life at Puget Sound, with Sketches of Travel.* Boston: Lee and Shepard, 1884.

Levenson, J. C., et al., eds. *The Letters of Henry Adams.* Harvard University Press, 1982.

Lewis, R. W. B., ed. *Herman Melville.* Dell, 1962.

Lewis, Sinclair. "Cheap and Contested Labor: The Picture of a Southern Mill Town." United Features Syndicate, 1929.

Licht, Walter. *Industrializing America: The Nineteenth Century.* Johns Hopkins University Press, 1995.

Limerick, Patricia Nelson. "The Adventures of the Frontier in the Twentieth Century." In *The Frontier in American Culture,* ed. J. Grossman. University of California Press, 1994.

Lincoln, Abraham. "Second Lecture on Discoveries and Inventions." In *Collected Works of Abraham Lincoln,* volume 3, ed. R. Brasler. Rutgers University Press, 1954.

Lindley, Harlow, ed. *Indiana as Seen by Early Travelers.* Indianapolis: Indiana Historical Society, 1916.

Logan, George. "Letters Addressed to the Yeomanry of the United States." *American Museum* 12 (1792), September-October. Reprinted in *The Philosophy of Manufactures,* ed. M. Folsom and S. Lubar (MIT Press, 1982).

Lopez, Barry. *Crossing Open Ground.* Vintage, 1989.

Loree, L. F. "Address of L. F. Loree at the Hotel Astor." April 23, 1923.

Loree, L. F. "Introduction of J. S. Alexander." April 23, 1923.

Lucic, Karen. *Charles Sheeler and the Cult of the Machine.* London: Reaktion, 1991.

Lumpkin, Grace. *To Make My Bread.* New York: Macauley, 1932.

Lyell, Charles. *A Second Visit to the United States.* London: John Murray, 1849.

Maas, Arthur, and Raymond Andersen. *. . . and the Desert Shall Rejoice.* MIT Press, 1978.

Macherey, Pierre. *A Theory of Literary Production*. Routledge & Kegan Paul, 1978.

Madison, James, Alexander Hamilton, and John Jay. *The Federalist*. New American Library, 1960.

Malone, Michael P. "The Historiography of the American West." In *The American West as Seen by Europeans and Americans*, ed. R. Kroes. Amsterdam: Free University Press, 1989.

Malthus, Thomas. *Essay on the Principles of Population*, fourth edition. London: Printed for J. Johnson by T. Bensley, 1807.

Marcy, Captain R. B. *The Prairie and Overland Traveler*. London: Sampson Low & Son, 1860.

Marmor, Theodore R. "Anti-Industrialism and the Old South: The Agrarian Perspective of J. C. Calhoun." *Comparative Studies in Society and History* 9 (1967): 377–406.

Marsh, George Perkins. *Man and Nature*. Scribner, 1864; reprint: Harvard University Press, 1965.

Marshall, Edward. "Thomas A. Edison on Immortality." *Columbian Magazine*, January 1911: 603–612.

Martin, Albro. *Railroads Triumphant: The Growth, Rejection, and Rebirth of a Vital American Force*. Oxford University Press, 1992.

Martin, John Frederick. *Profits in the Wilderness: Entrepreneurship and the Founding of New England Towns in the Seventeenth Century*. University of North Carolina Press, 1991.

Martin, Ronald E. *American Literature and the Universe of Force*. Duke University Press, 1981.

Martineau, Harriet. *Society in America*, ed. S. Lipset. Doubleday, 1962.

Marx, Leo. *The Machine in the Garden*. Oxford University Press, 1965.

McDermott, John Francis, ed. *Before Mark Twain*. Southern Illinois University Press, 1968.

McKelvey, Blake. *Rochester, the Water-Power City, 1812–1854*. Harvard University Press, 1945.

McNichol, Tim. "The New Red Menace." *Wired* 9 (2001), July: 140–147.

McWilliams, Carey. *Factories in the Field: The Story of Migratory Farm Labor in California*. University of California Press, 1999 (reprint of 1935 edition).

Meinig, D. W. *Continental America, 1800–1867*, volume 2. Yale University Press, 1993.

Meinig, D. W. *The Shaping of America: Atlantic America, 1492–1800*. Yale University Press, 1986.

Meinig, D. W. *Transcontinental America, 1850–1915*. Yale University, 1998.

Mercer, Henry C. "The Origin of Log Houses in the United States." Bucks County Historical Society, 1924.

Merchant, Carolyn. "Reinventing Eden." In *Uncommon Ground*, ed. W. Cronon. Norton, 1996.

Merchant, Carolyn. *Ecological Revolutions: Nature, Gender, and Science in New England*. University of North Carolina Press, 1989.

Merrick, George Byron. *Old Times on the Upper Mississippi*. Cleveland: Arthur H. Clark, 1909.

Miller, George H. *Railroads and the Granger Laws*. University of Wisconsin Press, 1971.

Miller, Perry. *Errand into the Wilderness*. Harvard University Press, 1956.

Miller, Perry. *Life of the Mind in America from the Revolution to the Civil War*. Harcourt, Brace and World, 1965.

Miller, William D. *Pretty Bubbles in the Air: America in 1919*. University of Illinois Press, 1991.

Miner, H. Craig. *The Corporation and the Indian: Tribal Sovereignty and Industrial Civilization in Indian Territory, 1865–1907*. University of Missouri Press, 1976.

Mitchell, Broadus, and George Sinclair Mitchel. *The Industrial Revolution in the South*. Johns Hopkins University Press, 1930.

Mollhausen, Baldwin. *Diary of a Journey from the Mississipi to the Coasts of the Pacific with a United States Government Expedition*. London: Longman, Brown, Green, 1858.

Moore, Dennis D., ed. *More Letters from the American Farmer: An Edition of the Essays in English Left Unpublished by Crèvecoeur*. University of Georgia Press, 1995.

Muir, John. "The American Forests." *The Atlantic* 80 (1897), August: 145–157.

Muir, John. *A Thousand-Mile Walk to the Gulf*. Houghton Mifflin, 1916; reprint, 1998.

Mulhall, Michael. "The Increase of Wealth." *North American Review* 140 (1885): 78–85.

Nash, Roderick. *Wilderness and the American Mind*, third edition. Yale University Press, 1982.

Nason, Frank Lewis. *The Vision of Elijah Berl*. Little, Brown, 1905.

Neihardt, John. *Black Elk Speaks*. University of Nebraska Press, 1961 (reprint of 1932 edition).

Neill, Charles P. *Report on Strike of Textile Workers in Lawrence, Mass, in 1912*. Government Printing Office, 1912.

Newell, Frederick Haynes. *Irrigation in the United States*. New York: Thomas Y. Crowell, 1902.

Noble, David W. "Frederick Jackson Turner and Henry Nash Smith Revisited, 1890–1950–1990." In *The American West as Seen by Europeans and Americans,* ed. R. Kroes. Amsterdam: Free University Press, 1989.

Nobles, Gregory. *Divisions throughout the Whole: Politics and Society in Hampshire County, Massachusetts, 1740–1775.* Cambridge University Press, 1983.

Norris, Frank. *The Octopus: A Story of California.* Bantam, 1958 (reprint of 1903 edition).

Norton, A. Banning. *Reminiscences of the Log Cabin and Hard Cider Campaign.* Cleveland: A. B. Norton 1888.

Novak, Barbara. *American Painting of the Nineteenth Century.* Praeger, 1969.

Novak, Barbara. *Nature and Culture: American Landscape and Painting, 1825–1875.* Oxford University Press, 1980.

Nye, David E. *American Technological Sublime.* MIT Press, 1994.

Nye, David E. *Consuming Power: A Social History of American Energies.* MIT Press, 1998.

Nye, David E. *Electrifying America: Social Meanings of a New Technology.* MIT Press, 1990.

Nye, David E. *Henry Ford: Ignorant Idealist.* Kennikat, 1979.

Nye, David E. *Narratives and Spaces: Technology and the Construction of American Culture.* Columbia University Press, 1998.

Nye, David E., ed. *Technologies of Landscape: Reaping to Recycling.* University of Massachusetts Press, 2000.

O'Brien, Jim. "A Beaver's Perspective on North American History." Reprinted in *Major Problems in American Environmental History,* ed. C. Merchant. Heath, 1993.

Official Guide, New York World's Fair, 1964/1965. Time Inc., 1964.

Official Guide. Chicago: A Century of Progress, 1933.

"Ollapondiana." *Knickerbocker* 8 (July 1836), p. 74.

"Ollapondiana." *Knickerbocker* 8 (September 1836), p. 348.

Olmsted, Frederick Law. *The Cotton Kingdom,* ed. A. Schlesinger Jr. Knopf, 1966.

Olson, Sherry H. *The Depletion Myth: A History of Railway Use of Timber.* Harvard University Press, 1971.

O'Malley, Michael. *Keeping Watch.* Viking, 1990.

Onuf, Peter S. "Liberty, Development, and Union: Visions of the West in the 1780s." *William and Mary Quarterly* 43 (1986), no. 2: 179–213.

Opie, John. *Ogallala: Water for a Dry Land,* second edition. University of Nebraska Press, 2000.

Page, Dorothy Mayra. *Gathering Storm.* New York: International Publishers, 1932.

Parker, Margaret Errell. *Lowell: A Study of Industrial Development.* Kennikat, 1970 (reprint of 1940 edition).

Parkman, Francis. "The Forests and the Census." *The Atlantic,* June 1885: 835–839.

Pioneers: Narratives of Noah Harris Letts and Thomas Allen Banning. Lakeside, 1972.

Pisani, Donald J. *From the Family Farm to Agribusiness: the Irrigation Crusade in California and the West, 1850–1931.* University of California Press, 1984.

Pisani, Donald J. *To Reclaim a Divided West: Water, Law, and Public Policy, 1848–1902.* University of New Mexico Press, 1992.

Plumb, Henry B. *History of Hanover Township.* Wilkes-Barre: R. Baur, 1885.

Pope-Hennessy, Una, ed. *Aristocratic Journey: Letters of Mrs. Basil Hall.* Putnam, 1934.

Porter, Robert P. *The West, from the Census of 1880.* Rand, McNally, 1882.

Powell, John Wesley. "The Irrigable Lands of the Arid Region." *Century* 39 (1890), March: 766–776.

Powell, John Wesley. *Report on the Lands of the Arid Region of the United States,* ed. W. Stegner. Harvard University Press, 1962 (reprint of 1879 edition).

Powell, Lyman P., ed. *Historic Towns of New England.* Putnam, 1898.

Prentice, Archibald. *A Tour of the United States.* London: Charles Gilpen, 1848.

Pursell, Carroll. *The Machine in America: A Social History of Technology.* Johns Hopkins University Press, 1995.

Raban, Jonathan. *Bad Land: An American Romance.* London: Picador, 1996.

Rajchman, John. *Constructions.* MIT Press, 1998.

Reisner, Marc. *Cadillac Desert: The American West and Its Disappearing Water.* Viking, 1986.

Reisner, Marc, and Sarah Bates. *Overtapped Oasis: Reform or Revolution for Western Water.* Island, 1990.

Renner, Rolf Günter. *Edward Hopper, 1882–1967, Transformer of the Real.* Köln: Taschen, 1993.

Reps, John V. *Cities of the American West: A History of Frontier Urban Planning.* Princeton University Press, 1979.

Reps, John V. *Panoramas of Progress: Pacific Northwest Cities and Towns on Nineteenth-Century Lithographs.* Washington State University Press, 1984.

Richardson, Albert D. *Beyond the Mississippi.* Hartford: American Publishing, 1867.

Robbins, Roy M. *Our Landed Heritage: The Public Domaine, 1776–1936.* University of Nebraska Press, 1962.

Robinson, Douglas. *American Apocalypses: The Image of the End of the World in American Litearature.* Johns Hopkins University Press, 1985.

Robinson, Harriet H. *Loom and Spindle, or Life Among the Early Mill Girls.* Press Pacifica, 1976.

Roosevelt, Franklin D. "Address at Dedication of Boulder Dam." In *The Public Papers and Addresses of Franklin D. Roosevelt*, volume 4. Random House, 1938.

Roosevelt, Theodore. "The Forest in the Life of a Nation." In *Proceedings of the American Forest Congress held at Washington, DC, 1905*. Washington: Suter, 1905.

Roske, Ralph J. *Everyman's Eden: A History of California*. Macmillan, 1968.

Ross, Steven J. "Struggles for the Screen: Workers, Radicals, and the Political Uses of Silent Film." *American Historical Review* 96 (1991), no. 2: 333–367.

Rowe, John Carlos, ed. *New Essays on The Education of Henry Adams*. Cambridge University Press, 1996.

Rowlandson, Mary. *A narrative of the captivity, sufferings and removes, of Mrs. Mary Rowlandson, who was taken prisoner by the Indians, with several others, and treated in the most barbarous and cruel manner by those vile savages*. Boston: Nathaniel Coverly, 1770.

Rozwenc, Edwin C., ed. *Ideology and Power in the Age of Jackson*. Doubleday, 1964.

Rusling, James F. *The Great West and Pacific Coast*. New York: Sheldon, 1877 (expansion of 1867 edition).

Russell, Carl P. *Firearms, Traps, and Tools of the Mountain Men*. Knopf, 1967.

Safire, William, ed. *Lend Me Your Ears: Great Speeches in History*. Norton, 1997.

Sakolski, A. M. *The Great American Land Bubble*. Harper, 1932.

Salinger, J. D. *The Catcher in the Rye*. Bantam, 1964.

Salmond, John A. *Gastonia 1929: The Story of the Loray Mill Strike*. University of North Carolina Press, 1995.

Samuels, Ernest. *Henry Adams*. Harvard University Press, 1989.

Samuels, Ernest. *Henry Adams: The Major Phase*. Harvard University Press, 1964.

Santa Fe Railroad. *Kansas in 1875: Strong and Impartial Testimony to the Wonderful Productiveness of the Cottonwood and Arkansas Valleys: What Over 200 Editors Think of Their Present and Future*. Topeka, 1875.

Savelle, Max. *Empires into Nations: Expansion in America, 1713–1824*. University of Minnesota Press, 1974,

Schivelbusch, Wolfgang. *The Railway Journey*. Leamington Spa: Berg, 1986.

Schumpeter, Joseph. *Business Cycles: A Theoretical, Historical, and Statistical Analysis of the Capitalist Process*. McGraw-Hill, 1939.

Schuyler, David. "The Sanctified Landscape: The Hudson River Valley, 1820–1850." In *Landscape in America*, ed. G. Thompson. University of Texas Press, 1995.

Scranton, Philip. "Varieties of Paternalism: Industrial Structures and Social Relations of Porduction in American Textiles." *American Quarterly* 36 (1984), no. 2: 235–257.

Scranton, Philip. *Figured Tapestry*. Cambridge University Press, 1989.

Sears, John. *Sacred Places: American Tourist Attractions in the Nineteenth Century.* Oxford University Press, 1989.

Shaw, Ronald E. *Canals for a Nation: The Canal Era in the United States, 1790–1860.* University of Kentucky Press, 1990.

Sheriff, Carol. *The Artificial River: The Erie Canal and the Paradox of Progress, 1817–1862.* Hill and Wang, 1996.

Sherow, James Earl. *Watering the Valley: Development along the High Plains Arkansas River, 1870–1950.* University of Kansas Press, 1990.

Shurtleff, Harold R. *The Log Cabin Myth.* Harvard University Press, 1939.

Sibley, Mulford Q. *Political Ideas and Ideologies: A History of Political Thought.* Harper & Row, 1970.

Silko, Leslie Marmon. "Landscape, History and the Pueblo Imagination." *The Ecocriticism Reader: Landmarks in Literary Ecology* ed. C. Glotfelty and H. Fromm. University of Georgia Press, 1996.

Simpson, John Warfield. *Visions of Paradise: Glimpses of Our Landscape's Legacy.* University of California Press, 1999.

Slotkin, Richard. *Regeneration through Violence: The Mythology of the American Frontier.* Wesleyan University Press, 1973.

Smith, Adam. *The Wealth of Nations,* ed. D. Raphae. Knopf, 1991.

Smith, Barbara Clark. "Food Rioters and the American Revolution." *William and Mary Quarterly* 51 (1994), no. 1: 3–38.

Smith, Barbara Herrnstein. "Narrative Versions, Narrative Theories" *Critical Inquiry* 7 (1980): 213–236.

Smith, Henry Nash. *Virgin Land.* Harvard University Press, 1950.

Smith, Lincoln. *The Power Policy of Maine.* University of California Press, 1951.

Smith, Merritt Roe, and Leo Marx, eds. *Does Technology Drive History? The Dilemma of Technological Determinism.* MIT Press, 1994.

Smith, William Prescott. *The Book of the Great Railway Celebrations of 1857.* Appleton, 1858.

Smythe, William E. *The Conquest of Arid America.* Macmillan, 1905.

Spencer, Herbert. *First Principles,* fourth edition. Appleton, 1896.

Stange, Maren. *Symbols of Ideal Life.* Cambridge University Press, 1989.

Starr, Kevin. *Material Dreams: Southern California through the 1920s.* Oxford University Press, 1990.

Starr, Kevin. *Endangered Dreams: The Great Depression in California.* Oxford University Press, 1996.

Steinbeck, John. *The Grapes of Wrath.* Viking, 1958 (reprint of 1939 edition).

Steinberg, Theodore. *Nature Incorporated: Industrialization and the Waters of New England.* Cambridge University Press, 1991.

Stilgoe, John. *Common Landscape of America, 1580–1845.* Yale University Press, 1980.

Stilgoe, John. *Metropolitan Corridor.* Yale University Press, 1983.

Stover, John F. *The Routeledge Historical Atlas of the American Railroads.* Routeledge, 1999.

Stowell, David O. *Streets, Railroads, and the Great Strike of 1877.* University of Chicago Press, 1999.

Stradling, David. *Smokestacks and Progressives: Environmentalists, Engineers, and Air Quality in America, 1881–1951.* Johns Hopkins University Press, 1999.

Street, Alfred B. "Song of the Axe." *Graham's Magazine,* April 1855: 7–8.

Stromquist, Selton. *A Generation of Boomers: The Pattern of Railroad Labor Conflict in Nineteenth-Century America.* University of Illinois Press, 1987.

Strong, Josiah. *Expansion under New World Conditions,* ed. R. Weber. Garland, 1971 (reprint of 1900 edition).

Strong, Josiah. *Our Country: Its Possible Future and Its Present Crisis.* Harvard University Press, 1963 (reprint of 1885 edition).

Tarr, Joel A. *The Search for the Ultimate Sink: Urban Pollution in Historical Perspective.* University of Akron Press, 1996.

Taylor, George Rogers, ed. *The Turner Thesis.* Heath, 1956.

Taylor, George Rogers. *The Transportation Revolution.* Rinehart, 1951.

Teele, Ray Palmer. *Irrigation in the United States.* Appleton, 1915.

Tenner, Edward. *Why Things Bite Back.* Vintage, 1996.

Terrie, Philip G. *Contested Terrain: A New History of Nature and People in the Adirondacks.* Syracuse University Press, 1997.

Tesla, Nikola. "The Problem of Increasing Human Energy with Special Reference to the Harnessing of the Sun's Energy." *Century Magazine,* May 1900: 175–211.

Thoreau, Henry David. "Paradise (to Be) Regained." *United States Magazine and Democratic Review* 13 (1843), November. Reprinted in *The Philosophy of Manufactures,* ed. M. Folsom and S. Lubar (MIT Press, 1982).

Thoreau, Henry David. *The Journal of Henry David Thoreau,* ed. B. Torrey and F. Allen. Dover, 1962.

Thoreau, Henry David. *The Maine Woods.* New York: Bramhall House, 1950.

Thoreau, Henry David. *Walden.* Holt, Rinehart and Winston, 1948.

Thurston, Robert H. "The Borderland of Science." *North American Review* 150 (1890): 67–79.

Thurston, Robert H. "Our Progress in Mechanical Engineering: The President's Annual Address." *Transactions of the American Society of Mechanical Engineers* 2 (1881): 5–17.

Thurston, Robert H. "The Mission of Science." In *Proceedings of the American Association for the Advancement of Science, 1884.* Salem: AAAS, 1885.

Thurston, Robert H. "The Trend of National Progress." *North American Review* 161 (1895), September: 297–313.

Tichi, Cecelia. *New World, New Earth: Environmental Reform in American Literature from the Puritans through Whitman.* Yale University Press, 1979.

Tichi, Cecelia. *Shifting Gears: Technology, Culture, Literature in Modernist America.* University of North Carolina Press, 1987.

Tocqueville, Alexis de. *Journey to America.* Yale University Press, 1960.

Tredgold, Thomas. *A Practical Treatise on Rail-Roads and Carriages.* New York: E. Bliss and E. White, 1825.

Truman, Benjamin C. *History of the World's Fair.* Arno, 1976 (reprint of 1893 edition).

Turner, Frederick Jackson. "The Hunter Type." In *Frederick Jackson Turner's Legacy,* ed. W. Jacobs. University of Nebraska Press, 1977.

Turner, Frederick Jackson. "The Significance of the Frontier." In *Annual Report of the American Historical Association for the Year 1983.* Government Printing Office, 1894.

Turner, Frederick Jackson. *Rise of the New West, 1819–1829.* Harper, 1906.

Turner, Orsamus. *Pioneer History of the Holland Purchase of Western New York.* Buffalo: Jewett, Thomas, 1850.

Tyler, Ron. *Visions of America: Pioneer Artists in a New Land.* Thames and Hudson, 1983.

Updike, John. *Rabbit at Rest.* Knopf, 1990.

U.S. Census Office. *Sixth Census, 1840.* Government Printing Office, 1842.

U.S. Department of the Interior. *The Colorado River.* Government Printing Office, 1946.

U.S. Department of State. *Sixth Census or Enumeration of the United States.* Washington: Blair and Rives, 1841.

Uzzell, Thomas. "The Future of Electricity." *Colliers National Weekly,* December 2, 1916: 7–8.

Vorse, Mary. *The Passaic Textile Strike, 1926–27.* Passaic: General Relief Committee of Textile Strikers, 1929.

Vorse, Mary. *Strike!* New York: Liveright, 1930.

Wade, Richard C. *The Urban Frontier: The Rise of Western Cities, 1790–1830.* Harvard University Press, 1959.

Walkowitz, Daniel J. *Worker City, Company Town.* University of Illinois Press, 1978.

Wallace, Anthony F. C. *Rockdale: The Growth of an American Village in the Early Industrial Revolution.* Norton, 1980.

Walsh, Margaret. *The Manufacturing Frontier: Pioneer Industry in Ante-bellum Wisconsin, 1830–1860.* State Historical Society of Wisconsin, 1972.

Ward, John William. *Andrew Jackson: Symbol for an Age.* Oxford University Press, 1955.

Ware, Norman. *The Industrial Worker, 1840–1860.* Ivan R. Dee, 1990 (reprint of 1924 edition).

Warner, A. G. "Railroad Problems in a Western State." *Political Science Quarterly* 6 (1891), no. 1: 66–89.

Waters, Frank. *People of the Valley.* Swallow/Ohio University Press, 1969 (reprint of 1941 edition).

Way, Peter. *Workers and the Digging of North American Canals, 1780–1860.* Johns Hopkins University Press, 1997.

Weslager, C. A. *The Log Cabin in America: From Pioneer Days to the Present.* Rutgers University Press, 1969.

Westfall, Richard S. *Force in Newton's Physics.* London: McDonald, 1971.

White, George S. *Memoir of Samuel Slater.* A. M. Kelley, 1967 (reprint of 1836 edition).

White, Hayden. *The Content of the Form: Narrative Discourse and Historical Representation.* Johns Hopkins University Press, 1987.

White, Hayden. *Tropics of Discourse: Essays in Cultural Criticism.* Johns Hopkins University Press, 1978.

White, Richard. "Discovering Nature in North America." *Journal of American History* 79 (1992): 874–891.

White, Richard. "Frederick Jackson Turner and Buffalo Bill." In *The Frontier in American Culture,* ed. J. Grossman. University of California Press, 1994.

Whitman, Walt. *Leaves of Grass.* Norton, 1973.

Whitman, Walt. *The Collected Writings of Walt Whitman, Prose Works.* New York University Press, 1964.

Whittier, John Greenleaf. "The City of a Day." In Whittier, *Prose Works,* volume 2. Boston: Ticknor and Fields, 1866.

Whittier, John Greenleaf. "Pawtucket Falls." In Whittier, *Prose Works,* volume 2. Boston: Ticknor and Fields, 1866.

Whittier, John Greenleaf. *A Stranger in Lowell.* Boston: Waite, Peirce and Co., 1845.

Whitworth, Joseph, and George Wallis. *The Industry of the United States in Machinery, Manufactures and Useful and Ornamental Arts.* Routledge, 1854.

Wicks, William S. *Log Cabins and Cottages: How to Build and Furnish Them.* New York: Forest and Stream, 1900.

Williams, Michael. *Americans and Their Forests.* Cambridge University Press, 1989.

Willoughby, Lynn. *Flowing through Time: A History of the Lower Chattahouchee River.* University of Alabama Press, 1999.

Winner, Langdon. *Autonomous Technology: Technics-out-of-Control as a Theme in Political Thought.* MIT Press, 1977.

Worster, Donald. *An Unsettled Country.* University of New Mexico Press, 1994.

Worster, Donald. "Comment: A Response to 'John Wesley Powell and the Unmaking of the West.'" *Environmental History* 2 (1997), no. 2: 216–219.

Worster, Donald. *Rivers of Empire: Water, Aridity, and the Growth of the American West.* Pantheon, 1985.

Wright, Benjamin. "The Village of Little Falls." *American Railroad Journal, and Advocate of Internal Improvements,* April 1833, p. 244.

Wright, Carroll D. *The Industrial Evolution of the United States.* Meadville, Pa.: Chautauqua Century Press, 1895.

Wright, Harold Bell. *The Winning of Barbara Worth.* Gretna, La.: Pelican, 1999 (reprint of 1911 edition).

Wright, Louis B., ed. *The Prose Works of William Byrd of Westover.* Harvard University Press, 1966.

Wycoff, William. *The Developer's Frontier: The Making of the Western New York Landscape.* Yale University Press, 1988.

Yates, Gayle Graham, ed. *Harriet Martineau on Women.* Rutgers University Press, 1985.

Zelinsky, Wilbur. *Exploring the Beloved Country: Geographical Forays into American Society and Culture.* University of Iowa Press, 1994.

Index